Food Chemistry: Concepts, Technology and Applications

Food Chemistry: Concepts, Technology and Applications

Editor: Lisa Jordan

MURPHY & MOORE
www.murphy-moorepublishing.com

www.murphy-moorepublishing.com

MURPHY & MOORE

Cataloging-in-Publication Data

Food chemistry : concepts, technology and applications / edited by Lisa Jordan.
p. cm.
Includes bibliographical references and index.
ISBN 978-1-63987-725-6
1. Food--Analysis. 2. Food--Composition. 3. Food. 4. Chemistry. I. Jordan, Lisa.
TX541 .F66 2023
664.07--dc23

Murphy & Moore Publishing
1 Rockefeller Plaza,
New York City,
NY 10020, USA

ISBN 978-1-63987-725-6

Contents

Preface

Over the recent decade, advancements and applications have progressed exponentially. This has led to the increased interest in this field and projects are being conducted to enhance knowledge. The main objective of this book is to present some of the critical challenges and provide insights into possible solutions. This book will answer the varied questions that arise in the field and also provide an increased scope for furthering studies.

Food chemistry is a discipline that examines the chemical composition of food in order to determine the nature of food, the causes of its deterioration, the principles underlying food processing, and the methods for improving food for public consumption. It involves the application of concepts from disciplines such as chemistry, engineering, microbiology, and nutrition in order to improve the safety, nutrition, and availability of food. Some of the major areas of study within this discipline are water, lipids and carbohydrates. Water is present in most food items, and offers a favorable environment for bacterial growth. Therefore, in order to increase the shelf life of a product, the amount of water can be altered using processes such as freezing, dehydration and refrigeration. This book attempts to explore the concepts, technology and applications of food chemistry. A number of latest researches have been included to keep the readers up-to-date with the latest developments in this area of study. It aims to serve as a resource guide for students and experts alike and contribute in the growth of the discipline.

I hope that this book, with its visionary approach, will be a valuable addition and will promote interest among readers. Each of the authors has provided their extraordinary competence in their specific fields by providing different perspectives as they come from diverse nations and regions. I thank them for their contributions.

Editor

1

The Toxic Impact of Honey Adulteration

Rafieh Fakhlaei [1], Jinap Selamat [1,2,*], Alfi Khatib [3,4], Ahmad Faizal Abdull Razis [2,5], Rashidah Sukor [2], Syahida Ahmad [6] and Arman Amani Babadi [7]

1 Food Safety and Food Integrity (FOSFI), Institute of Tropical Agriculture and Food Security, Universiti Putra Malaysia, Serdang 43400, Selangor, Malaysia; rafieh.fakhlaei@gmail.com
2 Department of Food Science, Faculty of Food Science and Technology, Universiti Putra Malaysia, Serdang 43400, Selangor, Malaysia; madfaizal@upm.edu.my (A.F.A.R.); rashidah@upm.edu.my (R.S.)
3 Pharmacognosy Research Group, Department of Pharmaceutical Chemistry, Kulliyyah of Pharmacy, International Islamic University Malaysia, Kuantan 25200, Pahang Darul Makmur, Malaysia; alfikhatib@iium.edu.my
4 Faculty of Pharmacy, Airlangga University, Surabaya 60155, Indonesia
5 Natural Medicines and Products Research Laboratory, Universiti Putra Malaysia, Serdang 43400, Selangor, Malaysia
6 Department of Biochemistry, Faculty of Biotechnology & Biomolecular Sciences, Universiti Putra Malaysia, Serdang 43400, Selangor, Malaysia; syahida@upm.edu.my
7 School of Energy and Power Engineering, Jiangsu University, Zhenjiang 212013, China; ar.amani65@gmail.com

* Correspondence: jinap@upm.edu.my.

Abstract: Honey is characterized as a natural and raw foodstuff that can be consumed not only as a sweetener but also as medicine due to its therapeutic impact on human health. It is prone to adulterants caused by humans that manipulate the quality of honey. Although honey consumption has remarkably increased in the last few years all around the world, the safety of honey is not assessed and monitored regularly. Since the number of consumers of honey adulteration have increased in recent years, their trust and interest in this valuable product has decreased. Honey adulterants are any substances that are added to the pure honey. In this regard, this paper provides a comprehensive and critical review of the different types of adulteration, common sugar adulterants and detection methods, and draws a clear perspective toward the impact of honey adulteration on human health. Adulteration increases the consumer's blood sugar, which can cause diabetes, abdominal weight gain, and obesity, raise the level of blood lipids and can cause high blood pressure. The most common organ affected by honey adulterants is the liver followed by the kidney, heart, and brain, as shown in several in vivo research designs.

Keywords: honey; adulteration; sugar adulterants; toxicity

1. Introduction

The standards of *Codex Alimentarius* [1] defines honey as the natural sweet substance from the nectar of plants or secretions of living parts of the plants that are stored and dehydrated by honey bees to improve its nutritional properties and become consumable for humans.

Honey, traditionally, is used for its anti-aging properties, enhancing the immune system, killing bacteria, treatment of bronchial phlegm, and relieving a sore throat, cough, and cold [2]. Moreover, according to literature, honey represents various pharmacological properties such as anti-inflammatory [3], antioxidant [4], anti-cancer activities against breast and cervical cancer [5], prostate cancer [6], and osteosarcoma [7]. The therapeutic effect of honey on human health can be either oral administration or topical application. In this regard, reference [8] revealed the therapeutic properties of oral administration of honey for the treatment of laryngitis, osteoporosis, gastrointestinal ulcers, anorexia, insomnia and constipation, and liver, cardiovascular and gastrointestinal problems. On the other hand, advantages of topical application of honey are prescribed for eczema, lip sores, sterile and infected wounds, genital lesions, burns, surgery scars, and athlete's foot [9].

The food industry is one of the critical and fast developing industries worldwide, owing to the tremendous growth of the human population and increased interest of consumers toward the consumption of high-quality products. Moreover, it has been proven that low-quality food products and junk foods may have an adverse impact on consumers' health [10]. Food adulteration will multiple this risk since the nature of food has been altered. "Food adulteration" is described as the act of intentionally decreasing the quality of food either by adding or swapping low-quality materials or eliminating various important integrant. When the cheaper and low-grade elements are added to an original product threaten the consumer's health, it is considered and declared "adulterated." Honey, as one of the most common foods worldwide, also has been subjected to adulteration [11]. Hence, its quality control and safety protocols have become the center of attention of many international committees.

Although honey is recognized as high-quality food, it is more vulnerable to adulteration, mislabeling, and unethical mixing with cheaper and low-grade honey, sugars, and other substances. Moreover, due to its limited availability, proved therapeutic and healing properties, and the increased population concerns regarding their health, there is a rising demand for the natural food product. This increased economic value would make honey a vulnerable adulteration target [12]. Moreover, while honey is a well-known high nutritional value food, it can also be toxic naturally by transferring plant toxins such as pyrrolizidine alkaloids, or because of adulterants that are added into the pure honey by mankind to gain economic profits [13]. Food adulteration has been a major concern for consumers, as it does not only decrease the quality of food products but also results in several adverse health effects. Authentic testing of food and the toxicology of adulterants is required for a value assessment to assure consumer protection against fraudulent activities. According to the regulation set by *Alimentarius* [1], consumers have the right to receive truthful information about the food that they are going to consume. It has also mentioned that honey should not have any added ingredients, any foreign matter, flavor, aroma, or taint absorbed from foreign substances during processing and storage, nor any removal of a particular constituent. Moreover, honey should not be heated or processed to such an extent that its essential composition is changed and its quality impaired. Although honey adulteration is a serious issue worldwide that requires several actions to be solved, there is currently a lack of an effective method to regulate the adulterated honey production [14]. In addition, honey adulteration is a key factor in the honey price fluctuation on the market. Several actions, locally and internationally, have been taken to detect fraud and solve the problem, but there is no actual solution to control the production of adulterated honey [15].

Reference [15] mentioned that sugars could be used in two different ways as adulterants; direct adulteration and indirect adulteration. During direct adulteration, a certain ratio of syrups is added to harvested honey to increase its sweet taste, while in indirect adulteration the bees were overfed with sugar syrups to increase the honey yield in hives. According to Se et al. [13], the most frequent sugar syrups for honey adulteration are high fructose corn syrup (HFCS), corn sugar syrup (COSS), inverted sugar syrup (ISS), and cane sugar syrup (CASS); there is a high preference towards HFCS (from simple isomerization of COSS) according to Se et al. [16].

Reference [17] mentioned that in Ethiopia, other adulterants are used for honey, such as water, banana, wheat, and maize syrup or flour. Furthermore, Damto [18] reported the water content, honey processing, and botanical origin as a direct honey adulteration, while admitting that pre-mature harvesting and veterinary drugs, especially an extra dosage, should be considered indirect methods.

According to Jaafar et al. [19], olive oil, milk, honey, saffron, orange juice, coffee, and apple juice are the seven most likely food ingredients to be targeted for economically-motivated adulteration of food (food fraud), as per their article published in the Journal of Food Science. Nowadays, honey adulteration has a major impact on economic loss, not only due to the declining honey quality but also because of the difficulty in the marketing of pure honey. According to Johnson et al. [12], the majority of honey sold on the market is caramelized sucrose that does not have any nutritional value. Since this adulterated material does not have the nutritional value that honey has, consumers have lost sight of the nutritional value of honey.

Some authors have also reported that honey adulteration may cause a reduction in the protein content of honey. In addition, honey adulteration, overheating, or prolonged storage of honey can reduce or eliminate its protein content, as seen in the paper by Lawal et al. [20]. As regards to nitrogen content, reference [21] claimed that there are some variations concerning the nitrogen content of pure and adulterated honey and also pure HFCS, as it ranges from 0.6 for pure honey, 0.3% for 1:1 adulterated honey/HFCS to 0.1% for pure HFCS. The authors mentioned that the nitrogen, proline, potassium, and sodium content of pure honey is much higher than adulterated honey. Abdel-Aal et al. [22] confirmed that the protein content is a reliable factor to investigate the honey adulteration in the samples with less than 30% added sugar. The amino acid content of honey is 50–300 mg/kg, while proline is the most abundant amino acid (50–85%). Particularly, the proline concentration ratio (180 mg/kg) is an indication value to differentiate between natural and adulterated honey [23], which is more accurate in comparison to the protein content [24,25].

Unfortunately, there would be a risk of detecting pure honey as the adulterated type if C4 plants are being used in honey production with other plants since it causes the product to exceed the $\delta^{13}C$ value [23]. Plants are classified into three groups, in accordance with the photosynthesis pathway: C3 plants and C4 plants can bio-fixate carbon dioxide into a 3-carbon compound by using the Calvin cycle and a 4-carbon compound via the Hatch-Slack cycle, respectively, while crassulacean acid metabolism (CAM) plants can use both cycles [26].

However, the $\delta^{13}C$ value might reduce by the addition of a sugar adulterant (below 23.5%) and mislead the detection of honey adulteration [27]. Therefore, both C isotope ratios of raw honey and protein fractions ($\delta^{13}C_{honey}$ and $\delta^{13}C_{protein}$) should be considered to detect the honey adulteration by the addition of sugar [28].

Adulterants are any substances that are added to the original and pure product. Honey can be adulterated directly (addition of adulterants) [28], indirectly (bee-feeding) [26,29], or by blending it with other cheap honey [30,31], which will be discussed further below. The quality of honey is closely related to its impurities and adulterants.

The adverse health impact of honey adulteration on consumers may lead to increased blood sugar, followed by the release of the insulin hormone and type II diabetes, abdominal weight gain and obesity, a rise in the blood lipid levels, and high blood pressure [32]. Furthermore, adulterants can affect internal organs, potentially causing a fatty liver [13], acute and chronic kidney injury [33] and elevate visceral fat pads and total body fat, which can lead to death [12,15].

Honey adulteration is due to various reasons, such as adding sugars to enhance the taste based on the consumer's preference or gaining financial profits by mixing cheap and low-quality honey to the expensive honey to increase the yield. As we have checked various research databases such as Scopus, the Web of Science, and Google Scholar, there is not any significant study on the toxicological effect of honey adulteration on human health up to this date. Hence, this review paper can be a foundation for scientists and researchers to investigate commercial honey adulterants and the disadvantages of adulterated honey on consumers' health. This review comprehensively covers the outline regarding different types of adulterants, methods of honey adulteration, and the toxic impact of honey adulteration on internal organs via various in vivo and in vitro studies.

1.1. Honey

Honey is defined as the excretions of insects sucking on the living parts of plants. Honeybees are the most well-known plant-sucking insects and can collect and transform honey, and deposit, dehydrate, store and leave honey in the honeycomb to ripen and mature. Honeybees collect pollen and nectar from a variety of flowering plants and convert it into the wax and honey [34]. Only worker honeybees forage for food, consuming as much nectar from each flower as they can. After foraging, worker honeybees return to the hive/comb and pass the collected nectar to the other worker honeybees. This worker holds the nectar on her tongue until the liquid evaporates, creating honey. The honey is then stored in a cell within the hive/comb.

1.2. Type of Honey

Reference [5] and Alvarez-Suarez et al. [35] classified honey according to its origin as follows:

(1) Blossom honey: the main source of this honey is the nectar of flowers such as linden, clover, citrus, cotton, thyme, and acacia honey.
(2) Honeydew honey: the source of this honey is the "honeydew" (*Rhynchota* genus insects pierce plant cells, ingest plant sap, and then secrete it again) collected by bees. A typical example of honeydew honey is pine, oak, fir, and leaf honey.
(3) Monofloral honey: named according to the plant that the bees that have produced the honey forage predominantly.
(4) Multifloral honey (polyfloral): the source of this honey is several botanical flowers, with none of them predominant. Meadow blossom honey and forest honey are classified in this category.

All these classifications indicate the quality and physicochemical properties of honey. The honey composition and quality varies according to the botanical origin, geographic area, and harvesting season [26]. Moreover, honey can be classified based on the bee species (stingless and honeybee) as mentioned in the next sections.

1.2.1. Stingless Bee Honey

Among all five genera of the stingless bee, there are the only two types of stingless bees that produce honey, namely, *Melipona* and *Trigona* [36]. As a result of the previous study, stingless bee honey, in comparison to sting bee honey, has been shown to be higher in moisture content, acidity, and lower in diastase activity [37]. Furthermore [5], stingless bees can be differentiated from sting bee in terms of morphology (absence of sting) (Figure 1a), collection of nectar, their short harvest distance, and their honeycomb-less hives.

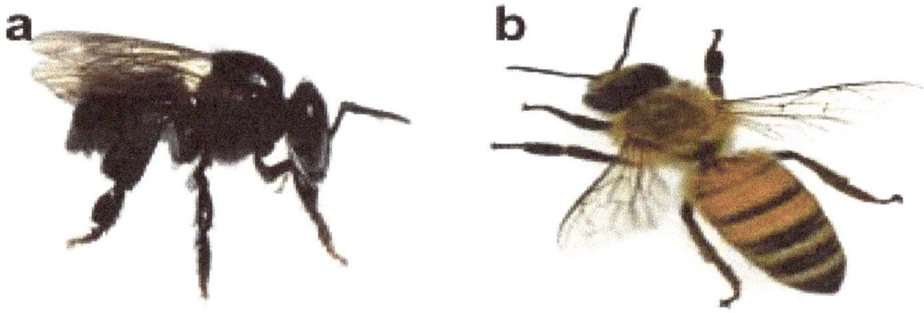

Figure 1. Different types of bees: (**a**) *Heterotrigona itama* bee, (**b**) *Apis mellifera* bee.

These species of bees are not dangerous to humans and are very active as compared to other bees. Among all these species, only two types, named *Heterotrigona itama* (*H. itama*) and *Geniotrigona thoracica*, are kept by beekeepers, as they produce a higher volume of honey as compared to the others. In this regard, Se et al. [16] stated that *H. itama* honey has significantly higher physicochemical values when it comes to moisture content, water activity, free acidity and color intensity, and better antioxidant properties in comparison with other types of honey. Furthermore, Bakar et al. [38] mentioned that *H. itama* honey is less concentrated, with a more sour taste and aroma in comparison with other types of honey. Moreover, stingless bee store honey in a honey pot as opposed to a honeycomb.

1.2.2. Honeybee (*Apis*)

Among all recognized *Apis* species, there are only two commercially used by mankind, namely *A. mellifera* and *A. cerana*. This is due to the behavioral limitation of other species, such as dwarf and giant honeybees, which practice open-air nesting and cannot be kept in manmade hives. A comparison between these two species, *A. mellifera* generally is more productive in producing honey compared to *A. cerana* [3]. Furthermore, honeybees (Figure 1b) store honey in a honeycomb.

Honey from honeybees is a viscous, high-nutrient food. Its most important contents are fructose and glucose (80% ± 2), water (16% ± 1), ash (0.2%), and amino acids (<0.1%), while enzymes, vitamins, phenolic compounds, and other substances are present in trace amounts [39].

1.3. Worldwide Honey Production and Consumption

Globally, the honey market was worth more than USD 7.5 billion in 2018, and it will reach USD 10.5 billion by 2025 [5]. As regards the geographical production distribution, the Asia Pacific had the biggest market share in 2018. China is the top producer, with a production volume of 490.84 K (27.5% of the global production), followed by Turkey with a production volume of 105.53 K (5.9% of the global production) and Iran with a production volume of 80.56 K (4.5% of the global production) [40]. This noticeable geographical production growth can be attributed to the regional rise in the supply and demand chain, and also to the growing awareness of consumers regarding the benefits of honey and its advantages over normal sugar. Furthermore, the United States placed the fourth among the top producers of honey, with a production volume of 73.43 K (4.1% global production) in 2016. The production of honey in the European Union (EU) has been increased moderately (250,000 tons/year in 2015), which established the EU as the next largest honey supplier after China [40]. The main producers in the EU are Romania, Spain and Germany followed by Hungary, France, Greece, and Poland [41].

1.4. Regulation Related to Honey

The authors in [40] established two criteria regarding the sugar content of honey: (1) the total concentration of fructose and glucose should not be less than 60% (*w*/*w*) for blossom honey, and not less than 45% (*w*/*w*) for honeydew honey and related blends; (2) the sucrose concentration should not exceed 5% (*w*/*w*) (Table 1). Concerning the above information, CODEX and INTERGOVERNMENTAL [14] reported that 14% of honey from EU and non-EU was considered adulterated by sugar.

In this regard, the Australian New Zealand food standard [41] proposed the regulation that a food that is prescribed as 'honey' must contain no less than 60% reducing sugars and no more than 21% moisture.

Based on the established China honey standard, Reference [42] mentioned there are additional requirements for cane sugar content in Eucalyptus, citrus, clover, Lychee, and wild osmanthus honey, namely, not less than 10 g/100 g, while in other honey it cannot be less than 5 g/100 g. Moreover, the zinc content in honey is also limited to ≤25 mg/kg.

Table 1. Essential composition and quality factors of pure honey (adopted from Wei et al. [43]).

Component		Honey	Value
1.	Moisture content	Honey that is not listed below	<20%
		Heather honey (*Calluna*)	<23%
2.	Sugar content		
a.	Fructose + Glucose content	Honey that is not listed below	>60 g/100 g
		Honeydew honey and its blends with blossom honey	>45 g/100 g
b.	Sucrose content	Honey that is not listed below	<5 g/100 g
		Alfalfa (*Medicago sativa*), *Citrus* spp., False Acacia (*Robinia pseudoacacia*), French Honeysuckle (*Hedysarum*), Menzies Banksia (*Banksia menziesii*), Red Gum (*Eucalyptus camaldulensis*), Leatherwood (*Eucryphia lucida*), *Eucryphia milligani*	<10 g/100 g
		Lavender (*Lavandula* spp.) and Borage (*Borago officinalis*)	<15 g/100 g
3.	Water-insoluble solid content	Honey that is not listed below	<0.1 g/100 g
		Pressed honey	<0.5 g/100 g

1.5. Nutritional Value of Honey

Based on the nutritional value of honey, this healthy natural food contains not only fructose, glucose, and water, but also trace amounts of valuable ingredients, such as flavorings, vitamins (B_1, B_2, B_3, B_5, B_6, B_9, C, and K), minerals (Na, Ca, K, Mg, P, Se, Cu, Fe, Mn, Cr, Zn), enzymes and antioxidant [14] (Tables 2 and 3).

Table 2. Chemical elements found in pure honey.

Minerals	Amount (mg/100 g)	Vitamins	Amount (mg/100 g)
Sodium (Na)	1.6–17	Thiamine (B1)	0.00–0.01
Calcium (Ca)	3–31	Riboflavin (B2)	0.010.02
Potassium (K)	40–3500	Niacin (B3)	0.10–0.20
Magnesium (Mg)	0.7–13	Pantothenic acid (B5)	0.02–0.11
Phosphorus (P)	2–15	Pyridoxine (B6)	0.01–0.32
Selenium (Se)	0.002–0.01	Folic acid (B9)	0.002–0.01
Copper (Cu)	0.02–0.6	Ascorbic acid (C)	2.2–2.5
Iron (Fe)	0.03–4	Phyllochinon (K)	0.025
Manganese (Mn)	0.02–2		
Chromium (Cr)	0.01–0.3		
Zink (Zn)	0.05–2		

Adopted from Ajibola, Chamunorwa, and Erlwanger [34].

Honey is one of humankind's oldest food products that preserve human health and shields them from various diseases, such as cancer, a cold, sore throat, etc. [34,44]. There is a study that shows that fructose in pure honey tends to lower blood glucose in animal models of diabetes [18]. The mechanism of reducing blood glucose is as follows: it reduces the rate of intestinal absorption [45,46], extends the gastric emptying time [47], and reduces food intake [48]. Fructose stimulates glucokinase in hepatocytes, which play an important role in the uptake and storage of glucose as glycogen by the liver. Hence, glucose, which is present in combination with fructose in honey, enhances the absorption of fructose and promotes its hepatic actions through its enhanced delivery to the liver [49].

Table 3. Nutritional composition of pure honey.

Nutrition	Blossom Honey	Honeydew Honey
	Range	Range
Water	15–20	
Total sugars		
Monosaccharides		
Fructose	30–45	28–40
Glucose	24–40	19–32
Disaccharides		
Sucrose	0.1–4.8	0.1–4.7
Others	2.0–8.0	1.0–6.0
Trisaccharides		
Erlose	0.5–6.0	0.1–6.0
Melezitose	NA	0.3–22
Others	0.5–1.0	0.1–6.0
Minerals	0.1–0.5	0.6–2.0
Amino acids, proteins	0.2–0.4	0.4–0.7
Acids	0.2–0.8	0.8–1.5
pH value	3.2–4.5	4.5–6.5

Data in g/100 g of honey, adopted from [48].

Consumption of honey as a food or a natural sweetener can boost the human body's energy level. Honey has high antioxidant activity and antimicrobial properties and can be used for healing wounds, treat obesity, diabetes, and cancers [34].

Several studies have revealed that honey has biological properties and health benefits on the human body, as discussed here. An in-depth study by [7] stated that natural honey has a significant effect on reducing cholesterol (7%), triglycerides (2%), C-reactive protein (7%), and homocysteine (6%).

The therapeutic effect of honey on human health can be either through oral administration or topical application. In this regard, Al-Waili et al. [50] revealed the therapeutic properties of oral administration of honey for the treatment of laryngitis, osteoporosis, gastrointestinal ulcers, anorexia, insomnia and constipation, and liver, cardiovascular and gastrointestinal problems. On the other hand, advantages of the topical application of honey are prescribed for eczema, lip sores, sterile and infected wounds, genital lesions, burns, surgery scars, and athlete's foot [9].

The therapeutic effect of honey can be tested on animals and then extrapolated to humans. Therefore, according to an in vivo study done by [10] on rats early mortality and histopathology result, several health modifications have been observed after the consumption of commercially adulterated honey: weight gain, an abnormal parameter in renal and hepatic function, a rise in the level of circulating triglycerides, cholesterol and glucose levels, fat deposition augmentation and severe organ toxicities. Moreover, this research is in agreement with the study done by Samat et al. [15] on rats, which showed that adulterated honey consumption may be harmful and cause liver and kidney dysfunction.

2. Honey Adulterants

Low-cost sugars and commercial syrups are common substances for honey adulteration. Ismail and Ismail [51] described well-known adulterants from sugar cane and sugar beet such as corn syrup (CS), HFCS, glucose syrup (GS), sucrose syrup (SS), inverted syrup (IS), and high fructose inulin syrup (HFIS). Adulteration of honey by sugars alters the chemical and biochemical properties of honey, such as the enzymatic activity, electrical conductivity, and specific compounds contents [13].

Honey adulterates are selected based on the following three factors: the specific region of origin, the economic benefits and the sugars or sweeteners accessibility. A well-known example is employing rice and wheat syrup extraction as adulterants in Turkey and France [13]. The plant syrups can be harvested from heating vegetable juices or partial enzymatic hydrolysis [31,43,52,53]. According to [54,55], European countries adulterate the honey with HFIS. In this section, common sweeteners that are used in commercial honey adulteration will be introduced and their health impact discussed. The chemical structures of these widely-used sugar adulterants are presented in Figure 2.

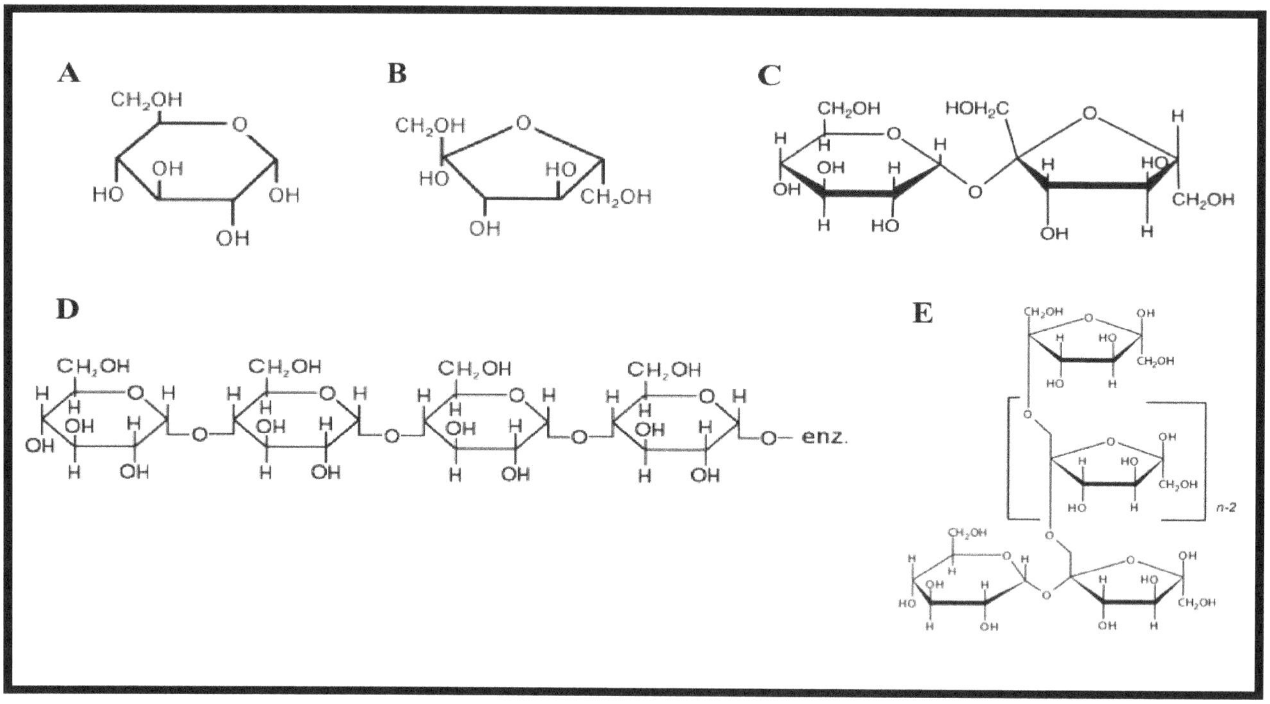

Figure 2. Chemical structures of widely used sugar adulterants in honey. (**A**) Glucose, (**B**) Fructose, (**C**) Sucrose, (**D**) Rice syrup, and (**E**) Inulin syrup.

2.1. *Cane Sugar*

Cane sugar is sucrose consisting of two sugar molecules monosaccharides (glucose and fructose). Glucose and fructose are monosaccharides with an identical chemical formula ($C_6H_{12}O_6$) but a different chemical conformation, linked together by a weak glycosidic bond to produce sucrose ($C_{12}H_{22}O_{11}$), a disaccharide. Generally, cane sugar is obtained by extracting juice from the sugar cane, a perennial C4 grass, followed by purification by chemical and physical means, evaporation to remove the water and separation of the sugar crystal [56]. Cane sugar originates from plants with a C4 metabolic pathway (Hatch-Slack cycle), while nectar originates from the C3 metabolic pathway (Calvin cycle).

In another study [57], cane sugar was used in both direct and indirect adulteration of honey. During direct adulteration, 10, 20, and 40% of syrup were added to the honey sample; for the indirect method, the bees were fed with syrup.

Concerning the toxicity of sugar, the lethal dose (LD_{50}) is a useful tool to measure the short-term toxicity and causes the death of 50% of a test animal population. Hence, the acute oral LD_{50} of cane sugar in rats is 29,700 mg/kg BW, almost 30 g/kg, which placed this sugar into a particularly safe scale [30].

2.2. *Corn Syrup*

Corn syrup, or high fructose corn syrup (HFCS), is a viscous, odorless, and colorless liquid that is much denser than water. Corn syrup is a liquid sweetener derived by cornstarch hydrolysis, which is used as a sweetener in foods [58]. Based on its fructose content, corn syrup is classified as: HFCS-42 (42% fructose), HFCS-55 (55% fructose), HFCS-90 (90% fructose) [59]. The fructose from high fructose corn syrup cannot be used directly to generate energy and has to be stored in the liver as fat or glycogen. Hence, the extreme amount of fructose from HFCS cannot be processed beneficially in the body [60]. The LD_{50} of rare sugar syrup, which is obtained from HFCSm is 15,000 mg/kg BW for rats with no abnormalities; in humans, the acute non-effect level, which caused diarrhea, was estimated as 0.9 g/Kg BW as a dry solid base [61].

2.3. Palm Sugar

Palm sugar is extracted from the flower buds of the palm. It is a natural sweetener undergoing minimum steps during the chemical-free procedure. One study [62] reported that the major carbohydrates in palm sugar were sucrose, followed by glucose and fructose. The significant advantage of palm sugar is the lack of a blood sugar spiking effect, owing to its low glycemic index (~35). The most popular honey adulterant in India is jaggery syrup, which is prepared by the evaporation of palm tree extraction the evaporation of the sap of palm trees [63]. While sucrose and glucose are the main sugar components of palm sugar, the LD_{50} for sucrose is 29,700 mg/kg BW and for glucose 25,800 mg/kg BW for a rat [64].

2.4. Invert Sugar

Invert sugar (IS) is produced by cleavage of the sucrose into its monosaccharides building blocks, fructose, and dextrose. The inversion procedure is usually performed by heating the sucrose syrup in the presence of acids, alkali, or invertase [58]. The sugar content of IS originates from beet and cane plants, mimicking the pure honey sugar profile [65]. Invert sugar has been widely used in beverages and food industries for making non-crystallized cream, jams, artificial honey, and liquid sugar [16].

Inverted beet syrup is one of the most well-known adulterants, which can be tailored to mimic the natural sucrose (glucose-fructose) profile of honey and, as beet is a C3 plant, it is usually difficult to detect. In one study, various quantities of inverted beet syrup were added to the pure honey samples of clover, orange, and buckwheat [66]. Invert sugar is a generally accepted safe substance and it does not present toxic effects. It is recommended to take precautions in patients that present diabetes mellitus and also in patients with the rare hereditary problems of fructose intolerance, glucose-galactose malabsorption, or sucrase-isomaltase insufficiency [67].

As inverted sugar is a mixture of fructose and glucose, the LD_{50} value for this sugar is adopted from its structural sugars; the LD_{50} for fructose and glucose are 25,800 mg/kg BW and 29,700 mg/kg BW, respectively [68].

2.5. Rice Syrup

Rice syrup (RS), a product of rice polysaccharide hydrolysis, originating from a C3 plant (similar to beet syrup), is one of the most popular honey adulterants in China [58].

Rice syrup contains three sugars: maltotriose (52%), maltose (45%), and glucose (3%). Since maltose is two molecules of glucose and maltotriose is three molecules of glucose, rice syrup acts like 100% glucose inside the body. Honey adulterated with RS has recently emerged on the honey market. Rice syrup is a C3 syrup adulterant that follows a similar Calvin cycle of photosynthesis as natural honey [28]. Thus, rice syrup as a honey adulterant is a critical issue that affects quality assurance and food safety [69].

The unethical substitute brown rice syrup with HFCS in some organic foods has raised researchers' concerns due to its high arsenic content [70]. Baby formulas that contain organic brown rice syrup (OBRS) have an increased arsenic level, above the drinking water standard as per the research led by [71], and there is no regulation to govern this particular scenario. Rice syrup acts as glucose inside the body; the LD_{50} of rice syrup is the same as glucose, which is 25,800 mg/kg BW [72].

2.6. Inulin Syrup

Inulin is naturally occurring polysaccharide, belonging to a class of fructans. These nutritional fibers are a chain of fructose residues linked to glucose at the end of the chain. The linkage arrangements of the fructose molecules determine the fructan type. For example, in the case of inulin, the chain of β2-1 linked fructose has been terminated by glucose. The common source of this polysaccharide are wheat, onion, bananas, garlic, asparagus, sunchoke, and chicory [58]. Furthermore, [70] added different proportions (5, 10, and 20%, *w/w*) of high fructose inulin syrup to a nectar honey sample to intentionally simulate honey adulteration.

Classical toxicology tests are difficult to apply to inulin, which is a micro ingredient. Although some high dose animal tests have been performed, none have revealed any toxic effects [56]. Thus, the LD_{50} values for fructose, glucose and sucrose are 25,800 mg/kg BW, 29,700 mg/kg BW and 29,700 mg/kg BW, respectively [73].

3. Adulteration Method

Commercial honey adulteration is typically classified as direct, indirect, and blending [58], as shown in Figure 2. As discussed earlier, the authors in one paper [60] categorized the honey adulteration into two types; direct and indirect. The direct addition of sugar syrups is a post-production procedure of adding certain ratios to increase honey sweetness [13]. Meanwhile, indirect adulteration occurs by overfeeding the bees during the main nectar period with honey, chemicals, and industrial sugars to recover more honey from hives [28]. Blending is another honey adulteration procedure, which can be explained as mixing pure and high-quality honey with cheap and low-quality honey [19,31]. However, the syrup or sugar residues of some reported studies are identical to the natural residues in the honey. Therefore, the detection of these adulterants have proved difficult, and scientists have to discover new methods to distinguish the differences between pure and adulterated honey [32].

3.1. Direct Adulteration

Direct adulteration of honey is commonly performed by the direct addition of a certain amount of sucrose syrup into the honey. The source of sucrose syrup could be sugar beet, HFCS, maltose syrup, or industrial sugar syrups (glucose and fructose) obtained from heat, enzyme, or acid treatment of starch [74]. Direct adulteration causes harm to consumers and pure honey producers [28].

According to a study done by [31], to make direct adulterated honey, pure honey was mixed with different concentrations (7%, 15%, and 30%) of date and inverted sugar syrups. The authenticity of the samples was estimated by applying a multivariate analysis (PCA and LDA) into the physicochemical and rheological analysis. Amiry et al. [75], on the other hand, performed direct adulteration of honey with synthesized sugar. Adulteration was done by mixing glucose powder with distilled water and then adding it to the pure honey in laboratory design experiments. The honey adulteration was detected by an optical microfiber sensor in this research.

Moreover, Irawati et al. [76] added the two rice syrups directly with a ratio of 1.2:1 to the pure honey to provide a mixture with a viscosity close to the honey. Furthermore, three-dimensional fluorescence spectra (3DFS) and multivariate calibrations were employed as a detection method for honey authenticity.

3.2. Indirect Adulteration

Indirect adulteration of honey is the incorporation of sugars into honey via bee-feeding [77]. In this manner, low-quality honey, chemicals, and industrial sugars were incorporated into the honey during a natural process that happened in the bee's digestive system [78]. During indirect honey adulteration, an extreme amount of sugar syrup was fed to the bee colonies in the main nectar flow period [31].

One paper [79] described the indirect adulteration as follows: the colonies were settled in empty beehives with bees, brood, and honey frames. Standard bee-feeding methods were applied in the early spring [77]. The sucrose syrup (1:1.5 *w/w*, water: sugar) as adulterants were presented to each colony to ensure the growth and strength of the forage worker bee population of colonies for the main nectar flow season. The honey and honeycomb frames of all colonies were taken at the end of the nectar flow season [80]. After settling the bees in the hives, no more syrup was provided to the colonies.

Furthermore, other researchers [81] developed a method analyzing the biochemical properties to distinguish indirect sucrose syrup (SS) adulterated honey from pure blossom honey. According to the result of this study, adulteration of honey by SS has no significant effect on the content of honey, since over 95% of the fed SS to the bees was converted to fructose and glucose. During the initial step of the bees' SS feeding, the sucrose content was hydrolyzed into its building blocks. However, after extending the intensive feeding duration, the bees start to convert and store the glucose into maltose and maltotriose oligosaccharides [82].

The amount of sucrose present in pure honey owes little to the activity of invertase enzymes present in pure honey. This is the enzyme in charge of breaking down the sucrose into its building blocks and reducing the sucrose content of pure honey. Hence, it can be concluded that high sucrose levels in a honey sample might be due to the adulteration of honey [83].

According to the experiment done by [18] on indirect adulteration of honey, prolonged SS feeding will have identical results as with artificial direct adulteration. Furthermore, the extended prolonged SS bee-feeding reduces the nutritional properties and quality of the honey in the same manner of direct honey adulteration [30].

3.3. Blending

In this method, high quality (pure and rare) honey is mixed with cheaper honey that has lower quality and nutrition. Adulteration of pure honey with synthetic honey has become much more prevalent in recent years [30].

In China and Venezuela, a well-known fraud in the honey industry consists of blending the costly acacia honey with rape honey (cheaper honey) to increase the marketing profit since the light amber color of rape honey is highly similar to that of the yellow-colored acacia honey [76]. Acacia honey originates from *Robinia pseudacacia* blossoms and is commonly consumed nectar honey because of its mild aromatic and clear aspect and light yellowish color, and because it does not crystallize [32]. In contrast, rape honey is sweeter, with a light amber color, and easily crystallizes [84]. Hence, as the collection of *M. favosa* honey has steadily increased since 1985, it is the perfect target for blending with the cheaper honey of *A. mellifera* [32]. In this study, acacia honey was blended with different concentrations (5–50%, *w*/*w*) of rape honey, and the simple method of liquid chromatography-electrochemical detection (LC-ECD) was used to detect the adulteration.

Since the honey adulterants and adulteration methods have been discussed earlier, detection methods to investigate the authenticity of honey become the priority in the food safety of honey. These detections methods must be capable of distinguishing between pure and adulterated honey to assist the authorities in regulating the market and protect the consumer's right.

4. Detection Method in Honey Adulteration

The addition of sugar to honey is the common fraud in honey adulteration. Sugar addition always has been coupled with thermal treatment to produce a homogenous mixture to sell it as pure honey to the consumers. During applying heat to the honey, HMF can be produced as an intermediate compound via the Maillard reaction. So, the amount of HMF is an indicator of honey adulteration. There is a possibility of HMF formation at low temperature and can be elevated by the increase of applied temperature or storage [85]. High-performance liquid chromatography (HPLC), gas chromatography with mass spectrometry (GC-MS), micellar electrokinetic capillary chromatography (MEKC), and voltammetry are among the methods that applied for the determination of HMF in adulterated honey [86]. A physiochemical analysis is another application to differentiate pure honey from adulterated honey. Honey adulterated with fructose and saccharide showed a lighter color (L values), while pure honey shows darker red (a values) and yellow (b values) color. Moreover, there is a decreasing trend in pH values and an increasing trend in water activity (a_w) with the addition of fructose and saccharides to the pure honey [87].

In addition to HMF, diastase enzymatic activity can also be used for detecting sugar adulteration [88]. Enzymatic activity (diastase and invertase) is one of the simplest and most applicable analytical methods, though it is not always conclusive and must be paired with other analytical methods to generate a reliable result. As a comparison between these two enzymatic activities, invertase is a better marker for the quality control of honey, since it degraded much faster during the heating up procedure of honey, compared to amylase [14]. Moreover, Differential Scanning Calorimeter (DSC) can be used to determine the thermal properties of honey influenced by the direct addition of sugar syrup adulterants [89]. Furthermore, the conventional polymerase chain reaction (PCR) can detect several genes of rice molasses, which has been used in the direct adulteration of honey. Furthermore, the standard curve attained from real-time PCR can quantify the rice molasses' DNA amount and calculate the exact level of adulterants [90].

Several detection techniques have been developed in recent studies with the aim of detection of direct honey adulterated with sugar, such as high-performance anion-exchange chromatography with pulsed amperometric detection (HPAEC-PAD) for the detection of CS and HFCS, HPLC, and stable carbon isotope ratio analysis (SCIRA) for the detection of HFS [91], GC-MS for the detection of HFCS [32], cavity ring-down spectroscopy (CRDS) and isotope ratio mass spectrometry (IRMS) for the detection of CS [92], three-dimensional fluorescence spectroscopy (3DFS) for the detection of RS [93], Raman spectroscopy for the detection of HFCS [77], matrix-assisted laser desorption/ionization mass spectrometry (MALDI-MS) for the detection of IS and CS [94] and headspace-gas chromatography coupled to ion mobility spectrometry (HS-GC-IMS) for the detection of CS [95].

However, ultrahigh-performance liquid chromatography coupled with quadrupole time-of-flight mass spectrometry (UHPLC/Q-TOF-MS) was applied for indirect adulteration of honey, through bee feeding with multi-class sugar syrups, such as HFCS, RS, and IS [96]. Hence, the liquid chromatography-electrochemical detection (LC-ECD) [69] and laser-induced breakdown spectroscopy (LIBS) [97] methods were used for the detection of acacia honey blended with rape honey.

The chemometric analysis is widely applied to the detection method as a helpful tool for the reduction of samples' complexity and classifies them into groups based on their similarities. The most commonly used chemometrics techniques are the principal component analysis (PCA) [98], orthogonal projection to latent structure discriminant analysis (OPLS-DA) [97], partial least square-linear discrimination analysis (PLS-LDA) [96], and partial least squares regression (PLSR) [94]. Table 4 represents the most recent detection methods for the identification of honey adulteration based on sugar adulterants.

Table 4. Summary of analytical approaches of articles related to the honey adulteration with sugar.

Adulterants	Targeted	Non-Targeted	Reference
Rice molasses	Conventional and real-time PCR	-	[98]
Sugar syrup	DSC	PCA	[91]
HFS	HPLC; SCIRA	-	[90]
CS and HFCS	HPAEC-PAD	-	[32]
RS	3DFS	PCA	[17]
CS	HS-GC-IMS	OPLS-DA	[77]
HFIS	GC-MS	-	[96]
HFCS	RAMAN spectroscopy	PLS-LDA	[92]
IS and CS	MALDI/MS	-	[94]
CS	CRDS; IRMS	-	[95]
RS, HFCS, CS, IS	UHPLC/Q-TOF-MS	-	[93]
Rape honey	LC-ECD	PCA	[69]
Rape honey and HFCS	LIBS	PLS	[97]

5. Adverse Health Impact of Honey Adulteration

The adverse health impacts of consuming adulterated honey on human health are not completely established yet due to an absence of systematic and scientific studies and lack of public awareness. Pure honey showed significantly lower toxicity due to containing simple sugar (glucose and fructose) and other essential nutrients such as proteins, antioxidants, and minerals [98]. While honey has an antibacterial effect, helping to fight common cold and some digestive problems, the mixture of inverted sugar or jaggery can sometimes restrict the antibacterial properties of honey and lead to stomach disorders [78]. Adulteration harms consumers' health, which may cause increased blood sugar followed by the release of the insulin hormone and type II diabetes, abdominal weight gain, and obesity, a raise in the level of blood lipid, and high blood pressure [99].

Consumption of glucose from sugar-adulterated honey may elevate insulin secretion. Insulin activates the plasma membrane enzyme system with the properties of NADPH-oxidase resulting in not only the production of H_2O_2 and fructose but also increases uric acid in humans and rodents [13]. Uric acid generation in the body has unfavorable effects, namely, the inability to scavenge lipophilic radicals and breaking the radical chain propagation within the lipid membrane [100]. Under the same circumstances, glucose and fructose of sugar produce ROS through various mechanisms in the body, which is detrimental toward human health and causes chronic diseases such as atherosclerosis, diabetes, obesity, hypertension, coronary artery diseases, and finally heart failure [101]. Glucose that is either naturally produced in the body or added to foodstuff can be detrimental to humans.

Furthermore, the authors of one paper [102] investigated the short-term (two weeks) and long-term (16 weeks) effects of two different brands of honey consumption in the Malaysian market using male Sprague Dawley rats. The finding was illustrated that a total of five rats from both adulterated honey groups showed early mortality and many abnormal signs developed compared to rats fed with natural honey (pineapple honey) as a control. The abnormalities in the adulterated honey group represent significant body weight, fat pads, and BMI, and serum lipid profile (triglycerides, cholesterol, and glucose level) drastically increases. Since glucose and fructose can instantly convert to energy inside the cells, they represent a lower glycemic index compared to sucrose [15]. These results could indicate that long-term consumption of adulterated honey has a harmful impact on human health, requiring local and international authorities' actions to control and regulate this matter.

The kidney serology and toxicology study of rats that consumed adulterated honey for 16 weeks showed kidney damage due to losing their capability to expel creatinine and urea from the serum [50]. In another in vivo study [15], it was proved that a long-term high sucrose diet may lead to the increment of both urea and creatinine. Meanwhile, Li et al. [103] stated that prolonged consumption of HFCS would lead to rat's glomerular filtration failure.

According to a recent study by Arise and Malomo [104] on rats, hypercholesterolemia, hypertriglyceridemia, and hyperinsulinemia are caused by prolonged feeding of any sugar syrup such as sucrose or fructose, which leads to animal death. Furthermore, Ajibola et al. [34] reported the synthesis of triacylglycerides in the liver as a result of increasing the gene expressions of lipogenesis enzymes such as acetyl CoA carboxylase and fatty acid synthase.

The morphological alteration in response to diseases and toxic compounds are often critical to our understanding of normal and abnormal function. Histological studies facilitate a new chapter of science into the architecture of tissue, which is biologically and clinically valuable. Larson-Meyer and Willis [105] observed a pyknotic nucleus with a reduced diameter in the liver of the tilapia fish—*Oreochromis niloticus*—exposed to different dilutions of sugar cane vinasse (1%, 2%, 5%, and 10%). These results could explain the effect of the consumption of the high amount of fructose in adulterated honey, which leads to a rise in the fasting serum uric acid level in the liver, followed by increased insulin resistance (caused diabetes), NAFLD score, lobular inflammation and steatosis grades [106]. On the other hand, fructose intake stimulates de novo lipogenesis (DNL) in the liver, which prevents fatty acid oxidation and leads to accumulation of fatty acid in the liver [107].

In other research [107], rats fed with sugar adulterated honey exhibited a pale reddish color, exceptional kidney size, and observed nodules outside the kidneys. There was also an abnormal difference in the appearance of the liver, with slight discoloration to brown, particularly in the middle of this organ's surface. In detail, some of the livers showed a roughly brown surface and their sizes were smaller, with whitish micronodules on the entire liver surface. Regardless of weight differentiation, the relative weight of the kidneys and lungs from rats fed with adulterated honey showed a significant increase compared to control rats. Moreover, the heart and brain from the rats fed with adulterated honey exhibited significant decreases [15]. These types of intensive investigations may be harmful toward the human body, and authorities should avoid it in the future.

Natural honey possesses some beneficial effects, such as lower total cholesterol and LDL in healthy overweight human subjects, while consumption of 3 to 20% of dietary fructose caused the elevation of total cholesterol and LDL by 9% and 11%, respectively [15]. In this regard, the addition of sugar to honey could be critical toward human health.

Although investigating the direct effect of adulterated honey on human health is an impossible process, it is a crucial procedure and needs ethical approval [108]. To our knowledge, no studies investigated the toxic effect of adulterated honey on humans. However, more animal studies need to be done on the toxic impact of adulterated honey in order to correlate the results to humans.

Table 5 represents the recent studies that investigated the in vivo effect of various adulterants on internal organs. Several sugars were tested on the internal organs in Sparague Dawley rats, such as commercial honey, sucrose, high fructose corn syrup, sugar syrup, and fructose; these caused liver and kidney damage [102], elevated blood urea and creatinine [15], kidney failure [103], elevated visceral fat pads and total body fat, which caused the animal death [104] and acute and chronic kidney injury leading to renal failure [34] and hyperglycemia [12], respectively. Moreover, sugar cane vinasse fed to *Oreochromis niloticus* fish caused liver damage [33]. Furthermore, feeding broiler chicken sugar cane extract resulted in a higher value of the intestinal villus height and area [106].

Table 5. Different in vivo studies of sugar adulterant impact on an internal organ.

Type of Sugar	Animal Type	Affected Organ	Reference
Inverted sugar	Human being	Stomach disorder	[109]
Commercial honey	Sparague Dawley rat	Increase body weight, serum lipid, liver, and kidney damage	[99]
Sucrose	Sparague Dawley rat	Increase urea and creatinine level	[15]
HFCS	Sparague Dawley rat	Kidney failure	[103]
Sugar syrup	Sparague Dawley rat	Hypercholesterolemia, hypertriglyceridemia, and hyperinsulinemia	[104]
Sugar cane vinasse	Oreochromis niloticus (tilapia fish)	Liver damage	[34]
Fructose	Rat	Renal failure	[106]
Sugar cane extract	Broiler chicken	Hypertrophied intestinal villi and epithelial cells	[12]
High fructose	Sparague Dawley rat	Hypertriglyceridemia	[109]

6. Conclusions

The significant impact of honey adulteration on market loss, reducing the quality of honey, shows the importance of studies to investigate different honey adulterants, adulteration methods, and detection methods. Moreover, this fraud has an adverse impact on the honey production industry and market by reducing the trust of consumers on this valuable product. According to this review, all six studied sugar adulterants (cane sugar, corn syrup, palm sugar, invert sugar, rice syrup, and inulin syrup) have health disadvantages toward human health based on their LD_{50} value and internal organ toxicology. The kidney and liver are the main organs that fail due to the consumption of sugar-adulterated honey. Diabetes, CKD, and AKI are the direct results of honey adulteration stated by in vivo histological examination. These diseases have a noticeable impact on human daily life and social health. Hence, the discussed detection methods can be used to identify honey adulteration via chemometric methods. While these techniques require specific laboratory knowledge and equipment, rapid and accurate detection methods, with ease of public accessibility to save customer's expenses and time, and guarantee their safety, are absent. As has been discussed in this review, more prompt actions must be taken by authorities to prevent the production, trading, and marketing of adulterated honey and discover the other harmful honey adulterants available on the market. The overall result demonstrates that honey adulteration is a threat to food safety, food security, and ecological sustainability of this important and valuable product.

Author Contributions: Conceptualization, J.S., R.F., A.A.B.; Data curation, R.F., A.A.B.; Formal analysis, R.F., A.K.; Funding acquisition, J.S.; Investigation, R.F., R.S.; Methodology, R.F., A.F.A.R., S.A.; Project administration, J.S.; Resources, J.S., A.K., A.F.A.R., R.S., S.A.; Software, A.K., R.F., S.A.; Supervision, J.S.; Validation, J.S.; Visualization, R.F., A.A.B.; Writing—original draft preparation, R.F., A.A.B.; Writing—review and editing, J.S., R.F., A.A.B.; All authors have read and agreed to the published version of the manuscript.

Acknowledgments: The authors also would like to thank the Ministry of Education Malaysia (MOE) for the HICoE rendered to the Institute of Tropical Agriculture and Food Security (ITAFoS), University Putra Malaysia (UPM).

References

1. Food and Agriculture Organization of the United Nations (FAO); World Health Organization (WHO). General Standard for Food Additives, CODEX STAN 192-1995. In *Codex Alimentarius Commission*; FAO: Rome, Italy; WHO: Geneva, Switzerland, 2018; Available online: http://www.fao.org/gsfaonline/docs/CXS_192e.pdf (accessed on 1 August 2020).
2. Akhmazillah, M.; Farid, M.; Silva, F. High-pressure processing (HPP) of honey for the improvement of nutritional value. *Innov. Food Sci. Emerg. Technol.* **2013**, *20*, 59–63. [CrossRef]
3. Saiful Yazan, L.; Zali, M.; Shyfiq, M.F.; Mohd Ali, R.; Zainal, N.A.; Esa, N.; Sapuan, S.; Ong, Y.S.; Tor, Y.S.; Gopalsamy, B.; et al. Chemopreventive properties and toxicity of Kelulut honey in Sprague Dawley rats induced with Azoxymethane. *BioMed Res. Int.* **2016**, *2016*. [CrossRef] [PubMed]
4. Kassim, M.; Yusoff, K.M.; Ong, G.; Sekaran, S.; Yusof, M.Y.B.M.; Mansor, M. Gelam honey inhibits lipopolysaccharide-induced endotoxemia in rats through the induction of heme oxygenase-1 and the inhibition of cytokines, nitric oxide, and high-mobility group protein B1. *Fitoterapia* **2012**, *83*, 1054–1059. [CrossRef] [PubMed]
5. Rao, P.V.; Krishnan, K.T.; Salleh, N.; Gan, S.H. Biological and therapeutic effects of honey produced by honey bees and stingless bees: A comparative review. *Rev. Bras. Farmacogn.* **2016**, *26*, 657–664. [CrossRef]
6. Fauzi, A.N.; Norazmi, M.N.; Yaacob, N.S. Tualang honey induces apoptosis and disrupts the mitochondrial membrane potential of human breast and cervical cancer cell lines. *Food Chem. Toxicol.* **2011**, *49*, 871–878. [CrossRef]
7. Samarghandian, S.; Farkhondeh, T.; Samini, F. Honey and health: A review of recent clinical research. *Pharmacogn. Res.* **2017**, *9*, 121.
8. Ghashm, A.A.; Othman, N.H.; Khattak, M.N.; Ismail, N.M.; Saini, R. Antiproliferative effect of Tualang honey on oral squamous cell carcinoma and osteosarcoma cell lines. *BMC Complement. Altern. Med.* **2010**, *10*, 49. [CrossRef]
9. Mahmoudi, R.; Ghojoghi, A.; Ghajarbeygi, P. Honey safety hazards and public health. *J. Chem. Health Risks* **2016**, *6*, 249–267.
10. Mijanur Rahman, M.; Gan, S.H.; Khalil, M. Neurological effects of honey: Current and future prospects. *Evid. Based Complement. Altern. Med.* **2014**. [CrossRef]
11. Fuhrman, J. The hidden dangers of fast and processed food. *Am. J. Lifestyle Med.* **2018**, *12*, 375–381. [CrossRef]
12. Johnson, R.J.; Fuggle, S.V.; Mumford, L.; Bradley, J.A.; Forsythe, J.L.; Rudge, C.J.; Kidney Advisory Group of NHS Blood and Transplant. A New UK 2006 National Kidney Allocation Scheme for deceased heart-beating donor kidneys. *Transplantation* **2010**, *89*, 387–394. [CrossRef] [PubMed]
13. Soares, S.; Amaral, J.S.; Oliveira, M.B.P.; Mafra, I. A comprehensive review on the main honey authentication issues: Production and origin. *Compr. Rev. Food Sci. Food Saf.* **2017**, *16*, 1072–1100. [CrossRef]

14. Food and Agriculture Organization of the United Nations (FAO); World Health Organization (WHO). Joint FAO/WHO Food Standard Programme Codex Alimentarius Commission Twenty-Fourth Session Geneva, 2–7 July 2001. In *Codex Alimentarius Commission*; ALINORM 01/34A; FAO: Rome, Italy; WHO: Geneva, Switzerland, 2001.
15. Samat, S.; Kanyan Enchang, F.; Nor Hussein, F.; Wan Ismail, W.I. Four-week consumption of Malaysian honey reduces excess weight gain and improves obesity-related parameters in high fat diet induced obese rats. *Evid. Based Complement. Altern. Med* **2017**, *2017*. [CrossRef] [PubMed]
16. Se, K.W.; Ghoshal, S.K.; Wahab, R.A.; Ibrahim, R.K.R.; Lani, M.N. A simple approach for rapid detection and quantification of adulterants in stingless bees (*Heterotrigona itama*) honey. *Food Res. Int.* **2018**, *105*, 453–460. [CrossRef]
17. Morales, V.; Corzo, N.; Sanz, M. HPAEC-PAD oligosaccharide analysis to detect adulterations of honey with sugar syrups. *Food Chem.* **2008**, *107*, 922–928. [CrossRef]
18. Damto, T. A Review on Effect of Adulteration on Honey Properties. *SSRN* **2019**. [CrossRef]
19. Jaafar, M.; Othman, M.; Yaacob, M.; Talip, B.; Ilyas, M.; Ngajikin, N.; Fauzi, N. A Review on Honey Adulteration and the Available Detection Approaches. *Int. J. Integr. Eng.* **2020**, *12*, 125–131.
20. Lawal, R.; Lawal, A.; Adekalu, J. Physico-chemical studies on adulteration of honey in Nigeria. *Pak. J. Biol. Sci. PJBS* **2009**, *12*, 1080–1084. [CrossRef]
21. De Almeida-Muradian, L.B.; Stramm, K.M.; Horita, A.; Barth, O.M.; da Silva de Freitas, A.; Estevinho, L.M. Comparative study of the physicochemical and palynological characteristics of honey from *Melipona subnitida* and *Apis mellifera*. *Int. J. Food Sci. Technol.* **2013**, *48*, 1698–1706. [CrossRef]
22. Abdel-Aal, E.M.; Ziena, H.; Youssef, M. Adulteration of honey with high-fructose corn syrup: Detection by different methods. *Food Chem.* **1993**, *48*, 209–212. [CrossRef]
23. Nisbet, C.; Kazak, F.; Ardalı, Y. Determination of quality criteria that allow differentiation between honey adulterated with sugar and pure honey. *Biol. Trace Elem. Res.* **2018**, *186*, 288–293. [CrossRef] [PubMed]
24. Da Silva, P.M.; Gauche, C.; Gonzaga, L.V.; Costa, A.C.O.; Fett, R. Honey: Chemical composition, stability and authenticity. *Food Chem.* **2016**, *196*, 309–323. [CrossRef]
25. Cotte, J.-F.; Casabianca, H.; Lhéritier, J.; Perrucchietti, C.; Sanglar, C.; Waton, H.; Grenier-Loustalot, M.-F. Study and validity of 13C stable carbon isotopic ratio analysis by mass spectrometry and 2H site-specific natural isotopic fractionation by nuclear magnetic resonance isotopic measurements to characterize and control the authenticity of honey. *Anal. Chim. Acta* **2007**, *582*, 125–136. [CrossRef] [PubMed]
26. Çinar, S.B.; Ekşi, A.; Coşkun, İ. Carbon isotope ratio (13C/12C) of pine honey and detection of HFCS adulteration. *Food Chem.* **2014**, *157*, 10–13. [CrossRef] [PubMed]
27. Förstel, H. The natural fingerprint of stable isotopes—Use of IRMS to test food authenticity. *Anal. Bioanal. Chem.* **2007**, *388*, 541–544. [CrossRef]
28. Wu, L.; Du, B.; Vander Heyden, Y.; Chen, L.; Zhao, L.; Wang, M.; Xue, X. Recent advancements in detecting sugar-based adulterants in honey—A challenge. *TrAC Trends Anal. Chem.* **2017**, *86*, 25–38. [CrossRef]
29. Gallardo-Velázquez, T.; Osorio-Revilla, G.; Zuñiga-de Loa, M.; Rivera-Espinoza, Y. Application of FTIR-HATR spectroscopy and multivariate analysis to the quantification of adulterants in Mexican honeys. *Food Res. Int.* **2009**, *42*, 313–318. [CrossRef]
30. Cordella, C.; Militao, J.S.; Clément, M.-C.; Drajnudel, P.; Cabrol-Bass, D. Detection and quantification of honey adulteration via direct incorporation of sugar syrups or bee-feeding: Preliminary study using high-performance anion exchange chromatography with pulsed amperometric detection (HPAEC-PAD) and chemometrics. *Anal. Chim. Acta* **2005**, *531*, 239–248. [CrossRef]
31. Guler, A.; Kocaokutgen, H.; Garipoglu, A.V.; Onder, H.; Ekinci, D.; Biyik, S. Detection of adulterated honey produced by honeybee (*Apis mellifera* L.) colonies fed with different levels of commercial industrial sugar (C3 and C4 plants) syrups by the carbon isotope ratio analysis. *Food Chem.* **2014**, *155*, 155–160. [CrossRef]
32. Wang, S.; Guo, Q.; Wang, L.; Lin, L.; Shi, H.; Cao, H.; Cao, B. Detection of honey adulteration with starch syrup by high performance liquid chromatography. *Food Chem.* **2015**, *172*, 669–674. [CrossRef]
33. Shapiro, A.; Mu, W.; Roncal, C.; Cheng, K.-Y.; Johnson, R.J.; Scarpace, P.J. Fructose-induced leptin resistance exacerbates weight gain in response to subsequent high-fat feeding. *Am. J. Physiol. Regulat. Integr. Compar. Physiol.* **2008**, *295*, R1370–R1375. [CrossRef] [PubMed]
34. Ajibola, A.; Chamunorwa, J.P.; Erlwanger, K.H. Dietary supplementation with natural honey promotes growth and health of male and female rats compared to cane syrup. *Sci. Res. Essays* **2013**, *8*, 543–553.

35. Alvarez-Suarez, J.M.; Gasparrini, M.; Forbes-Hernández, T.Y.; Mazzoni, L.; Giampieri, F. The composition and biological activity of honey: A focus on Manuka honey. *Foods* **2014**, *3*, 420–432. [CrossRef] [PubMed]
36. Ratiu, I.A.; Al-Suod, H.; Bukowska, M.; Ligor, M.; Buszewski, B. Correlation Study of Honey Regarding their Physicochemical Properties and Sugars and Cyclitols Content. *Molecules* **2020**, *25*, 34. [CrossRef] [PubMed]
37. Gupta, R.K. Taxonomy and distribution of different honeybee species. In *Beekeeping for Poverty Alleviation and Livelihood Security*; Springer: Dordrecht, The Netherlands, 2014; pp. 63–103.
38. Bakar, N.A.; Sata, N.S.A.M.; Ramlan, N.F.; Ibrahim, W.N.W.; Zulkifli, S.Z.; Abdullah, C.A.C.; Ahmad, S.; Amal, M.N.A. Evaluation of the neurotoxic effects of chronic embryonic exposure with inorganic mercury on motor and anxiety-like responses in zebrafish (*Danio rerio*) larvae. *Neurotoxicol. Teratol.* **2017**, *59*, 53–61. [CrossRef] [PubMed]
39. Machado De-Melo, A.A.; Almeida-Muradian, L.B.d.; Sancho, M.T.; Pascual-Maté, A. Composition and properties of *Apis mellifera* honey: A review. *J. Apic. Res.* **2018**, *57*, 5–37. [CrossRef]
40. García, N.L. The current situation on the international honey market. *Bee World* **2018**, *95*, 89–94. [CrossRef]
41. Aries, E.; Burton, J.; Carrasco, L.; De Rudder, O.; Maquet, A. *Scientific Support to the Implementation of a Coordinated Control Plan with a View to Establishing the Prevalence of Fraudulent Practices in the Marketing of Honey*; European Commission: Brussels, Belgium, 2016; pp. 1–38.
42. Bogdanov, S.; Lüllmann, C.; Martin, P.; von der Ohe, W.; Russmann, H.; Vorwohl, G.; Oddo, L.P.; Sabatini, A.-G.; Marcazzan, G.L.; Piro, R. Honey quality and international regulatory standards: Review by the International Honey Commission. *Bee World* **1999**, *80*, 61–69. [CrossRef]
43. Wei, G.-X.; Huang, J.-K.; Jun, Y. Honey safety standards and its impacts on China's honey export. *J. Integr. Agric.* **2012**, *11*, 684–693. [CrossRef]
44. Schievano, E.; Stocchero, M.; Morelato, E.; Facchin, C.; Mammi, S. An NMR-based metabolomic approach to identify the botanical origin of honey. *Metabolomics* **2012**, *8*, 679–690. [CrossRef]
45. Erejuwa, O.O.; Sulaiman, S.A.; Ab Wahab, M.S. Honey: A novel antioxidant. *Molecules* **2012**, *17*, 4400–4423. [CrossRef] [PubMed]
46. Kwon, S.; Kim, Y.J.; Kim, M.K. Effect of fructose or sucrose feeding with different levels on oral glucose tolerance test in normal and type 2 diabetic rats. *Nutr. Res. Pract.* **2008**, *2*, 252–258. [CrossRef] [PubMed]
47. Kellett, G.L.; Brot-Laroche, E.; Mace, O.J.; Leturque, A. Sugar absorption in the intestine: The role of GLUT2. *Annu. Rev. Nutr.* **2008**, *28*, 35–54. [CrossRef] [PubMed]
48. Bobiş, O.; Dezmirean, D.S.; Moise, A.R. Honey and diabetes: The importance of natural simple sugars in diet for preventing and treating different type of diabetes. *Oxid. Med. Cell. Longev.* **2018**, *2018*. [CrossRef]
49. Meireles, M.; Martel, F.; Araújo, J.; Santos-Buelga, C.; Gonzalez-Manzano, S.; Duenas, M.; de Freitas, V.; Mateus, N.; Calhau, C.; Faria, A. Characterization and modulation of glucose uptake in a human blood—Brain barrier model. *J. Membr. Biol.* **2013**, *246*, 669–677. [CrossRef]
50. Al-Waili, N.; Salom, K.; Al-Ghamdi, A.; Ansari, M.J. Antibiotic, pesticide, and microbial contaminants of honey: Human health hazards. *Sci. World J.* **2012**, *2012*. [CrossRef]
51. Ismail, M.M.; Ismail, W.I.W. Development of stingless beekeeping projects in Malaysia. In Proceedings of the CSSPO International Conference 2018: Towards Inclusive & Sustainable Agriculture—Harmonizing Environmental, Social and Economic Dimensions: Is it Possible? Sarawak, Malaysia, 9–11 July 2018; p. 5. [CrossRef]
52. Tosun, M. Detection of adulteration in honey samples added various sugar syrups with 13C/12C isotope ratio analysis method. *Food Chem.* **2013**, *138*, 1629–1632. [CrossRef]
53. Spiteri, M.; Dubin, E.; Cotton, J.; Poirel, M.; Corman, B.; Jamin, E.; Lees, M.; Rutledge, D. Data fusion between high resolution 1 H-NMR and mass spectrometry: A synergetic approach to honey botanical origin characterization. *Anal. Bioanal. Chem.* **2016**, *408*, 4389–4401. [CrossRef]
54. Corradini, C.; Cavazza, A.; Bignardi, C. High-performance anion-exchange chromatography coupled with pulsed electrochemical detection as a powerful tool to evaluate carbohydrates of food interest: Principles and applications. *Int. J. Carbohydr. Chem.* **2012**, *2012*, 1–13. [CrossRef]
55. Low, N.; Hammond, D. Detection of high fructose syrup from inulin in apple juice by capillary gas chromatography with flame ionization detection. *Fruit Process.* **1996**, *4*, 135–141.
56. Ruiz-Matute, A.I.; Soria, A.C.; Martínez-Castro, I.; Sanz, M.L. A new methodology based on GC-MS to detect honey adulteration with commercial syrups. *J. Agric. Food Chem.* **2007**, *55*, 7264–7269. [CrossRef] [PubMed]

57. Li, S.; Zhang, X.; Shan, Y.; Su, D.; Ma, Q.; Wen, R.; Li, J. Qualitative and quantitative detection of honey adulterated with high-fructose corn syrup and maltose syrup by using near-infrared spectroscopy. *Food Chem.* **2017**, *218*, 231–236. [CrossRef] [PubMed]
58. European Chemical Agency. *Evaluation of New Scientific Evidence Concerning the Restrictions Contained in Annex XVII to Regulation (EC) No. 1907/2006 (REACH): Review of New Available Information for Di-'isononyl'Phthalate (DIN)*; European Chemical Agency: Helsinki, Finland, 2010.
59. Olivares-Pérez, A.; Mejias-Brizuela, N.; Grande-Grande, A.; Fuentes-Tapia, I. Corn syrup holograms. *Optik* **2012**, *123*, 447–450. [CrossRef]
60. Zábrodská, B.; Vorlová, L. Adulteration of honey and available methods for detection—A review. *Acta Vet. Brno* **2015**, *83*, 85–102. [CrossRef]
61. Malik, V.S.; Popkin, B.M.; Bray, G.A.; Després, J.-P.; Willett, W.C.; Hu, F.B. Sugar-sweetened beverages and risk of metabolic syndrome and type 2 diabetes: A meta-analysis. *Diabetes Care* **2010**, *33*, 2477–2483. [CrossRef]
62. Yamada, T.; Iida, T.; Takamine, S.; Hayashi, N.; Okuma, K. Safety evaluation of rare sugar syrup: Single-dose oral toxicity in rats, reverse mutation assay, chromosome aberration assay, and acute non-effect level for diarrhea of a single dose in humans. *Shokuhin Eiseigaku Zasshi. J. Food Hyg. Soc. Jpn.* **2015**, *56*, 211–216. [CrossRef]
63. Luis, G.; Rubio, C.; Gutiérrez, A.; Hernández, C.; González-Weller, D.; Revert, C.; Castilla, A.; Abreu, P.; Hardisson, A. Palm tree syrup; nutritional composition of a natural edulcorant. *Nutr. Hosp.* **2012**, *27*, 548–552.
64. Mishra, S.; Kamboj, U.; Kaur, H.; Kapur, P. Detection of jaggery syrup in honey using near-infrared spectroscopy. *Int. J. Food Sci. Nutr.* **2010**, *61*, 306–315. [CrossRef]
65. Gehlawat, J. New Technology for Invert Sugar and High Fructose Syrups from Sugarcane. *Indian J. Chem. Technol.* **2001**, *8*, 28–32.
66. Veana, F.; Flores-Gallegos, A.C.; Gonzalez-Montemayor, A.M.; Michel-Michel, M.; Lopez-Lopez, L.; Aguilar-Zarate, P.; Ascacio-Valdés, J.A.; Rodríguez-Herrera, R. Invertase: An Enzyme with Importance in Confectionery Food Industry. In *Enzymes in Food Technology*; Springer: Singapore, 2018; pp. 187–212.
67. Paradkar, M.; Irudayaraj, J. Discrimination and classification of beet and cane inverts in honey by FT-Raman spectroscopy. *Food Chem.* **2002**, *76*, 231–239. [CrossRef]
68. White, J.S.; Foreyt, J.P.; Melanson, K.J.; Angelopoulos, T.J. High-fructose corn syrup: Controversies and common sense. *Am. J. Lifestyle Med.* **2010**, *4*, 515–520. [CrossRef]
69. Du, B.; Wu, L.; Xue, X.; Chen, L.; Li, Y.; Zhao, J.; Cao, W. Rapid screening of multiclass syrup adulterants in honey by ultrahigh-performance liquid chromatography/quadrupole time of flight mass spectrometry. *J. Agric. Food Chem.* **2015**, *63*, 6614–6623. [CrossRef]
70. Siddiqui, A.J.; Musharraf, S.G.; Choudhary, M.I. Application of analytical methods in authentication and adulteration of honey. *Food Chem.* **2017**, *217*, 687–698. [CrossRef]
71. Lai, T.N.H.; André, C.; Rogez, H.; Mignolet, E.; Nguyen, T.B.T.; Larondelle, Y. Nutritional composition and antioxidant properties of the sim fruit (*Rhodomyrtus tomentosa*). *Food Chem.* **2015**, *168*, 410–416. [CrossRef]
72. Jackson, B.P.; Taylor, V.F.; Karagas, M.R.; Punshon, T.; Cottingham, K.L. Arsenic, organic foods, and brown rice syrup. *Environ. Health Perspect.* **2012**, *120*, 623–626. [CrossRef]
73. Drabińska, N.; Jarocka-Cyrta, E.; Markiewicz, L.H.; Krupa-Kozak, U. The effect of oligofructose-enriched inulin on faecal bacterial counts and microbiota-associated characteristics in celiac disease children following a gluten-free diet: Results of a randomized, placebo-controlled trial. *Nutrients* **2018**, *10*, 201. [CrossRef]
74. Mehryar, L.; Esmaiili, M. Honey and honey adulteration detection: A review. In Proceedings of the 11th International Congress on Engineering and Food, Athens, Greece, 22–26 May 2011; pp. 1713–1714.
75. Amiry, S.; Esmaiili, M.; Alizadeh, M. Classification of honeys adulterated with date and invert syrups. *Food Chem.* **2017**, *1*, 390–397. [CrossRef]
76. Irawati, N.; Isa, N.M.; Mohamed, A.F.; Rahman, H.A.; Harun, S.W.; Ahmad, H. Optical microfiber sensing of adulterated honey. *IEEE Sens. J.* **2017**, *17*, 5510–5514. [CrossRef]
77. Chen, Q.; Qi, S.; Li, H.; Han, X.; Ouyang, Q.; Zhao, J. Determination of rice syrup adulterant concentration in honey using three-dimensional fluorescence spectra and multivariate calibrations. *Spectrochim. Acta Part. A Mol. Biomol. Spectrosc.* **2014**, *131*, 177–182. [CrossRef] [PubMed]
78. Bogdanov, S.; Jurendic, T.; Sieber, R.; Gallmann, P. Honey for nutrition and health: A review. *J. Am. Coll. Nutr.* **2008**, *27*, 677–689. [CrossRef] [PubMed]

79. Oddo, L.P.; Piana, L.; Bogdanov, S.; Bentabol, A.; Gotsiou, P.; Kerkvliet, J.; Martin, P.; Morlot, M.; Valbuena, A.O.; Ruoff, K. Botanical species giving unifloral honey in Europe. *Apidologie* **2004**, *35*, S82–S93. [CrossRef]
80. Sammataro, D.; Avitabile, A. *The Beekeeper's Handbook*; Cornell University Press: Ithaca, NY, USA, 1998.
81. Guler, A. The effects of the shook swarm technique on honey bee (*Apis mellifera* L.) colony productivity and honey quality. *J. Apic. Res.* **2008**, *47*, 27–34. [CrossRef]
82. Guler, A.; Bakan, A.; Nisbet, C.; Yavuz, O. Determination of important biochemical properties of honey to discriminate pure and adulterated honey with sucrose (*Saccharum officinarum* L.) syrup. *Food Chem.* **2007**, *105*, 1119–1125. [CrossRef]
83. Dumté, M.E.J. Development of a Method for the Quantitative Detection of Honey in Imported Products. University of Waikato. Ph.D. Thesis, University of Waikato, Hamilton, New Zealand, 2010.
84. Kenjerić, D.; Mandić, M.L.; Primorac, L.; Bubalo, D.; Perl, A. Flavonoid profile of *Robinia* honeys produced in Croatia. *Food Chem.* **2007**, *102*, 683–690. [CrossRef]
85. Vit, P.; Pedro, S.R.; Roubik, D. *Pot-Honey: A Legacy of Stingless Bees*; Springer: Berlin/Heidelberg, Germany, 2013.
86. Başar, B.; Özdemir, D. Determination of honey adulteration with beet sugar and corn syrup using infrared spectroscopy and genetic-algorithm-based multivariate calibration. *J. Sci. Food Agric.* **2018**, *98*, 5616–5624. [CrossRef]
87. Hošťálková, A.; Klingelhöfer, I.; Morlock, G.E. Comparison of an HPTLC method with the Reflectoquant assay for rapid determination of 5-hydroxymethylfurfural in honey. *Anal. Bioanal. Chem.* **2013**, *405*, 9207–9218. [CrossRef]
88. Yilmaz, M.T.; Tatlisu, N.B.; Toker, O.S.; Karaman, S.; Dertli, E.; Sagdic, O.; Arici, M. Steady, dynamic and creep rheological analysis as a novel approach to detect honey adulteration by fructose and saccharose syrups: Correlations with HPLC-RID results. *Food Res. Int.* **2014**, *64*, 634–646. [CrossRef]
89. Baglio, E. Overheating Indexes and Honey Quality. In *Chemistry and Technology of Honey Production*; Springer: Cham, Switzerland, 2018; pp. 23–40.
90. Sobrino-Gregorio, L.; Vargas, M.; Chiralt, A.; Escriche, I. Thermal properties of honey as affected by the addition of sugar syrup. *J. Food Eng.* **2017**, *213*, 69–75. [CrossRef]
91. Sobrino-Gregorio, L.; Vilanova, S.; Prohens, J.; Escriche, I. Detection of honey adulteration by conventional and real-time PCR. *Food Control.* **2019**, *95*, 57–62. [CrossRef]
92. Ruiz-Matute, A.I.; Rodríguez-Sánchez, S.; Sanz, M.L.; Martínez-Castro, I. Detection of adulterations of honey with high fructose syrups from inulin by GC analysis. *J. Food Compos. Anal.* **2010**, *23*, 273–276. [CrossRef]
93. Mantha, M.; Urban, J.R.; Mark, W.A.; Chernyshev, A.; Kubachka, K.M. Direct comparison of cavity ring down spectrometry and isotope ratio mass spectrometry for detection of sugar adulteration in honey samples. *J. AOAC Int.* **2018**, *101*, 1857–1863. [CrossRef]
94. Li, S.; Shan, Y.; Zhu, X.; Zhang, X.; Ling, G. Detection of honey adulteration by high fructose corn syrup and maltose syrup using Raman spectroscopy. *J. Food Compos. Anal.* **2012**, *28*, 69–74. [CrossRef]
95. Qu, L.; Jiang, Y.; Huang, X.; Cui, M.; Ning, F.; Liu, T.; Gao, Y.; Wu, D.; Nie, Z.; Luo, L. High-Throughput Monitoring of Multiclass Syrup Adulterants in honey based on the oligosaccharide and polysaccharide profiles by MALDI mass spectrometry. *J. Agric. Food Chem.* **2019**, *67*, 11256–11261. [CrossRef]
96. Arroyo-Manzanares, N.; García-Nicolás, M.; Castell, A.; Campillo, N.; Viñas, P.; López-García, I.; Hernández-Córdoba, M. Untargeted headspace gas chromatography–Ion mobility spectrometry analysis for detection of adulterated honey. *Talanta* **2019**, *205*, 120123. [CrossRef]
97. Wang, J.; Xue, X.; Du, X.; Cheng, N.; Chen, L.; Zhao, J.; Zheng, J.; Cao, W. Identification of acacia honey adulteration with rape honey using liquid chromatography–electrochemical detection and chemometrics. *Food Anal. Methods* **2014**, *7*, 2003–2012. [CrossRef]
98. Peng, J.; Xie, W.; Jiang, J.; Zhao, Z.; Zhou, F.; Liu, F. Fast Quantification of Honey Adulteration with Laser-Induced Breakdown Spectroscopy and Chemometric Methods. *Foods* **2020**, *9*, 341. [CrossRef]
99. Awasthi, S.; Jain, K.; Das, A.; Alam, R.; Surti, G.; Kishan, N. Analysis of food quality and food adulterants from different departmental & local grocery stores by qualitative analysis for food safety. *IOSR JESTFT* **2014**, *8*, 22–26.
100. Afroz, R.; Tanvir, E.; Paul, S.; Bhoumik, N.C.; Gan, S.H.; Khalil, M.I. DNA damage inhibition properties of sundarban honey and its phenolic composition. *J. Food Biochem.* **2016**, *40*, 436–445. [CrossRef]

101. Muraoka-Cook, R.; Shin, I.; Yi, J.; Easterly, E.; Barcellos-Hoff, M.; Yingling, J.; Zent, R.; Arteaga, C. Activated type I TGF β receptor kinase enhances the survival of mammary epithelial cells and accelerates tumor progression. *Oncogene* **2006**, *25*, 3408–3423. [CrossRef]
102. Samat, S.; Enchang, F.K.; Abd Razak, A.; Hussein, F.N.; Ismail, W.I.W. Adulterated honey consumption can induce obesity, increase blood glucose level and demonstrate toxicity effects. *Sains Malays.* **2018**, *47*, 353–365.
103. Li, L.; Zhao, Z.; Xia, J.; Xin, L.; Chen, Y.; Yang, S.; Li, K. A long-term high-fat/high-sucrose diet promotes kidney lipid deposition and causes apoptosis and glomerular hypertrophy in bama minipigs. *PLoS ONE* **2015**, *10*, e0142884. [CrossRef]
104. Arise, R.; Malomo, S. Effects of ivermectin and albendazole on some liver and kidney function indices in rats. *Afr. J. Biochem. Res.* **2009**, *3*, 190–197.
105. Larson-Meyer, D.E.; Willis, K.S. Vitamin D and athletes. *Curr. Sports Med. Rep.* **2010**, *9*, 220–226. [CrossRef] [PubMed]
106. Marinho, J.F.U.; Correia, J.E.; de Castro Marcato, A.C.; Pedro-Escher, J.; Fontanetti, C.S. Sugar cane vinasse in water bodies: Impact assessed by liver histopathology in tilapia. *Ecotoxicol. Environ. Saf.* **2014**, *110*, 239–245. [CrossRef]
107. Jensen, T.; Abdelmalek, M.F.; Sullivan, S.; Nadeau, K.J.; Green, M.; Roncal, C.; Nakagawa, T.; Kuwabara, M.; Sato, Y.; Kang, D.-H. Fructose and sugar: A major mediator of non-alcoholic fatty liver disease. *J. Hepatol.* **2018**, *68*, 1063–1075. [CrossRef]
108. Yaghoobi, N.; Al-Waili, N.; Ghayour-Mobarhan, M.; Parizadeh, S.; Abasalti, Z.; Yaghoobi, Z.; Yaghoobi, F.; Esmaeili, H.; Kazemi-Bajestani, S.; Aghasizadeh, R.; et al. Natural honey and cardiovascular risk factors; effects on blood glucose, cholesterol, triacylglycerole, CRP, and body weight compared with sucrose. *Sci. World J.* **2008**, *8*, 463–469. [CrossRef]
109. Khambualai, O.; Yamauchi, K.-E.; Ruttanavut, J.; Incharoen, T.; Kashimura, J. Effect of sugar cane extract, commercial probiotic and their mixture on growth performance and intestinal histology in broiler chickens. *Am. J. Anim. Vet. Sci.* **2010**, *5*, 132–138. [CrossRef]

2

Phytochemical Profile and Biological Properties of *Colchicum triphyllum* (Meadow Saffron)

Biancamaria Senizza [1], Gabriele Rocchetti [1,*], Murat Ali Okur [2], Gokhan Zengin [2], Evren Yildiztugay [3], Gunes Ak [2], Domenico Montesano [4,*] and Luigi Lucini [1]

1 Department for Sustainable Food Process, Università Cattolica del Sacro Cuore, Via Emilia Parmense 84, 29122 Piacenza, Italy; biancamaria.senizza@unicatt.it (B.S.); luigi.lucini@unicatt.it (L.L.)
2 Department of Biology, Science Faculty, Selcuk University, Campus, 4213 Konya, Turkey; murataliokur@hotmail.com (M.A.O.); gokhanzengin@selcuk.edu.tr (G.Z.); akguneselcuk@gmail.com (G.A.)
3 Department of Biotechnology Science Faculty, Selcuk University, Campus, 4213 Konya, Turkey; eytugay@gmail.com
4 Department of Pharmaceutical Sciences, Food Science and Nutrition Section, University of Perugia, Via S. Costanzo 1, 06126 Perugia, Italy
* Correspondence: gabriele.rocchetti@unicatt.it (G.R.); domenico.montesano@unipg.it (D.M.)

Abstract: In this work, the phytochemical profile and the biological properties of *Colchicum triphyllum* (an unexplored Turkish cultivar belonging to Colchicaceae) have been comprehensively investigated for the first time. Herein, we focused on the evaluation of the in vitro antioxidant and enzyme inhibitory effects of flower, tuber, and leaf extracts, obtained using different extraction methods, namely maceration (both aqueous and methanolic), infusion, and Soxhlet. Besides, the complete phenolic and alkaloid untargeted metabolomic profiling of the different extracts was investigated. In this regard, ultra-high-performance liquid chromatography coupled with quadrupole time-of-flight mass spectrometry (UHPLC-QTOF-MS) allowed us to putatively annotate 285 compounds when considering the different matrix extracts, including mainly alkaloids, flavonoids, lignans, phenolic acids, and tyrosol equivalents. The most abundant polyphenols were flavonoids (119 compounds), while colchicine, demecolcine, and lumicolchicine isomers were some of the most widespread alkaloids in each extract analyzed. In addition, our findings showed that *C. triphyllum* tuber extracts were a superior source of both total alkaloids and total polyphenols, being on average 2.89 and 10.41 mg/g, respectively. Multivariate statistics following metabolomics allowed for the detection of those compounds most affected by the different extraction methods. Overall, *C. triphyllum leaf* extracts showed a strong in vitro antioxidant capacity, in terms of cupric reducing antioxidant power (CUPRAC; on average 96.45 mg Trolox Equivalents (TE)/g) and ferric reducing antioxidant power (FRAP) reducing power (on average 66.86 mg TE/g). Interestingly, each *C. triphyllum* methanolic extract analyzed (i.e., from tuber, leaf, and flower) was active against the tyrosinase in terms of inhibition, recording the higher values for methanolic macerated leaves (i.e., 125.78 mg kojic acid equivalent (KAE)/g). On the other hand, moderate inhibitory activities were observed against AChE and α-amylase. Strong correlations ($p < 0.01$) were also observed between the phytochemical profiles and the biological activities determined. Therefore, our findings highlighted, for the first time, the potential of *C. triphhyllum* extracts in food and pharmaceutical applications.

Keywords: meadow saffron; metabolomics; UHPLC-QTOF-mass spectrometry; extraction methods; bioactive compounds; antioxidants

1. Introduction

Colchicum triphyllum Kunze is a spring/autumn-flowering species belonging to the Colchicaceae family, widely distributed in Turkey and Balkans [1]. It is also known as "autumn crocus" or "meadow saffron". Plants belonging to Colchicaceae are mainly used in pharmaceutical applications, thanks to therapeutic, anti-inflammatory, and antitumoral activities [2] attributed to the presence of colchicinoids (alkaloids), such as colchicine and demecolcine. In this regard, colchicine is used in the treatment of gout [3] and Behcet's disease [4], while demecolcine, together with trimethyl-colchicine acid methyl ester, demonstrated anti-neoplastic activity and is particularly suitable for the treatment of leukemia [5]. Besides, bioactive compounds characterizing plants belonging to the Colchicaceae family, such as alkaloids (e.g., colchicine), have been characterized and have been widely studied because of their beneficial effects for the treatment of cirrhosis, psoriasis, and amyloidosis [6]. Interestingly, less toxic derivatives of colchicine have also been studied as anticancer and antitumoral agents. *Colchicum* spp. also contains a considerable distribution of bioactive compounds, such as polyphenols. In particular, according to the literature [7], the most abundant (poly)-phenolic compounds are lignans, flavonoids, phenolic acids, and tannins.

Notably, *Colchicum* leaves share morphological similarities (mainly looking at leaves) with other plant species, such as *Allium ursinum* L. (wild garlic); fortunately, poisoning is rare, although some accidents (with also lethal outcomes) caused by the ingestion of toxic *Colchicum* alkaloids have been described in the scientific literature [8]. In addition, some *Colchicum* species (mainly *C. autumnale* L.) may be confused for the same reasons as Crocus spp. (mainly *Crocus sativus*); despite their morphological similarity, *Colchicum* flowers are typically larger with six stamens, while Crocus flowers are smaller with three longer stamens. Another issue related to saffron (*Crocus sativus*) is the lack of knowledge by consumers about the correct shape and botanical characteristics of the products, thus leading to potential episodes of adulteration and counterfeiting procedures on the market. In fact, besides morphological analogies, some plant-tissues from *Colchicum* spp. are extremely toxic, thus potentially affecting human health.

There are previous works based on the description and characterization of *Colchicum* spp. and its alkaloid distribution (focused mainly on colchicine and demecolcine) in several Jordanian *Colchicum* species. Recently, Rocchetti and co-authors [9] profiled, for the first time, flowers, leaves, and tubers of *Colchicum szovitsii* subsp. *szovitsii*, showing a great abundance of flavonols, phenolic acids, and total alkaloids that were the main components responsible for biological activities detected. However, to the best of our knowledge, there are no comprehensive studies based on a detailed characterization of both polyphenols and alkaloids characterizing different parts (i.e., flower, leaves, tuber) of *C. triphyllum* and based on untargeted metabolomics (i.e., ultra-high-performance liquid chromatography coupled with quadrupole time-of-flight (UHPLC-QTOF) mass spectrometry). Besides, considering that, to date, no efficient methods to synthesize *Colchicum* alkaloids have been found; colchicine and other alkaloids are mainly obtained from plant sources by different extractions techniques. Therefore, in this work, infusion, maceration (using methanol and water) together with Soxhlet extraction techniques were used to promote the extraction of both polyphenols and alkaloids from *C. triphyllum*, aiming to identify the most discriminant markers of each extraction technique used. Finally, in vitro antioxidant and enzyme inhibitory assessments were carried out to investigate the biological and pharmaceutical potential of this plant species.

2. Materials and Methods

2.1. Plant Material

The plant materials of *Colchicum triphyllum*, namely flowers, leaves, and tubers, were collected at Konya in Turkey in 2019 (Konya, around Silla Dam Lake, steppes 1200 m; Collection date: 03.02.2019). The plant materials were collected and identified by botanist Dr. Evren Yildiztugay (Selcuk University, Department of Biotechnology, Konya, Turkey, Voucher number: EY-2968). In the sampling, about twenty plants were collected in the same population. The plant materials were cleaned (first, washing with tap water and then rising by distilled water), and soil and other contaminants were removed. The plant parts, namely flowers, leaves, and tubers, were carefully separated, and these plants were dried in a shaded and well-ventilated environment at the Department of Biology, Selcuk University. After drying (about 10 days), the plant materials were powdered by using a laboratory mill (Retsch, SM-200), and the powdered materials were used to obtain extracts in the same week. The powdered plant materials were stored in well-ventilated conditions in the dark (about 20 °C). We performed the analysis in about one month after sampling.

2.2. Extraction Methods

In this work, three different extraction methods using different solvents were tested. In this regard, to obtain extracts, we performed infusion, maceration, and Soxhlet extraction techniques. Regarding infusion, the plant materials (5 g) were kept in 100 mL of boiling water for 20 min and then filtered. In the maceration technique, the plant materials (5 g) were mixed with 100 mL of both methanol and water for 24 h at room temperature. In the Soxhlet technique, the plant materials (5 g) were extracted with 100 mL methanol by using a Soxhlet apparatus for 6 h. Final extracts were obtained by using a vacuum evaporator and lyophilization. Finally, each obtained extract was stored in a refrigerator until further analyses.

2.3. UHPLC-QTOF Profiling of Polyphenols and Alkaloids

The untargeted phytochemical profile of the different *C. triphyllum* extracts was investigated through ultra-high-pressure liquid chromatography (Agilent 1290 HPLC liquid chromatograph; Agilent Technologies, Santa Clara, CA, USA) coupled to a quadrupole-time-of-flight mass spectrometer (Agilent 6550 iFunnel; Agilent Technologies, Santa Clara, CA, USA). The experimental conditions for the analysis of plant extracts using untargeted metabolomics were optimized in previous works from our research group [9–11]. The mass spectrometer acquired ions in the range 50–1200 m/z in positive (ESI+) scan mode. Three technical replications were considered, with an injection volume of 6 μL. An in-house database built, combining Phenol-Explorer 3.6 with some of the most important alkaloids reported in the literature on Colchicaceae, was then used for annotation purposes, exploiting the entire isotopic profile (i.e., combining monoisotopic accurate mass, isotopic ratios, and spacing) with a mass accuracy below 5 ppm. Therefore, the approach used was based on a Level 2 of identification (i.e., putatively annotated compounds), as set out by the COSMOS Metabolomics Standards Initiative [12–14]. Afterward, Agilent Profinder B.06 software was used for post-acquisition data filtering, retaining only those compounds identified within 100% of replications in at least one condition. Thereafter, to provide more quantitative information on the different annotated compounds, polyphenols were first ascribed into classes and subclasses and then quantified using standard solutions (80/20, *v/v* methanol/water) of pure standard compounds analyzed with the same method [9]. The following phenolic classes were targeted: anthocyanins (quantified as cyanidin equivalents), flavones (quantified as luteolin equivalents), flavonols (quantified as catechin equivalents), lignans (quantified as sesamin equivalents), low-molecular-weight phenolics (quantified as tyrosol equivalents), and phenolic acids (quantified as ferulic acid equivalents). Finally, a calibration curve of sanguinarine (Sigma grade, Sigma–Aldrich, S. Louis, MO, USA) was used to estimate the total alkaloid content. The results were finally expressed as mg equivalents/g dry matter.

2.4. In Vitro Antioxidant Capacity and Inhibitory Potential

For in vitro antioxidant capacity, different test systems, including radical quenching, reducing power, phosphomolybdenum, and ferrous ion chelating, were employed. The details of the methods are described in our earlier papers [15,16]. The results were reported as mg Trolox Equivalents (TE)/g extract and ethylenediaminetetraacetic acid (EDTA) equivalents (for ferrous ion chelating; mg EDTAE/g extract). For enzyme inhibitory activities, key enzymes for global health problems were selected, namely, α-amylase and α-glucosidase, acetylcholinesterase (AChE), butyrylcholinesterase (BChE), and tyrosinase and the inhibitory activities were compared to standard drugs (acarbose for amylase and glucosidase; galantamine for AChE and BChE; kojic acid for tyrosinase). All assays were performed considering three technical replications.

2.5. Statistical Analysis and Chemometrics

A one-way analysis of the variance (ANOVA) was performed considering data from each assay and using the software PASW Statistics 26.0 (SPSS Inc., Chicago, IL. USA), followed by a Duncan's post hoc test ($p > 0.05$). Pearson's correlations ($p < 0.05$; two-tailed) were also calculated using PASW Statistics 26.0. Afterward, the metabolomics-based dataset exported from Mass Profiler Profession B.12.06 (Agilent Technologies, Santa Clara, CA, USA) was elaborated into a second software, namely SIMCA 13 (Umetrics, Malmo, Sweden) for supervised orthogonal projections to latent structures discriminant analysis (OPLS-DA), as previously reported in previous works from our research group [9]. Two OPLS-DA models were built; the first one highlighted the differences in the phytochemical profiles as imposed by the extraction methods, while the second model showed the differences between the three plant-organs under investigation. Finally, the variables selection method VIP (i.e., variables' importance in projection) was used to evaluate those compounds mostly affected by the different extraction methods, together with those better discriminating the plant-organs. In particular, polyphenols and alkaloids showing a VIP score > 1 have been considered as marker compounds.

3. Results and Discussion

3.1. Phytochemical Profiling of the Different Extracts

To characterize the polyphenol and alkaloid composition of the different *Colchicum triphyllum* extracts we used untargeted metabolomics based on UHPLC-QTOF mass spectrometry. According to this approach, 285 compounds were putatively identified in the different matrix extracts, mainly including alkaloids, flavonoids (such as anthocyanins, flavonols, and flavones), lignans, and low-molecular-weight phenolic acids. Each compound annotated is provided in Supplementary Table S1 together with its abundance and composite mass spectra. Overall, the most abundant compounds detected when considering the metabolomic dataset were alkaloids (such as colchicine, demecolcine, and y-lumicolchicine), anthocyanins (such as petunidin 3-*O*-rutinoside and cyanidin 3-*O*-sophoroside), flavones (mainly apigenin and luteolin glucosides), and flavonols (mainly isomeric forms of kaempferol and quercetin). Regarding the class of lignans, the most abundant annotated compounds were secoisolariciresinol, pinoresinol, and its isomer, matairesinol (Supplementary Table S1). Afterward, a semi-quantitative approach based on standard compounds was used to evaluate the concentration of phytochemicals in the studied plant matrices. The results of this semi-quantitative approach are reported in Table 1.

Table 1. Semi-quantitative values for the main phenolic sub-classes and total alkaloids by ultra-high-performance liquid chromatography quadrupole time-of-flight (UHPLC-QTOF) mass spectrometry of the tested extracts together with extraction yields. Values are reported as the mean ± standard deviation ($n = 3$). The results are expressed as mg equivalents (Eq.)/g dry matter. Different superscript letters in the same column indicate significant differences ($p < 0.05$), as determined by Duncan's post-hoc test. nd = not detected.

Parts	Methods	Extraction Yield (%)	Total Alkaloids (mg Eq./g)	Anthocyanins (mg Eq./g)	Flavones (mg Eq./g)	Flavonols (mg Eq./g)	Phenolic Acids (mg Eq./g)	Lignans (mg Eq./g)	Tyrosols (mg Eq./g)
Flowers	Infusion	35.12	1.49 ± 0.49^{ab}	0.57 ± 0.05^{d}	1.98 ± 0.2^{c}	1.48 ± 0.21^{c}	0.11 ± 0.02^{a}	2.27 ± 0.29^{ab}	0.72 ± 0.35^{ab}
	Maceration-MeOH	42.6	1.77 ± 0.44^{abc}	0.78 ± 0.01^{f}	1.99 ± 0.94^{c}	1.71 ± 0.20^{d}	0.20 ± 0.00^{c}	2.67 ± 1.36^{abc}	1.09 ± 0.24^{bc}
	Maceration-Water	26.65	2.08 ± 0.02^{bcd}	0.06 ± 0.00^{b}	0.47 ± 0.08^{a}	0.14 ± 0.00^{a}	0.19 ± 0.01^{bc}	3.01 ± 0.28^{abc}	0.44 ± 0.03^{a}
	Soxhlet-MeOH	51.25	1.39 ± 0.01^{ab}	0.67 ± 0.01^{e}	2.52 ± 0.14^{c}	1.67 ± 0.01^{d}	0.15 ± 0.03^{abc}	1.28 ± 1.30^{a}	0.42 ± 0.01^{a}
Average value			1.68	2.08	1.74	1.25	0.16	2.31	0.67
Tubers	Infusion	13.44	1.86 ± 1.00^{abc}	0.005 ± 0.00^{a}	0.15 ± 0.00^{a}	nd ^{a}	0.50 ± 0.02^{g}	4.57 ± 0.03^{bc}	1.22 ± 0.05^{cd}
	Maceration-MeOH	28.41	3.19 ± 0.03^{cd}	nd ^{a}	1.06 ± 0.39^{b}	nd ^{a}	0.45 ± 0.05^{fg}	10.37 ± 0.72^{d}	1.84 ± 0.05^{e}
	Maceration-Water	44.02	3.46 ± 0.74^{d}	nd ^{a}	0.39 ± 0.24^{a}	nd ^{a}	0.41 ± 0.01^{e}	4.84 ± 1.41^{bc}	1.60 ± 0.01^{de}
	Soxhlet-MeOH	12.68	3.08 ± 1.14^{cd}	nd ^{a}	$0.45 \pm 0.02d^{a}$	nd ^{a}	0.35 ± 0.10^{de}	10.60 ± 2.50^{d}	2.85 ± 0.72^{f}
Average value			2.89	<0.01	0.51	nd	0.43	7.59	1.88
Leaves	Infusion	25.58	2.26 ± 0.49^{abcd}	0.06 ± 0.00^{b}	0.33 ± 0.01^{a}	0.07 ± 0.01^{a}	0.21 ± 0.01^{c}	4.77 ± 1.25^{d}	3.43 ± 0.07^{g}
	Maceration-MeOH	31.48	1.10 ± 0.54^{a}	0.05 ± 0.01^{b}	0.22 ± 0.04^{a}	0.11 ± 0.00^{a}	0.30 ± 0.04^{d}	1.65 ± 0.57^{a}	0.59 ± 0.09^{a}
	Maceration-Water	22.88	2.28 ± 0.04^{abcd}	0.13 ± 0.00^{c}	1.10 ± 0.01^{b}	0.43 ± 0.00^{b}	0.49 ± 0.06^{fg}	4.99 ± 2.75^{c}	2.48 ± 0.01^{f}
	Soxhlet-MeOH	33.61	2.79 ± 1.70^{bcd}	0.07 ± 0.01^{b}	0.56 ± 0.00^{ab}	0.09 ± 0.01^{a}	0.12 ± 0.00^{ab}	3.11 ± 0.15^{abc}	1.45 ± 0.04^{cde}
Average value			2.11	0.08	0.55	0.17	0.28	3.63	1.87

As can be observed, the three plant matrices were predominantly rich ($p < 0.05$) in alkaloids and lignans when compared to the other classes of compounds, while flavonols and anthocyanins showed the lower ($p < 0.05$) concentration; it is also interesting to notice that, when considering *C. triphyllum* tuber extracts, neither anthocyanins nor flavonols were detected. Besides, looking at *C. triphyllum* flower extracts, the maceration-water extraction method was found to promote the highest recovery of lignans and alkaloids (3.01 and 2.08 mg/g, respectively), while maceration-MeOH extraction encourages the recovery of anthocyanins, flavonols, phenolic acids, and tyrosols. Interestingly, Soxhlet extraction promoted the highest recovery of flavones (2.52 mg/g). *Colchicum* flowers are reported to be very similar from a morphological point of view to those of *Crocus Sativus L*. Overall, both plant species can be considered as a good source of (poly)-phenolic compounds In this regard, the saffron flower has been described as rich in flavonoids (such as flavonols and flavones), hydroxycinnamic acids, and lignans [10,17,18]. Regarding saffron alkaloids, Amin Mir and co-authors [19] found this class of compounds in both water and methanolic extracts of flowers. Another study by Hosseinzadeh et al. [20] based on the phytochemical screening of different *Crocus* extracts, highlighted the distribution of flavonoids (including anthocyanins) and tannins in both aqueous and ethanolic petal extracts, while alkaloids and saponins characterized the aqueous and ethanolic stigmas extracts. However, our data are difficult to compare with previously cited works, considering that in this work, both polyphenols and alkaloids were evaluated, targeting the whole flower. Recently, Jadouali and co-authors [21] evaluated the total phenolic content of different flower parts of Moroccan *Crocus sativus* L., showing a value of 54.59 mg gallic acid equivalents (GAE)/g for the whole saffron flower.

Considering the lack in the literature of similar comprehensive phytochemical screening on *C. triphyllum* extracts, we compared our findings with a previous work focused on a different species, namely *C. szovitsii*. In particular, the most abundant polyphenols characterizing *C. szovitsii* plant extracts were flavonols, phenolic acids, and tyrosols equivalents [9], while anthocyanins and flavanols were found to be less abundant. Another interesting result was obtained when comparing the extraction efficiency of polyphenols between the two *Colchicum* species and using Soxhlet-MeOH extractions. In particular, Soxhlet-MeOH promoted a better extraction of flavonols and flavones in tubers of *C. szovitsii* when compared to *C. triphyllum*, being 1.58 and 0.81 mg/g vs. not detectable values and 0.45 mg/g, respectively. Interestingly, the same extraction method (i.e., Soxhlet-MeOH) promoted a better recovery of lignans and tyrosols in *C. tryphillum*, being 10.60 and 2.85 mg/g, respectively. Regarding the alkaloids putatively annotated in *C. triphyllum* extracts, tubers obtained by water maceration showed the highest content (3.46 mg/g). This aspect is worthy of interest, considering that the colchicinoids (mainly colchicine and derivatives) are among the highly poisonous water-soluble alkaloids detected in flowers and seeds of *Colchicum* genus [22]. Regarding *C. szovitsii* leaf extracts, flavonols were better extracted by exploiting Soxhlet (21.95 mg/g), while phenolic acids and tyrosol equivalents by infusion (3.52 and 3.68 mg/g, respectively). On the other hand, *C. triphyllum* was revealed to be a great source of lignans (possessing a potential estrogenic activity), recording an average value of 3.63 mg/g when considering all the extraction methods tested (Table 1). Finally, concerning the total alkaloid content, *C. triphyllum* and *C. szovitsii* leaf extracts showed a comparable content, being 2.79 and 2.65 mg/g when considering Soxhlet and water-maceration, respectively.

3.2. Multivariate Statistical Discrimination of the Different Extraction Methods

Considering specifically the extraction type, a supervised multivariate statistical approach was carried out to find the compounds allowing the discrimination between the different methods. In more detail, an OPLS-DA (orthogonal projection to latent structures discriminant analysis) was depicted, followed by the variables importance in projection (VIP) method, to select those compounds mostly affected by the different extraction methods. As can be observed in Figure 1, a clear separation based on alkaloids and polyphenols content was achieved. In particular, each extraction method provided a differential phytochemical profile, although water-maceration extracts were different from the others. The model parameters were more than acceptable, being R^2Y (cum) (goodness-of-fit) = 0.97 and Q^2 (cum) (goodness-of-prediction) = 0.91. Besides, the model was cross-validated and showed neither suspect nor strong outliers. Afterward, the variables selection method VIP was exploited to find those metabolites mostly influenced by the extraction method used. Hence, 26 compounds were those possessing a VIP score > 1.2, including flavonoids (7 compounds), followed by low-molecular-weight phenolics (7 compounds), phenolic acids (6 compounds), and lignans (5 compounds). The list containing the remaining VIP compounds (1.2 < VIP score < 1) is reported in Supplementary Table S1. Interestingly, only one alkaloid, namely colchiceine, possessed a VIP score > 1.2 (i.e., 1.27). Regarding polyphenols, the higher VIP score was recorded for the lignan 1-acetoxypinoresinol, mainly found in tubers and leaves but not in flowers, followed by the caffeoylquinic acid isomers (VIP score = 1.39), whose presence was only detected in flowers treated with mac-MeOH. Furthermore, the lignans sesaminol, sesamolin, and episesaminol (VIP score = 1.36) were also better extracted with the infusion method. Some studies established that ethanol and methanol plant and fruit extracts can provide a better recovery of polyphenols when compared to water as the solvent, but the efficiency is strictly related to the extraction time [20]. Moreover, changes in temperature and the solvent-mixtures chosen could enhance and/or reduce the extraction efficiency. The Soxhlet method is a well-established technique requiring a smaller quantity of solvent compared to maceration, but according to literature, it is not suggested to promote the extraction of thermolabile compounds [23].

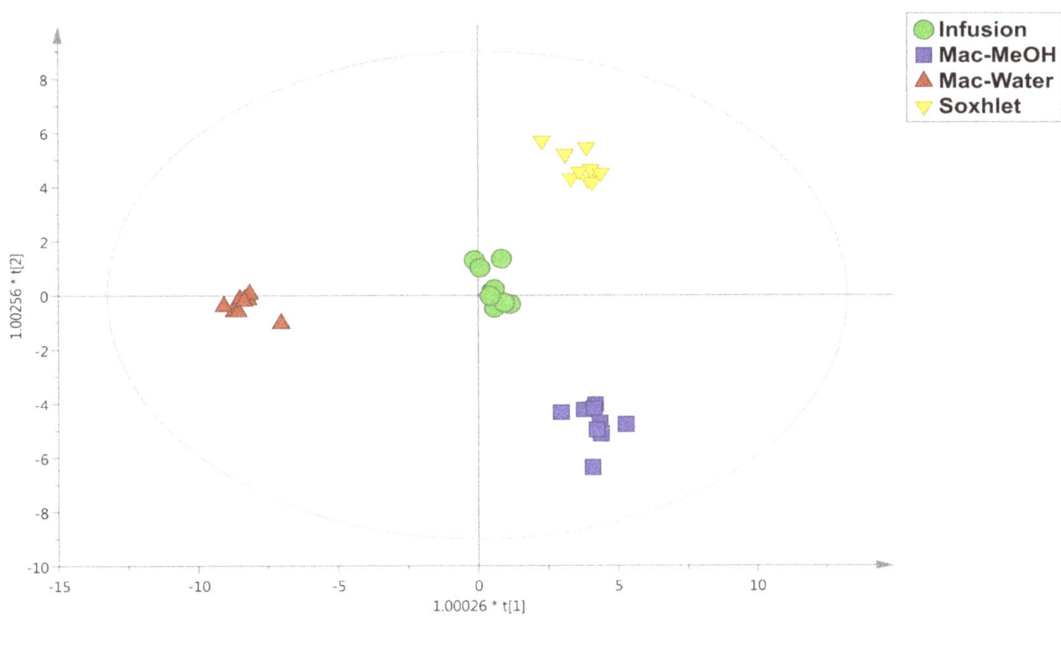

Figure 1. Orthogonal projections to latent structures discriminant analysis (OPLS-DA) score plot built according to polyphenol and alkaloid profiling and considering the different extraction methods as class membership criteria.

Thereafter, a second OPLS-DA model was built to check the major differences between the organs under investigation. The OPLS-DA score plot is reported in Figure 2.

Foods **2020**, *9*, 457

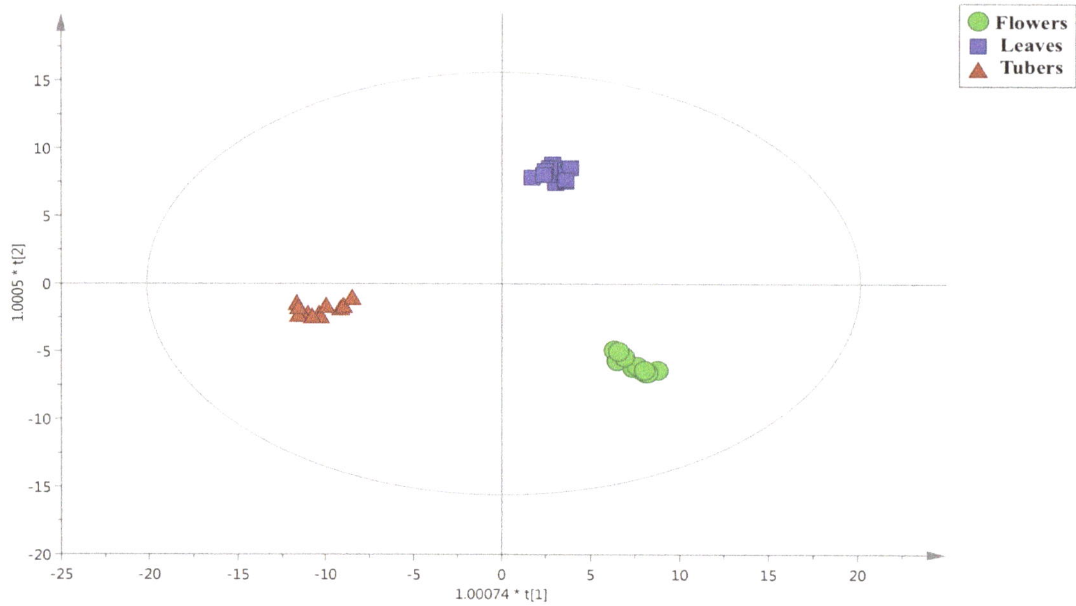

Figure 2. Orthogonal projections to latent structures discriminant analysis (OPLS-DA) score plot built according to polyphenol and alkaloid profiling and considering the different plant organs (i.e., flowers, leaves, and tubers) as class membership criteria.

As can be observed from Figure 2, the second latent vector t[2] provided a clear discrimination between tubers and the other *C. triphyllum* extracts, while the first latent vector t[1] revealed a more exclusive phytochemical profile for the leaf extracts. The goodness of the prediction model built on phenolics and alkaloids was confirmed by inspecting the goodness of fit and prediction, being 0.97 and 0.91, respectively. Finally, to check those compounds allowing the discrimination of the different organs, the VIP method was exploited, and the VIP markers can be found in Supplementary Table S1. Overall, 78 compounds were characterized by a VIP score > 1, being 34 flavonoids, 17 alkaloids, 11 phenolic acids, 11 lower-molecular-weight phenolics, and 5 lignans. In particular, the highest VIP scores were recorded for three alkaloids, namely androbine (1.51), autumnaline (1.46), and colchicoside (1.42). Looking at polyphenols, the most discriminant compounds highlighted by the VIP selection method were genistin (1.39), phlorin (1.35), and pelargonidin 3-*O*-glucoside (1.34). Overall, androbine was a specific marker of leaf extracts; autumnaline was detected in both tubers and leaves), while colchicoside was a marker of both tubers and flowers. Regarding polyphenols, genistin and pelargonidin 3-*O*-glucoside were detected mainly in flower extracts, while phlorin was a specific marker detected in tubers (Supplementary Table S1). Overall, our findings revealed a higher discrimination potential of alkaloids when compared to polyphenols. To date, more than 150 structurally elucidated alkaloids have been described for Colchicaceae genera [24]. In particular, eight distinct structural types of alkaloids characterize the Colchicaceae family, being phenethylisoquinolines (e.g., autumnaline), homoproaporphines (e.g., jolantine), homoaporphines (e.g., merobustine), androcymbines (e.g., szovitsidine), colchicines (e.g., colchicine), allocolchicines (e.g., jerusalemine), lumicolchicines (e.g., γ-lumicolchicine), and homoerythrinans (e.g., taxodine). Therefore, our findings (Supplementary Table S1) are completely following with the typical alkaloid composition reported for Colchicaceae plants. Regarding the polyphenols annotated, few comprehensive works based on high-resolution mass spectrometry are available in the literature. In previous work, Toplan et al. [1] reviewed the importance of *Colchicum* species in modern therapy together with its significance in Turkey, showing that these plants are very abundant in alkaloids, followed by polyphenols (mainly phenolic acids and flavonoids). However, looking to the different *Colchicum* species, few compounds are reported, namely ferulic acid, vanillin, luteolin, coumaric acid, caffeic acid, and 3,4-dihiydroxibenzaldehyde. Therefore, further works are required to better explore and give more insight on the phenolic composition of different *Colchicum* species.

3.3. In Vitro Antioxidant Capacity of the Tested Extracts

To date, the scientific evidence on the biological and pharmacological activities of extracts and bioactive compounds deriving from plants is very strong [25–31]. In the last years, many studies have explored the association between the ingestion of bioactive compounds and decreased risk of non-communicable diseases. Besides, considering that many non-nutrients with putative health benefits are reducing agents, hydrogen donors, singlet oxygen quenchers, or metal chelators, measurement of antioxidant activity using in vitro assays have become very popular over recent decades. In the current study, we investigated the antioxidant capacities of *C. triphyllum* flower, tuber, and leaf extracts by measuring their total in vitro antioxidant capacity (phosphomolybdenum assay), DPPH (2,2-diphenyl-1-picrylhydrazyl) and [2,2′-azinobis-(3-ethylbenzothiazoline-6-sulfonate)] (ABTS) scavenging capacity, ferric and cupric reduction activity (FRAP and CUPRAC), and metal chelating activity. However, as widely suggested in the recent scientific literature [32], these colorimetric and in vitro methods present many pitfalls and should be used as screening tools, thus supported by a comprehensive LC-MS characterization and quantification of those antioxidant compounds likely responsible of the activity observed. Based on the experimental findings (Table 2), it was observed that all of the studied extracts were characterized by potential health-promoting properties. Radical scavengers can prevent free radical-induced macromolecules or tissue damage by directly neutralizing free radicals and accepting or donating electron(s) to eliminate the unpaired condition of the radical [33]. Among the different *C. triphyllum* extracts tested in this study, we found that leaf extracts were the most active DPPH scavengers, recording an average value (when considering all the extraction method) of 45.22 mg TE/g. On the other hand, the methanolic tuber extracts obtained using the maceration method showed a similar activity, being 46.17 mg TE/g (Table 2). Regarding ABTS assay, the maceration method (based on methanol as extraction solvent) was found to promote the highest activity in tubers and leaves (being 50.93 and 52.55 mg TE/g, respectively). Overall, intriguing differences were outlined in the different organ-specific extracts when considering the cupric reduction activity (CUPRAC). In this regard, *C. triphyllum* leaf methanolic extracts showed higher values than flowers and tubers, recording 109.63 mg TE/g (using maceration method) and 123.34 mg TE/g (using Soxhlet method). Finally, regarding the other assays, FRAP ranged from 25.02 (infused flowers) up to 70.80 mg TE/g (leaves extracted using Soxhlet), metal chelating from 1.89 (tubers extracted using Soxhlet) up to 33.88 mg EDTAE/g (infused tubers), and total in vitro antioxidant capacity (phosphomolybdenum method) from 0.33 (infused tubers) up to 1.52 mmol TE/g (water macerated leaves). Therefore, taken together, our findings demonstrate that *C. triphyllum* leaf extracts were the most active as radical scavengers. This finding was in line with a previous work [9] focusing on another novel cultivar from Turkey, namely *Colchicum szovitsii subsp. szovitsii*.

Table 2. In vitro antioxidant activities of the tested extracts. Values are reported as the mean ± S.D. TE: Trolox equivalent; EDTAE: ethylenediaminetetraacetic acid equivalent. Different superscript letters in the same column indicate significant differences ($p < 0.05$), as determined by Duncan's post-hoc test.

Parts	Methods	Phosphomolybdenum (mmol TE/g)	DPPH (mg TE/g)	ABTS (mg TE/g)	CUPRAC (mg TE/g)	FRAP (mg TE/g)	Metal Chelating (mg EDTAE/g)
Flowers	Infusion	0.51 ± 0.04 ^{b}	31.69 ± 0.45 ^{c}	27.47 ± 1.57 ^{c}	35.00 ± 0.60 ^{a}	25.02 ± 0.34 ^{a}	20.90 ± 3.06 ^{e}
	Maceration-MeOH	0.46 ± 0.07 ^{ab}	33.99 ± 0.79 ^{cd}	20.53 ± 1.91 ^{a}	41.52 ± 1.13 ^{c}	25.71 ± 1.78 ^{a}	14.40 ± 0.55 ^{c}
	Maceration-water	1.10 ± 0.15 ^{d}	24.34 ± 0.60 ^{b}	19.26 ± 0.64 ^{a}	37.00 ± 0.22 ^{ab}	38.88 ± 0.67 ^{d}	21.53 ± 1.41 ^{ef}
	Soxhlet-MeOH	0.74 ± 0.09 ^{c}	34.63 ± 0.27 ^{cd}	24.48 ± 1.15 ^{b}	58.31 ± 0.89 ^{f}	30.72 ± 0.64 ^{b}	16.04 ± 0.41 ^{cd}
Average value		0.70	31.16	22.93	42.96	30.08	18.22
Tubers	Infusion	0.33 ± 0.02 ^{a}	12.14 ± 10.18 ^{a}	30.38 ± 0.38 ^{d}	38.38 ± 0.63 ^{b}	28.92 ± 0.74 ^{b}	33.88 ± 0.39 ^{g}
	Maceration-MeOH	0.72 ± 0.02 ^{c}	46.17 ± 1.15 ^{f}	50.93 ± 0.90 ^{h}	63.12 ± 1.52 ^{g}	46.16 ± 0.52 ^{e}	7.48 ± 0.27 ^{b}
	Maceration-water	0.77 ± 0.05 ^{c}	29.12 ± 1.05 ^{bc}	45.27 ± 0.60 ^{f}	44.75 ± 0.32 ^{d}	35.34 ± 0.71 ^{c}	23.99 ± 1.14 ^{f}
	Soxhlet-MeOH	0.85 ± 0.07 ^{c}	38.19 ± 0.91 ^{de}	38.79 ± 0.31 ^{e}	50.65 ± 2.06 ^{e}	35.85 ± 0.06 ^{c}	1.89 ± 0.25 ^{a}
Average value		0.67	31.40	41.34	49.22	36.57	16.81
Leaves	Infusion	1.25 ± 0.09 ^{e}	44.91 ± 0.35 ^{f}	47.90 ± 0.62 ^{g}	68.13 ± 0.45 ^{g}	60.66 ± 0.66 ^{f}	32.89 ± 2.20 ^{g}
	Maceration-MeOH	1.38 ± 0.05 ^{e}	45.49 ± 0.74 ^{f}	52.55 ± 1.65 ^{h}	109.63 ± 2.06 ^{l}	70.04 ± 3.09 ^{h}	22.26 ± 0.45 ^{ef}
	Maceration-water	1.52 ± 0.02 ^{f}	42.85 ± 1.09 ^{ef}	43.95 ± 1.34 ^{f}	84.71 ± 1.74 ^{i}	65.92 ± 1.39 ^{g}	15.74 ± 2.33 ^{cd}
	Soxhlet-MeOH	1.35 ± 0.10 ^{e}	47.65 ± 0.38 ^{f}	55.73 ± 0.36 ^{i}	123.34 ± 2.24 ^{m}	70.80 ± 0.88 ^{h}	18.25 ± 0.29 ^{d}
Average value		1.37	45.22	50.03	96.45	66.67	22.28

3.4. *Enzyme Inhibitory Activity of the Tested Extracts*

To investigate the enzyme inhibitory capacity of the different *C. triphyllum* extracts, the cholinesterases AChE and BChE, together with the enzymes tyrosinase, α-amylase, and α-glucosidase were considered. The results obtained using the above-mentioned enzymes are shown in Table 3.

Table 3. Enzyme inhibitory effects of the tested extracts. Values are reported as the mean ± S.D. Different superscript letters in the same column indicate significant differences ($p < 0.05$), as determined by Duncan's post-hoc test. GALAE: Galatamine equivalent; KAE: Kojic acid equivalent; ACAE: Acarbose equivalent; nd: not detected.

Parts	Methods	AChE Inhibition (mg GALAE/g)	BChE Inhibition (mg GALAE/g)	Tyrosinase Inhibition (mg KAE/g)	α-amylase Inhibition (mmol ACAE/g)	α-glucosidase Inhibition (mmol ACAE/g)
Flowers	Infusion	nd ^{a}	nd ^{a}	nd ^{a}	0.14 ± 0.01 ^{ab}	nd ^{a}
	Maceration-MeOH	3.41 ± 0.12 ^{d}	nd ^{a}	104.09 ± 3.66 ^{c}	0.66 ± 0.01 ^{ef}	2.69 ± 0.02 ^{b}
	Maceration-Water	nd ^{a}	nd ^{a}	nd ^{a}	0.27 ± 0.00 ^{c}	nd ^{a}
	Soxhlet-MeOH	4.09 ± 0.13 ^{f}	nd ^{a}	102.52 ± 0.56^{c}	0.61 ± 0.06 ^{d}	2.73 ± 0.01^{c}
Average value		1.87	nd	51.65	0.42	1.35
Tubers	Infusion	nd ^{a}	nd ^{a}	nd ^{a}	0.13 ± 0.00 ^{a}	nd ^{a}
	Maceration-MeOH	4.80 ± 0.02 ^{h}	3.91 ± 0.30^{b}	118.70 ± 0.85 ^{d}	0.73 ± 0.01 ^{g}	nd ^{a}
	Maceration-Water	nd ^{a}	nd ^{a}	nd ^{a}	0.18 ± 0.01 ^{b}	nd ^{a}
	Soxhlet-MeOH	5.11 ± 0.05 ^{i}	7.42 ± 1.48 ^{c}	118.61 ± 1.41 ^{d}	0.64 ± 0.01 ^{de}	nd ^{a}
Average value		2.48	2.83	59.32	0.42	nd
Leaves	Infusion	2.45 ± 0.16 ^{b}	nd ^{a}	nd^{a}	0.14 ± 0.002 ^{ab}	nd ^{a}
	Maceration-MeOH	3.62 ± 0.16 ^{e}	nd ^{a}	125.78 ± 0.13 ^{f}	0.71 ± 0.05 ^{fg}	2.75 ± 0.01 ^{d}
	Maceration-Water	2.95 ± 0.08 ^{c}	nd ^{a}	4.10 ± 0.34 ^{b}	0.24 ± 0.00 ^{c}	nd ^{a}
	Soxhlet-MeOH	4.56 ± 0.09 ^{g}	nd ^{a}	121.78 ± 0.57 ^{e}	0.73 ± 0.02 ^{g}	2.76 ± 0.01 ^{d}
Average value		3.39	nd	62.91	0.45	1.37

According to literature, the enzyme cholinesterase is a significant therapeutic target to alleviate the deterioration of cholinergic neurons in the brain and the loss of neurotransmission, i.e., one of the major causes of Alzheimer's disease [34]. Overall, AChE inhibition activity was detected in each *C. triphyllum* leaf extract analyzed (Table 3). In particular, the highest activity values were recorded for tuber extracts (obtained by maceration-MeOH) and leaf extracts (obtained through Soxhlet-MeOH), being 4.80 and 4.56 mg galatamine equivalent (GALAE)/g, respectively. No activity against AchE was observed when considering infusion and maceration-water techniques for flowers and tubers extracts. Considering BChE, only methanolic maceration (3.91 mg GALAE/g) and Soxhlet-MeOH of *C. triphyllum* tubers (7.42 mg GALAE/g) were able to produce effective inhibitory activity against this enzyme. Regarding the enzyme tyrosinase, it plays an important role in the melanogenesis and enzymatic browning, so its inhibitors can be considered as anti-browning compounds in food and agriculture industries and as depigmentation agents in the cosmetic and medicinal industries [35]. In our experimental conditions, we found that both Soxhlet and maceration (using methanol as extraction solvent) were able to produce extracts characterized by a strong tyrosinase inhibitory ability. In particular, the highest activity was recorded for *C. triphyllum* leaves using methanolic maceration (i.e., 125.78 mg KAE/g). Finally, regarding two of the most important enzymes from a nutritional standpoint, namely α-amylase and α-glucosidase, we found a low α-amylase inhibition potential for all the tested extracts, from 0.13 (for infused tubers) up to 0.73 mmol acarbose equivalent (ACAE)/g (for leaves and tubers, using Soxhlet and methanolic maceration, respectively). Interestingly, the α-glucosidase inhibition activity was observed only for flower and leaf extracts, with the maximum average activity recorded for leaves (i.e., 1.37 mmol ACAE/g). These two enzymes are potential targets in producing lead compounds for the management of diabetes. Therefore, further research on *C. triphyllum* extracts could open a new perspective for the management of health-related and metabolic problems.

Regarding a possible comparison between different *Colchicum* species, there are few works available in the literature to make a realistic discussion. That is why we decided to compare our findings with a previous work focused on the chemical characterization of *Colchicum szovitsii* [9]. Overall, the authors showed that *C. szovitsii* possessed a strong tyrosinase inhibitory action, mainly when considering methanolic macerated leaf extracts (i.e., 116.92 mg KAE/g), thus confirming our findings. Similar trends were observed for the remaining tested enzymes. Interestingly, when considering the existing literature on saffron (*Crocus sativus*), similar activities have been reported on different saffron extracts [36], with crocetin, dimethylcrocetin, and safranal the mainly responsible for acetylcholinesterase inhibitions, as revealed by both molecular docking and in vitro enzymatic studies. Besides, Menghini and co-authors [37] showed that saffron stigmas are characterized by both AChE and BChE inhibitions (i.e., 2.51 and 3.44 mg GALAE/g, respectively) and inhibition towards amylolytic enzymes, such as α-amylase (0.44 mmol ACAE/g) and α-glucosidase (6.34 mmol ACAE/g). The same authors also showed that these values are lower when considering the combination of *C. sativus* tepals and anthers.

3.5. Correlations

Pearson's correlation coefficients were then calculated to check the contribution of polyphenols and alkaloids to the antioxidant and enzymatic related properties observed when considering the different *C. triphyllum* extracts. A summarizing table reporting all of the correlation coefficients for each part analyzed (i.e., flowers, tubers, and leaves) is reported in Supplementary Table S1. Overall, a positive and significant ($p < 0.05$) correlation coefficient was found between total alkaloids and in vitro total antioxidant capacity (phosphomolybdenum method) when considering *C. triphyllum* tuber extracts, with a correlation coefficient of 0.673. Besides, no significant correlations coefficients were recorded for leaves, flowers, and tubers when considering total alkaloids and enzymatic assays. Regarding polyphenols, we found significant correlations mainly between anthocyanins characterizing *C. triphyllum* flower extracts and enzymatic inhibitory assays. In particular, the highest correlation coefficient was measured between anthocyanins and tyrosinase inhibition (0.740), followed by α-glucosidase (0.737), AChE (0.714), and α-amylase (0.601) inhibitions. These results are consistent with recent papers reporting that flavonoids (mainly anthocyanins, flavonols, and flavones) are one of the most important classes of natural enzyme inhibitors [38–43]. Interestingly, negative and significant correlation coefficients were found when considering enzymatic assays and anthocyanins characterizing tuber extracts. In this regard, only two anthocyanins were putatively annotated in tuber extracts, namely Vitisin A and Petunidin 3,5-O-diglucoside, likely characterized by no inhibitory potential. Regarding leaf extracts, no significant correlation coefficients were observed (Supplementary Table S1). Therefore, our findings suggested that a clear matrix-effect mainly due to specific anthocyanins is conceivable, when looking at the correlation coefficients observed. Further works based on in silico studies are strongly required to confirm our findings.

Regarding the in vitro antioxidant potential, anthocyanins, flavonols, and flavones characterizing *C. triphyllum* flower extracts showed significant correlations with DPPH radical scavenging ability, recording values of 0.974, 0.972, and 0.880, respectively. Looking at the tuber extracts, we found that lignans, anthocyanins, and flavones were the most interesting classes in terms of correlation potential (Supplementary Table S1). In this regard, flavones were strongly ($p < 0.01$) correlated to FRAP (0.882), CUPRAC (0.870), ABTS (0.758), and DPPH (0.744), while lignans recorded the highest correlation coefficient for CUPRAC (0.775). Interestingly, the anthocyanins characterizing tuber extracts showed negative correlations coefficients ($p < 0.01$) with each antioxidant assay (Supplementary Table S1). Overall, it is also important to emphasize the significant correlation coefficients recorded with enzymatic assays, when considering lignans and tyrosols extracted from tubers. Regarding lignans, *C. triphyllum* tubers were mainly characterized by isomeric forms of matairesinol (Supplementary Table S1); in the last years, the interest on lignans has risen because of their estrogenic-like effects but also when considering the potential inhibitory activity on several enzymes, as revealed by in silico and molecular docking studies [44]. Finally, looking at the correlation results for *C. triphyllum* leaf extracts, few significant and positive correlation coefficients were recorded (Supplementary Table S1); overall, the phenolic profile of leaf extracts (in terms of flavonoids and phenolic acids), showed a good degree of correlation with the total in vitro antioxidant capacity ($0.01 < p < 0.05$). Clearly, the observed correlation values between phytochemical profiles (by UHPLC-QTOF mass spectrometry) and antioxidant/biological assays may be explained by considering a different exposure of abiotic and biotic stress factors for each plant part, thus leading to the production of different levels of secondary metabolites. However, looking to the wide diversity in both phenolics and alkaloids of *C. triphyllum* extracts, further studies based on in silico and targeted compound evaluations appear to be worthwhile, to confirm the potential of this plant for food and pharmaceutical purposes.

4. Conclusions

In this work, we focused the attention on the unexplored species *Colchicum triphyllum* Kunze, belonging to the *Colchicum* genus, considering the scarcity of comprehensive studies on its phytochemical profiling and biological properties. Therefore, this study reports the biochemical characterization of different *C. triphyllum* extracts, according to their phenolic and alkaloid compositions, together with the evaluation of their in vitro antioxidant and enzyme inhibitory activities. In this regard, UHPLC-QTOF-MS allowed us to identify 289 compounds (mainly alkaloids and flavonoids). Regarding the plant parts studied, *C. triphyllum* tuber extracts were the most important source of both alkaloids and polyphenolic compounds, being on average 2.89 and 10.41 mg/g, respectively. Multivariate statistics following metabolomics showed a higher impact of the different extraction methods (i.e., maceration, infusion, and Soxhlet) on the polyphenolic profile rather than alkaloids. Overall, *C. triphyllum leaf* extracts showed the stronger in vitro antioxidant capacity (as CUPRAC and FRAP), while each *C. triphyllum* methanolic extract analyzed was active against the enzyme tyrosinase. Strong correlations ($p < 0.01$) were also observed between the phytochemical profiles (mainly lignans and tyrosol equivalents) and the activities determined. Therefore, although characterized by the presence of toxic alkaloids (such as colchicine), this work sustained the utilization of different extraction methods to produce rich extracts in terms of (poly)-phenols and alkaloids, largely contributing to both antioxidant and other pharmacological properties, making *C. triphyllum* a promising source of drugs and whitening agents for both food, pharmaceutical, and cosmetic industries. However, further studies are strongly recommended to understand better the toxicity and bioavailability of the putatively identified phytochemicals, aimed at replacing synthetic antioxidants and enzyme inhibitors.

Author Contributions: Conceptualization, G.R., G.Z., and M.A.O.; Methodology, G.R., B.S., and G.Z.; Resources, G.Z., L.L., E.Y., and D.M.; Writing—original draft preparation, B.S., G.R., G.Z., and G.A.; Writing—review and editing, G.R., D.M., G.Z., and L.L. All authors have read and agreed to the published version of the manuscript.

References

1. Toplan, G.G.; Gurer, C.; Mat, A. Importance of Colchicum species in modern therapy and its significance in Turkey. *J. Fac. Pharm.* **2016**, *46*, 129–144.
2. Alali, F.Q.; Tawaha, K.; El-Elimat, T. Determination of (–)-demecolcine and (–)-colchicine content in selected Jordanian Colchicum species. *Int. J. Pharm. Sci. Res.* **2007**, *62*, 739–742.
3. Terkeltaub, R.A.; Furst, D.E.; Bennett, K.; Kook, K.A.; Crockett, R.S.; Davis, M.W. High versus low dosing of oral colchicine for early acute gout flare: Twenty-four-hour outcome of the first multicenter, randomized, double-blind, placebo-controlled, parallel-group, dose-comparison colchicine study. *Arthritis Rheum.* **2010**, *62*, 1060–1068. [CrossRef] [PubMed]
4. Sakane, T.; Takeno, M. Novel approaches to Behcet's disease. *Expert Opin. Investig. Drugs* **2000**, *9*, 1993–2005. [CrossRef]
5. Cifuentes, M.; Schilling, B.; Ravindra, R.; Winter, J.; Janik, M.E. Synthesis and biological evaluation of B-ring modi ed colchicine and iso colchicine analogs. *Bioorg. Med. Chem. Lett.* **2006**, *16*, 2761–2764. [CrossRef]
6. Cocco, G.; Chu, D.C.; Pandolfi, S. Colchicine in Clinical Medicine. A Guide for Internists. *Eur. J. Intern. Med.* **2010**, *21*, 503–508. [CrossRef]
7. Ondra, P.; Válka, I.; Vičar, J.; Sütlüpinar, N.; Šimánek, V. Chromatographic determination of constituents of the genus Colchicum (Liliaceae). *J. Chromatogr. A* **1995**, *704*, 351–356. [CrossRef]
8. Brvar, M.; Ploj, T.; Kozelj, G.; Mozina, M.; Noc, M.; Bunc, M. Case report: Fatal poisoning with *Colchicum autumnale*. *Crit. Care* **2004**, *8*, R56–R59. [CrossRef]
9. Rocchetti, G.; Senizza, B.; Zengin, G.; Okur, M.A.; Montesano, D.; Yildiztugay, E.; Lobine, D.; Mahomoodally, M.F.; Lucini, L. Chemical Profiling and Biological Properties of Extracts from Different Parts of *Colchicum Szovitsii* Subsp. *Szovitsii*. *Antioxidants* **2019**, *8*, 632. [CrossRef]
10. Senizza, B.; Rocchetti, G.; Ghisoni, S.; Busconi, M.; De Los Mozos Pascual, J.; Fernandez, A.; Lucini, L.; Trevisan, M. Identification of phenolic markers for saffron authenticity and origin: An untargeted metabolomics approach. *Food Res. Int.* **2019**, *126*, 108584. [CrossRef]
11. Mohamed, M.B.; Rocchetti, G.; Montesano, D.; Ali, S.B.; Guasmi, F.; Grati-Kamoun, N.; Lucini, L. Discrimination of Tunisian and Italian extra-virgin olive oils according to their phenolic and sterolic fingerprints. *Food Res. Int.* **2018**, *106*, 920–927. [CrossRef]
12. Fellah, B.; Rocchetti, G.; Senizza, B.; Giuberti, G.; Bannour, M.; Ferchichi, A.; Lucini, L. Untargeted metabolomics reveals changes in phenolic profile following *in vitro* large intestine fermentation of non-edible parts of *Punica granatum* L. *Food Res. Int.* **2020**, *128*, 108807. [CrossRef] [PubMed]
13. Salek, R.M.; Neumann, S.; Schober, D.; Hummel, J.; Billiau, K.; Kopka, J.; Correa, E.; Reijmers, T.; Rosato, A.; Tenori, L.; et al. COordination of Standards in MetabOlomicS (COSMOS): Facilitating integrated metabolomics data access. *Metabolomics* **2015**, *11*, 1587–1597. [CrossRef]
14. Schrimpe-Rutledge, A.C.; Codreanu, S.G.; Sherrod, S.D.; McLean, J.A. Untargeted metabolomics strategies-Challenges and emerging directions. *J. Am. Soc. Mass Spectrom.* **2016**, *27*, 1897–1905. [CrossRef]
15. Zengin, G. A study on *in vitro* enzyme inhibitory properties of *Asphodeline anatolica*: New sources of natural inhibitors for public health problems. *Ind. Crops Prod.* **2016**, *83*, 39–43. [CrossRef]

16. Tufa, T.; Damianakos, H.; Zengin, G.; Graikou, K.; Chinou, I. Antioxidant and enzyme inhibitory activities of disodium rabdosiin isolated from *Alkanna sfikasiana* Tan, Vold and Strid. *S. Afr. J. Bot.* **2019**, *120*, 157–162. [CrossRef]
17. Mykhailenko, O.; Kovalyov, V.; Goryacha, O.; Ivanauskas, L.; Georgiyants, V. Biologically active compounds and pharmacological activities of species of the genus Crocus: A review. *Phytochemistry* **2019**, *162*, 56–89. [CrossRef]
18. Moratalla-López, N.; Bagur, M.J.; Lorenzo, C.; Martínez-Navarro, M.E.; Salinas, M.R.; Alonso, G.L. Bioactivity and Bioavailability of the Major Metabolites of Crocus sativus L. Flower. *Molecules* **2019**, *24*, 2827. [CrossRef] [PubMed]
19. Amin Mir, M.; Parihar, K.; Tabasum, U.; Kumari, E. Estimation of alkaloid, saponin and flavonoid, content in various extracts of *Crocus sativa*. *J. Med. Plant Res.* **2016**, *4*, 171–174.
20. Hosseinzadeh, H.; Younesi, H.M. Antinociceptive and anti-inflammatory effects of *Crocus sativus* L. stigma and petal extracts in mice. *BMC Pharmacol.* **2002**, *2*, 7.
21. Jadouali, S.M.; Atifi, H.; Mamouni, R.; Majourhat, K.; Bouzoubaâ, Z.; Laknifli, A.; Faouzi, A. Chemical characterization and antioxidant compounds of flower parts of Moroccan crocus sativus L. *J. Saudi Soc. Agric. Sci.* **2019**, *18*, 476–480. [CrossRef]
22. Suica-bunghez, I.; Ion, R.; Teodorescu, I.; Sorescu, A.; Stirbescu, R.; Stirbescu, N. Fitochemical and antioxidant characterization of autumn crocus (*Colchicum autumnale*) flowers and tubers plant extracts. *J. Sci. Arts* **2017**, *3*, 539–546.
23. Lapornik, B.; Prošek, M.; Golc Wondra, A. Comparison of extracts prepared from plant by-products using different solvents and extraction time. *J. Food Eng.* **2005**, *71*, 214–222. [CrossRef]
24. Larsson, S.; Rønsted, N. Reviewing Colchicaceae Alkaloids – Perspectives of Evolution on Medicinal Chemistry. *Curr. Top. Med. Chem.* **2014**, *14*, 274–289. [CrossRef]
25. Borrelli, F.; Borbone, N.; Capasso, R.; Montesano, D.; De Marino, S.; Aviello, G.; Aprea, G.; Masone, S.; Izzo, A.A. Potent relaxant effect of a *Celastrus paniculatus* extract in the rat and human ileum. *J. Ethnopharmacol.* **2009**, *122*, 434–438. [CrossRef] [PubMed]
26. Montesano, D.; Rocchetti, G.; Putnik, P.; Lucini, L. Bioactive profile of pumpkin: An overview on terpenoids and their health-promoting properties. *Curr. Opin. Food Sci.* **2018**, *22*, 81–87. [CrossRef]
27. Rocchetti, G.; Pellizzoni, M.; Montesano, D.; Lucini, L. Italian *Opuntia ficus-indica* cladodes as rich source of bioactive compounds with health-promoting properties. *Foods* **2018**, *7*, 24. [CrossRef]
28. Pandey, K.B.; Rizvi, S.I. Plant polyphenols as dietary antioxidants in human health and disease. *Oxid. Med. Cell. Logev.* **2009**, *2*, 897484. [CrossRef]
29. Saucedo-Pompa, S.; Torres-Castillo, J.A.; Castro-López, C.; Rojas, R.; Sánchez-Alejo, E.J.; Ngangyo-Heya, M.; Martínez-Ávila, G.C.G. Moringa plants: Bioactive compounds and promising applications in food products. *Food Res. Int.* **2018**, *111*, 438–450. [CrossRef]
30. de Morais Cardoso, L.; Pinheiro, S.S.; Martino, H.S.D.; Pinheiro-Sant'Ana, H.M. Sorghum (Sorghum bicolor L.): Nutrients, bioactive compounds, and potential impact on human health. *Crit. Rev. Food Sci. Nutr.* **2017**, *57*, 372–390. [CrossRef]
31. Borbone, N.; Borrelli, F.; Montesano, D.; Izzo, A.A.; De Marino, S.; Capasso, R.; Zollo, F. Identification of a new sesquiterpene polyol ester from *Celastrus paniculatus*. *Planta Med.* **2007**, *73*, 792–794. [CrossRef]
32. Granato, D.; Shahidi, F.; Wrolstad, R.; Kilmartin, P.; Melton, L.D.; Hidalgo, F.J.; Miyashita, K.; Camp, J.V.; Alasalvar, C.; Ismail, A.B.; et al. Antioxidant activity, total phenolics and flavonoids contents: Should we ban *in vitro* screening methods? *Food Chem.* **2018**, *264*, 471–475. [CrossRef]
33. Ahmadinejad, F.; Geir Møller, S.; Hashemzadeh-Chaleshtori, M.; Bidkhori, G.; Jami, M.-S. Molecular mechanisms behind free radical scavengers function against oxidative stress. *Antioxidants* **2017**, *6*, 51. [CrossRef]
34. Silman, I.; Sussman, J.L. Acetylcholinesterase: 'classical' and 'non-classical' functions and pharmacology. *Curr. Opin. Pharmacol.* **2005**, *5*, 293–302. [CrossRef]
35. Zolghadri, S.; Bahrami, A.; Hassan Khan, M.T.; Munoz-Munoz, J.; Garcia-Molina, F.; Garcia-Canovas, F.; Saboury, A.A. A comprehensive review on tyrosinase inhibitors. *J. Enzym. Inhib. Med. Chem.* **2019**, *34*, 279–309. [CrossRef] [PubMed]

36. Geromichalos, G.D.; Lamari, F.N.; Papandreou, M.A.; Trafalis, D.T.; Margarity, M.; Papageorgiou, A.; Sinakos, Z. Saffron as a source of novel acetylcholinesterase inhibitors: Molecular docking and *in vitro* enzymatic studies. *J. Agric. Food Chem.* **2012**, *60*, 6131–6138. [CrossRef] [PubMed]
37. Menghini, L.; Leporini, L.; Vecchiotti, G.; Locatelli, M.; Carradori, S.; Ferrante, C.; Zengin, G.; Recinella, L.; Chiavaroli, A.; Leone, S.; et al. *Crocus sativus* L. stigmas and byproducts: Qualitative fingerprint, antioxidant potentials and enzyme inhibitory activities. *Food Res. Int.* **2018**, *109*, 91–98. [CrossRef] [PubMed]
38. Rocchetti, G.; Giuberti, G.; Busconi, M.; Marocco, A.; Trevisan, M.; Lucini, L. Pigmented sorghum polyphenols as potential inhibitors of starch digestibility: An *in vitro* study combining starch digestion and untargeted metabolomics. *Food Chem.* **2020**, *312*, 126077. [CrossRef]
39. Rocchetti, G.; Giuberti, G.; Gallo, A.; Bernardi, J.; Marocco, A.; Lucini, L. Effect of dietary polyphenols on the *in vitro* starch digestibility of pigmented maize varieties under cooking conditions. *Food Res. Int.* **2018**, *108*, 183–191. [CrossRef] [PubMed]
40. Tan, Y.; Chang, S.K.C. Digestive enzyme inhibition activity of the phenolic substances in selected fruits, vegetables and tea as compared to black legumes. *J. Funct. Foods* **2017**, *38*, 644–655. [CrossRef]
41. Martinez-Gonzalez, A.I.; Díaz-Sánchez, Á.G.; de la Rosa, L.A.; Bustos-Jaimes, I.; Alvarez-Parrilla, E. Inhibition of α-amylase by flavonoids: Structure activity relationship (SAR). *Spectrochim. Acta A* **2019**, *206*, 437–447. [CrossRef] [PubMed]
42. Xiao, J.; Kai, G.; Yamamoto, K.; Chen, X. Advance in Dietary Polyphenols as α-Glucosidases Inhibitors: A Review on Structure-Activity Relationship Aspect. *Crit. Rev. Food Sci. Nutr.* **2013**, *53*, 818–836. [CrossRef] [PubMed]
43. Takahama, U.; Hirota, S. Interactions of flavonoids with α-amylase and starch slowing down its digestion. *Food Funct.* **2018**, *9*, 677–687. [CrossRef] [PubMed]
44. Mocan, A.; Zengin, G.; Crisan, G.; Mollica, A. Enzymatic assays and molecular modeling studies of *Schisandra chinensis* lignans and phenolics from fruit and leaf extracts. *J. Enzym. Inhib. Med. Chem.* **2016**, *31*, 200–210. [CrossRef] [PubMed]

3

Magnetic Stirring Assisted Demulsification Dispersive Liquid–Liquid Microextraction for Preconcentration of Polycyclic Aromatic Hydrocarbons in Grilled Pork Samples

Jitlada Vichapong [1,*], Yanawath Santaladchaiyakit [2], Rodjana Burakham [3] and Supalax Srijaranai [3]

1 Creative Chemistry and Innovation Research Unit, Department of Chemistry and Center of Excellence for Innovation in Chemistry, Faculty of Science, Mahasarakham University, Maha Sarakham 44150, Thailand
2 Department of Chemistry, Faculty of Engineering, Rajamangala University of Technology Isan, Khon Kaen Campus, Khon Kaen 40000, Thailand; sanyanawa@gmail.com
3 Materials Chemistry Research Center, Department of Chemistry and Center of Excellence for Innovation in Chemistry, Faculty of Science, Khon Kaen University, Khon Kaen 40002, Thailand; rodjbu@kku.ac.th (R.B.); supalax@kku.ac.th (S.S.)
* Correspondence: jitlada.v@msu.ac.th

Abstract: A simple microextraction method, magnetic stirring assisted demulsification dispersive liquid–liquid microextraction, for preconcentration of five polycyclic aromatic hydrocarbons (fluorene, phenanthrene, anthracene, fluoranthrene, and pyrene) was investigated prior to analysis by high performance liquid chromatography. In this method, a mixture of extraction solvent and disperser solvent was rapidly injected into sample solution. The magnetic stirrer agitator aided the dispersion of the extraction solvent into the sample solution. After the formation of an emulsion, the demulsifier was added, resulting in the rapid separation of the mixture into two phases. No centrifugation step was required. Several parameters affecting the extraction efficiency of the proposed method were studied, including addition of salt, kind and volume of extraction solvent, volume of demulsifier solvent, and extraction times. Under the optimum conditions, high enrichment factor, low limit of detections (LODs) and good precision were gained. The proposed method was successfully applied to analysis of polycyclic aromatic hydrocarbon residues in grilled pork samples.

Keywords: magnetic stirring assisted demulsification dispersive liquid–liquid microextraction; polycyclic aromatic hydrocarbons; grilled pork; high performance liquid chromatography

---

1. Introduction

Polycyclic aromatic hydrocarbons (PAHs) constitute of a large class of organic materials which are formed of two or more fused aromatic rings [1–3]. These compounds are generally drawn from incomplete combustion or high-temperature pyrolysis of organic materials, such as coal, petrol, wood, garbage, tobacco, meats or other organic foods [4]. Both the European Union (EU) and the United States Environmental Protection Agency (USEPA) have their own list of 16 PAHs as "priority organic pollutants", due to their high toxicity to the human health [5]. PAHs are non-polar, very hydrophobic compounds, with low water solubility. Therefore, in the aquatic environment they commonly exist in relatively low concentrations [6]. Consequently, powerful analytical methods are required to extract, separate, and identify these target analytes in the environment [7].

Several chromatographic methods including gas chromatography [8,9] and high-performance liquid chromatography [10,11] have been regularly used for separation and quantification of PAHs in various samples. Although these are sensitive and selective methods, PAHs usually occur in ppb level or lower in complex mediums containing various interfering compounds [12]. Therefore, sample preparation, matrix removal and preconcentration of the target analytes are needed before analysis in order to obtain sensitive and accurate results.

Conventional sample preparation methods such as solid-phase extraction (SPE) [13,14] and liquid–liquid extraction (LLE) [15,16] have been used for preconcentration and clean-up before analysis of PAHs. Unfortunately, these methods are tedious, time-consuming, and require large amounts of samples and toxic organic solvents [2]. Therefore, much effort has been made to develop a simple, sensitive and environmentally friendly sample preparation method termed dispersive liquid–liquid microextraction (DLLME) [17]. This method is based on the use of high density of extraction solvents such as chlorinated solvent. However, these are toxic and harmful to human health and environment. The other DLLME mode is based on the low density of extraction solvent. However, the main drawback of DLLME modes was the requirement of a centrifugation step. Recently, a new DLLME technique without a centrifugation step was introduced, the low-density based demulsification DLLME [18,19]. After the mixture of extraction solvent and disperser solvent was added into the aqueous solution, the cloudy solution was immediately separated by adding demulsifiers. No centrifugation step was required.

There are various agitators, such as vortex [20], ultrasound [21], and in-syringe [22], that have been used to enhance the dispersion and to accelerate the formation of fine droplets of extraction solvent. Zhang et al. [23] proposed a simple magnetic stirring assisted dispersive liquid–liquid microextraction (MSA-DLLME) method. After the injection of extraction solvent and disperser solvent into an aqueous solution, the sample was magnetically stirred. Consequently, the centrifugal step was not required. In this present study, the magnetic stirring assisted demulsified dispersive liquid–liquid microextraction combined with HPLC was investigated for extraction, preconcentration and simultaneous determination of PAHs. The magnetic stirrer was used to increase the dispersion and mass transfer between two phases. Special attention was paid to parameters providing the highest extraction efficacy of the extraction procedure. We carefully evaluated salt addition, kind and volume of extraction solvent, kind and volume of disperser solvent, volume of demulsifier solvent and extraction times. The proposed method was successfully applied to the determination of PAHs in grilled pork samples. To our knowledge, this was the first time that the simple magnetic stirring assisted demulsified dispersive liquid–liquid microextraction technique was used in the determination of trace levels of PAHs.

2. Materials and Methods

2.1. Chemicals and Reagents

All chemicals and reagents used were of analytical reagent grade or better. The analytical standards of polycyclic aromatic hydrocarbon including fluorene, phenanthrene, anthracene, fluoranthrene and pyrene were purchased from Sigma-Aldrich (Darmstadt, Germany). The stock standard solution ($1000\ mg \cdot L^{-1}$) was prepared by dissolving in methanol (MeOH). The working solution of standard was prepared daily by dilution with deionized water. Deionized water was prepared using RiOs™ Type I Simplicity 185 (Millipore Waters, Newford, MA, USA) with the resistivity of 18.2 MΩ.cm. Acetonitrile (ACN) and MeOH (HPLC grade) were obtained from Merck (Darmstadt, Germany). All solvents for HPLC were filtered through 0.45 μm filters (Millipore Corp., Newford, MA, USA) and degassed in an ultrasonic bath. NaCl, anhydrous Na_2CO_3 and anhydrous Na_2SO_4 were purchased from Ajax Finechem (Auckland, New Zealand), and CH_3COONa (Carlo Erba, France) was used. 1-Dodecanol and 1-octanol were provided by Merck (Darmstadt, Germany).

2.2. Chromatographic Conditions

The HPLC system was comprised of a Waters 1525 binary HPLC pump (Newford, MA, USA), a Rheodyne injector and photodiode array detector (PDA). The Empower 3 software was chosen for data acquisition. A LiChrospher® 100 RP-8 endcapped (4.6 × 150 mm, 5.0 μm) column (Merck, Darmstadt, Germany) was employed for the separation of all analytes, and was maintained at room temperature. The mobile phase was comprised of 67% acetonitrile in water with isocratic elution at a flow rate of 1 mL·min^{-1}. The injection volume was 20 μL. The detection of all target analytes was set at 254 nm. Five polycyclic aromatic hydrocarbons were separated within 10 min with the elution order of fluorene (t_R = 6.87 min), phenanthrene (t_R = 7.03 min), anthracene (t_R = 7.98 min), fluoranthrene (t_R = 8.98 min) and pyrene (t_R = 9.46 min).

2.3. *Sample Preparation of Grilled Pork Samples*

Grilled pork samples were collected from market in Maha Sarakham province, Northeastern Thailand. One gram of spiked and non-spiked were weighed and transferred to a centrifuge tube. Then, 10 mL of extraction solvent (2 mol·L^{-1} KOH in ethanol) was added to hydrolyze the sample. This sample was ultrasounded with fixed power for 5 min at 45 °C, followed by centrifugation (15 min, 4,000 rpm). The supernatant was collected into another centrifuge tube. The solid residue was extracted again with 3 mL of extraction solvent. All volumes of supernatant were gathered, the funnel was frozen for 1 h at −18 °C to precipitate the fat of the solution. Approximately 5 mL of solution were kept and followed by centrifugation. After that, an aliquot was applied for magnetic stirring assisted demulsified dispersive liquid–liquid microextraction under the selected conditions and analysis by a HPLC system.

2.4. *Magnetic Stirring Assisted Demulsified Dispersive Liquid–Liquid Microextraction*

A 10-mL aliquot of standard solution of each polycyclic aromatic hydrocarbon or grilled pork sample was mixed with 20% (*w*/*v*) of NaCl and then mixture solution containing 75 μL of extraction solvent (1-dodecanol) and 500 μL dispersive solvent (ACN) was quickly injected into the sample solution. The solution was then stirred at 1500 rpm to increase the mass transfer between two phases. An emulsion (water/extraction solvent/dispersive solvent) was formed. After that, 500 μL of ACN (as de-emulsifier solvent) was injected into the solution to break down the emulsion. The extraction was found to float to the top of the solution immediately. The extraction phase was kept and injected to HPLC for analysis.

2.5. *Method Validation*

The method validations such as linearity of calibration graph, limit of detection (LOD), limit of quantitation (LOQ), precision and enrichment factors (EFs) were studied. The linear range of standard calibration was conducted between 0.0005 and 1 mg·L^{-1} of PAH. LODs and LOQs were defined as the lowest detectable concentration with a signal to noise ratio of 3:1 and 10:1, respectively. EFs were defined as the concentration ratio of the analytes in the settled phase after performing microextraction methods and its initial concentration in the aqueous phase. The relative standard deviation (RSD, %) for intra-days and inter-day of extraction of PAHs were determined at three different concentration levels (0.10, 0.25 and 0.35 mg·L^{-1} of each PAHs) using 5 injections. The enrichment factor (EF) was defined as the concentration ratio of the analytes in the settled phase (Cset) and in the aqueous sample (Co).

2.6. *Statistical Analysis*

Data results are given as the mean ± standard deviation (SD) of three measurements (n = 3). In all graphs, a linear regression analysis was conducted using Microsoft Excel 2013 software (London, EC1V 2NX, United Kingdom).

3. Results and Discussion

3.1. Optimization of Magnetic Stirring Assisted Demulsified Dispersive Liquid–Liquid Microextraction Condition

In order to improve monitoring of PAH using magnetic stirring assisted demulsified dispersive liquid–liquid microextraction coupled with HPLC analysis, various experimental parameters including addition of salt, kind and volume of extraction solvent, volume of demulsifier solvent and extraction times were evaluated. The aqueous solution (10.00 mL) containing 100 ng·mL^{-1} of each PAH was used for optimization. All the experiments were performed in triplicate and the mean of the results were used for optimization.

Generally, the addition of salt decreases the solubility of analytes in aqueous solution [24,25] and therefore increases their partitioning into the organic phase for liquid phase microextraction. In this study the addition of various salts (1.0 g) such as sodium chloride (NaCl), sodium sulphate (Na_2SO_4), sodium carbonate (Na_2CO_3), sodium acetate (CH_3COONa) were considered and compared to a control, without salt addition. A comparison of extraction efficiency in terms of peak area with various kinds of salt is shown in Figure 1. Without salt addition, phase separation did not complete. The results demonstrated that the addition of NaCl and Na_2SO_4 provided no significant difference in the peak area of all PAHs except fluorene and phenanthrene. Therefore, NaCl was added in varying concentrations for further study. The effects of the amount of NaCl range from 0.5 to 2.5 g on the extraction efficiency of target analytes were also studied. The result (shown in Figure 2), was that the extraction efficacy of the analytes slightly increased as NaCl increased from 0.5 to 2.0 g, and trended to be decrease when the amount of NaCl was greater than 2.0 g. Finally, 2.0 g was chosen as the amount of NaCl in following experiments.

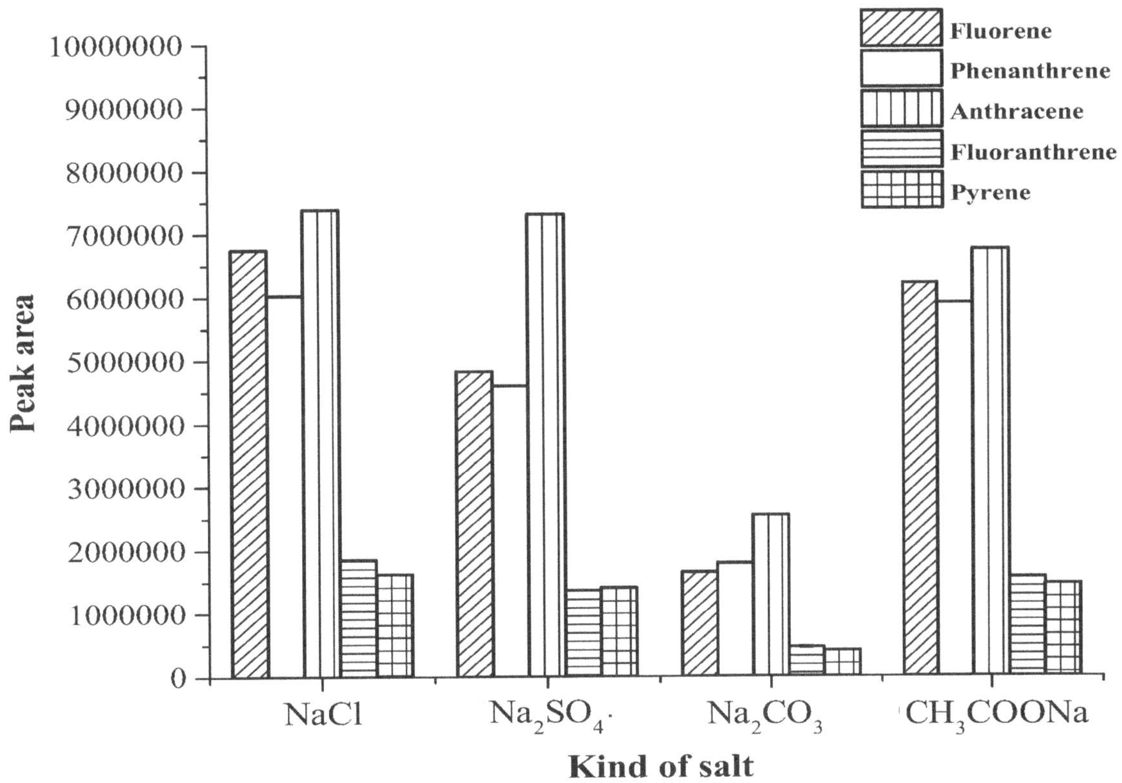

Figure 1. Effect of addition of salt on the extraction efficacy.

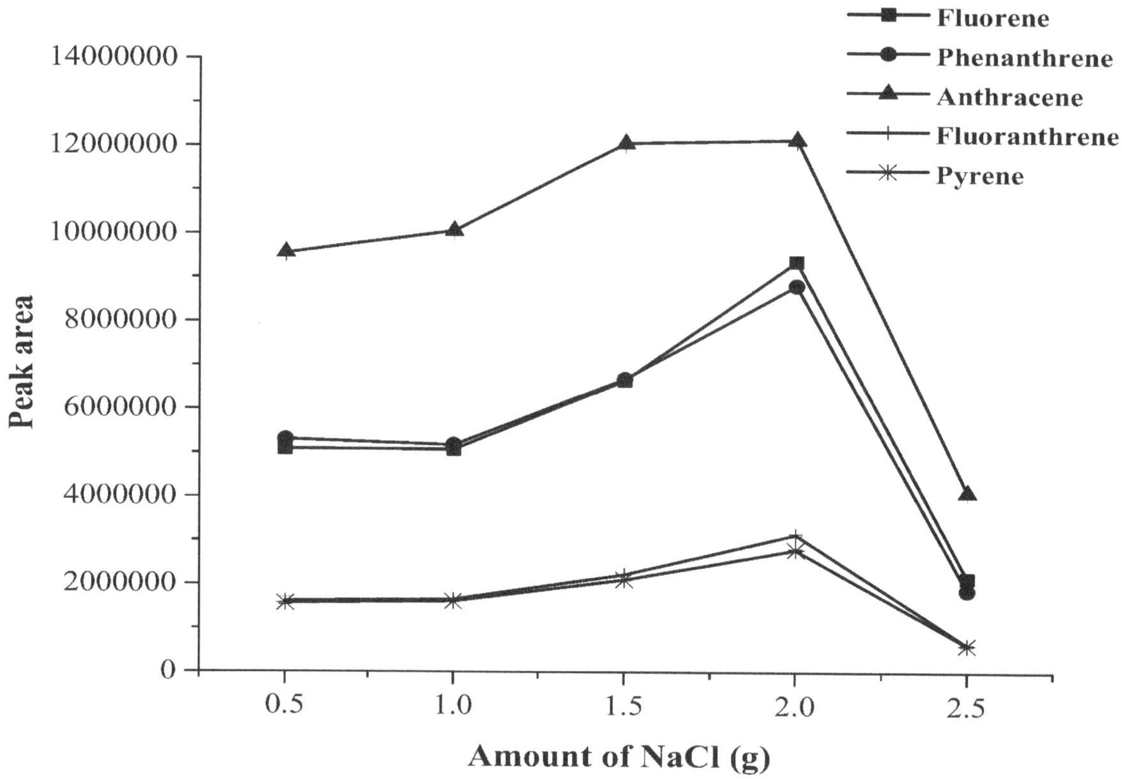

Figure 2. Effect of NaCl amount on the extraction efficacy.

Choosing a suitable extraction solvent is important for obtaining an efficient extraction procedure [26]. In this study, low density immiscible solvents (density = *d*) were investigated, 1-octanol ($d = 0.8270$ g·mL^{-1}), 1-undecanol ($d = 0.8298$ g·mL^{-1}), and 1-dodecanol ($d = 0.8309$ g·mL^{-1}). The cloudy solution was observed after a quick injection of a mixture of extraction solvent and dispersive solvent into the sample solution, the solution was then mixed using the magnetic stirrer. While using 1-octanol as an extraction solvent, it was found that phase separation did not occur. Thus, it was not suitable to be used for extraction solvent. In comparison of extraction efficacy between 1-undecanol and 1-dodecanol, it was found that 1-dodecanol provided high extraction efficacy in terms of peak area (data not shown). Thus, 1-dodecanol was chosen as the extraction solvent for further study. The effect of the 1-dodecanol volume was investigated in the range of 50–150 μL. It was found that, 50 μL of 1-dodecanol cannot complete phase separation. The peak area of all the target compounds decreased as the extraction solvent increased beyond 75 μL. Consequently, 1-dodecanol 75 μL was used as an optimum extraction solvent.

The selection of disperser solvent is important parameter in DLLME technique. This solvent should be miscible in both phases (extraction solvent and aqueous phase), moreover it should disperse the extraction solvent into the aqueous solution to form a cloudy state. To simplify the choosing process, the dispersive solvent should also be used as the demulsifier to break up the oil-in-water emulsion [27]. Various kinds of solvent such as ethanol, acetone, acetonitrile and methanol, were studied. The results are shown in Figure 3. It was found that acetonitrile as the disperser and demulsified solvent provided highest peak area. Thus, acetonitrile was chosen as disperser and demulsified solvent.

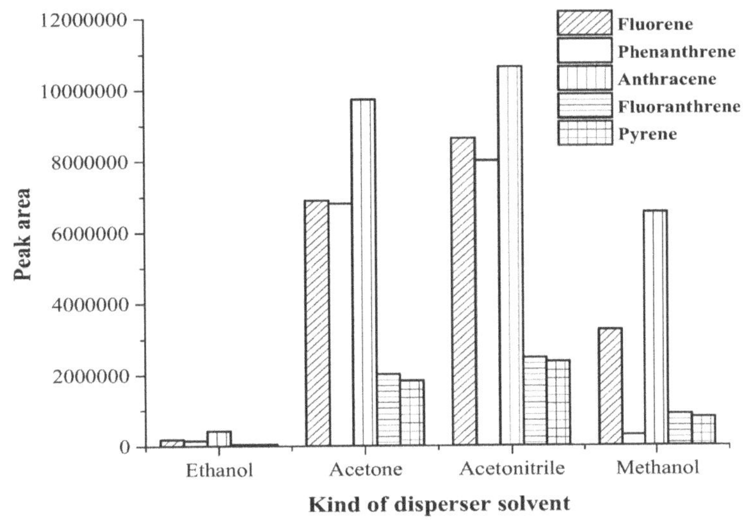

Figure 3. Effect of kind of disperser solvent on the extraction efficacy.

The volume of disperser solvent was studied within the range of 250–1,000 μL. It was found that the extraction efficiency slightly increased with increasing the volume of disperser solvent up to 500 μL and then decreased (as shown in Figure 4). Therefore, 500 μL of acetonitrile was used as disperser solvent. Moreover, volume of demulsified solvent was varied in the range 250–1,250 μL. It can be observed that a volume of 500 μL of demulsified solvent provided highest extraction efficiency in terms of peak area (as shown in Figure 5). Thus, acetonitrile 500 μL was selected as demulsified solvent.

In the two immiscible phase system, the mass transfer between the aqueous solution and extraction solvent depended on the extraction time and agitation [28]. The extraction time was defined as time that the sample was magnetic stirring agitated [29]. Extraction times on the extraction efficiency were evaluated for 1–10 minutes (data not shown) at 1500 rpm. The maximum peak area with the least standard deviation was obtained at four minutes, therefore it was selected as the optimum extraction time.

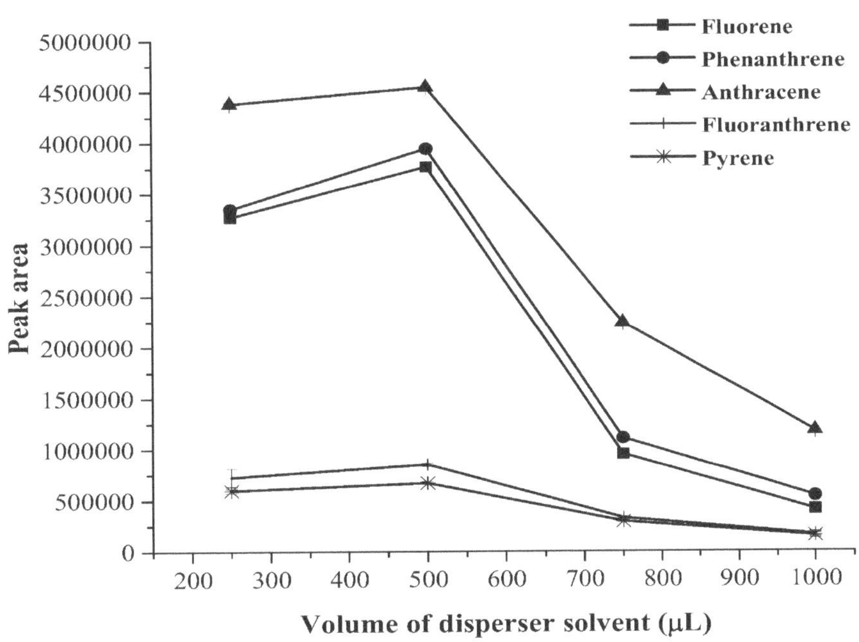

Figure 4. Effect of volume of disperser solvent on the extraction efficacy.

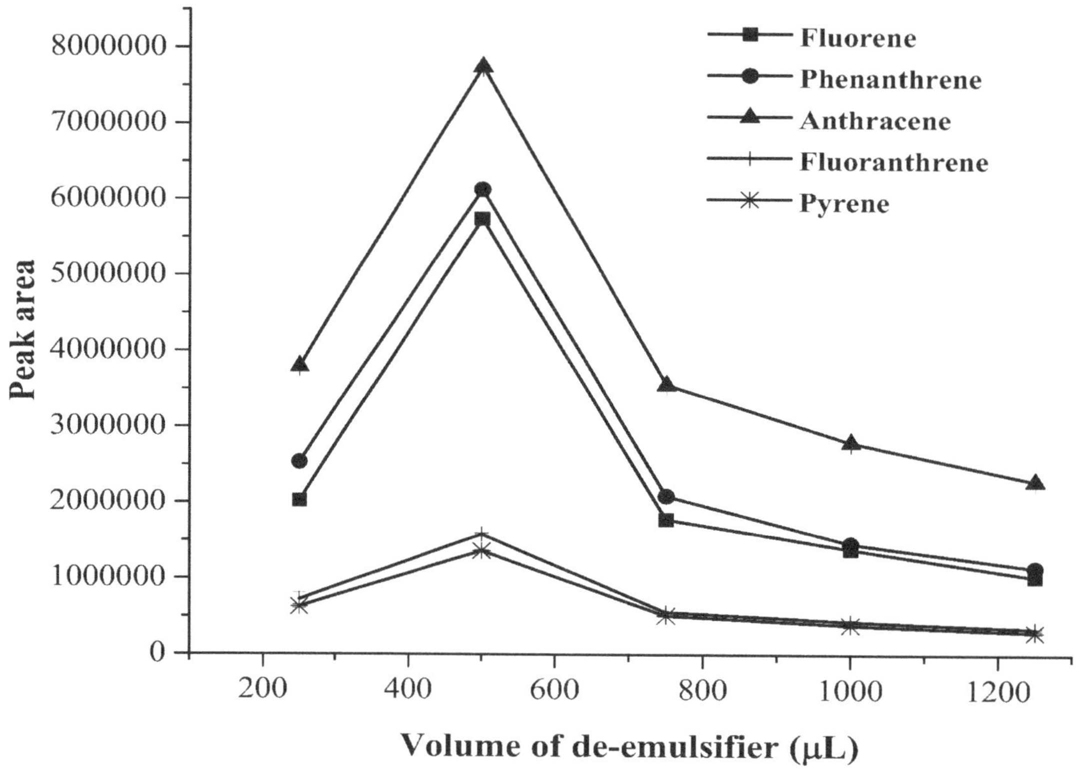

Figure 5. Effect of volume of de-emulsifier solvent on the extraction efficacy.

In summary, the optimum extraction conditions were sample solution 10.00 mL, 2.0 g NaCl, 1-dodecanol 75 μL used as extraction solvent, acetonitrile used as disperser (500 μL) and de-emulsified solvent (500 μL), and an extraction time of four minutes.

Table 1. Analytical performance of magnetic stirring assisted demulsified dispersive liquid–liquid microextraction method for determination of polycyclic aromatic hydrocarbons.

Analyte	Linear range (μg·mL^{-1})	Linear equation	R^2	EF	LOD (μg·mL^{-1})	LOQ (μg·mL^{-1})	Intra-day [a] (n = 5)		Inter-day (n = 3 × 5)	
							t_R	Area	t_R	Area
Fluorene	0.0005–1 (0.03–1) [b]	$y = (7 \times 10^6 x) + 87{,}587$	0.9992	65	0.0001	0.0003	0.15 (0.17)	2.18 (2.51)	0.39 (0.50)	6.72 (7.44)
Phenanthrene	0.0005–1 (0.03–1)	$y = (7 \times 10^6 x) + 87{,}587$	0.9973	64	0.0001	0.0003	0.17 (0.19)	2.79 (3.08)	0.22 (0.29)	2.73 (3.19)
Anthracene	0.0005–1 (0.03–1)	$y = 401{,}767x + 32{,}354$	0.9993	60	0.0003	0.0005	0.24 (0.39)	2.58 (3.70)	0.37 (0.60)	5.68 (6.25)
Fluoranthrene	0.0005–1 (0.03–1)	$y = (2 \times 10^6 x) + 20{,}850$	0.9989	66	0.0003	0.0005	0.20 (0.49)	3.55 (4.34)	0.33 (0.48)	5.51 (7.05)
Pyrene	0.0005–1 (0.03–1)	$y = (4 \times 10^6 x) + 30{,}813$	0.9997	67	0.0001	0.0003	0.32 (1.06)	3.89 (5.40)	0.48 (1.48)	7.22 (8.12)

[a] Precision was studied at standard concentration of 0.1 μg·mL^{-1}; [b] the values in parentheses are gained from direct HPLC analysis.

3.2. *Analytical Performance of the Proposed Method*

The analytical performances of the proposed method for analysis of polycyclic aromatic hydrocarbons were validated by obtaining linear range, coefficient of determination (R^2), Limit of detections (LODs), limit of quantitations (LOQs,) and enrichment factors (EFs). Table 1 summarizes the analytical performance of the proposed method. LODs and LOQs ranged from 0.0001 to 0.0003 mg·L^{-1} and 0.0003 to 0.0005 mg·L^{-1}, respectively. The calibration curve exhibited linearity over the range of 0.0005–1 mg·L^{-1} with R^2 greater than 0.99. RSDs were in the range of 0.22–1.48% and 2.73–7.22% for retention time and peak area, respectively. High EFs (60–67) were also gained. Figure 6 depicts typical chromatogram comparing analyses of (a) a mixture standard of polycyclic aromatic hydrocarbons with direct injection by HPLC and (b) a mixture of standard of polycyclic aromatic hydrocarbons after magnetic stirring assisted demulsified dispersive liquid–liquid microextraction technique.

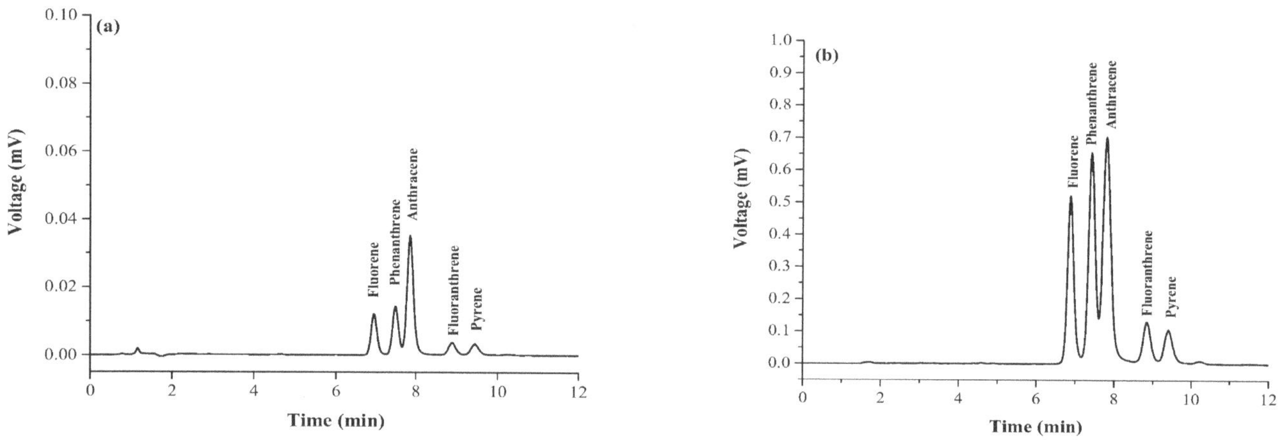

Figure 6. Chromatogram comparing analyses of (**a**) a mixture standard of polycyclic aromatic hydrocarbons with direct injection by HPLC and (**b**) a mixture of standard of polycyclic aromatic hydrocarbons after magnetic stirring assisted demulsified dispersive liquid–liquid microextraction technique: concentration of all standards was 100 ng·mL^{-1}.

3.3. *Grilled Pork Analysis*

To eliminate the matrix effect in real sample analysis, matrix-match calibration was used. A high degree of linearity was observed in the range of 0.01–0.5 mg·g^{-1} with R^2 greater than 0.995. LODs of the target analytes in real samples were studied with 3.3 × (SD y-Intercept/Average Slope) and LOQs were investigated using 10 × (SD y-Intercept/Average Slope) [24]. The obtained LODs and LOQs were in the range 0.001–0.005 mg·kg^{-1} and 0.004–0.010 mg·kg^{-1}, respectively. These are below the established EU maximum level (12 µg·kg^{-1}). For this analysis, all studied polycyclic aromatic hydrocarbons were found in the range of 0.30–1.00 mg·kg^{-1}. The results are summarized in Table 2. To test the accuracy of the proposed method, recovery was investigated by spiking standard solution at different levels (0.01, 0.05, 0.10 mg·kg^{-1}) before magnetic stirring assisted demulsified dispersive liquid–liquid microextraction. Figure 7 shows the chromatograms of grilled pork sample (untreated) (Figure 7a) as well as grilled pork samples spiked at concentrations of 0.01 mg·kg^{-1} (Figure 7b), 0.05 mg·kg^{-1} (Figure 7c), and 0.10 mg·kg^{-1} (Figure 7d). The chromatograms show that the sensitivity increases with increasing PAH concentration. The results (Table 3) show good analytical performance of the proposed method with the average recoveries for all studied analytes ranging from 82% to 99%, and good precision with relative standard deviation (RSD) less than 9%.

Table 2. The determination of polycyclic aromatic hydrocarbons in the studied grilled pork samples (n = 3).

Samples	Amount found ± SD, mg·kg^{-1}				
	Fluorene	Phenanthrene	Anthracene	Fluoranthrene	Pyrene
Grilled pork I (n = 3)	0.50 ± 0.10	1.00 ± 0.10	0.50 ± 0.10	0.80 ± 0.20	0.50 ± 0.01
Grilled pork II (n = 3)	-	0.70 ± 0.20	0.30 ± 0.20	0.50 ± 0.01	0.40 ± 0.01
Grilled pork III (n = 3)	0.30 ± 0.02	-	0.50 ± 0.20	0.70 ± 0.01	0.30 ± 0.20
Grilled pork IV (n = 3)	0.40 ± 0.10	0.40 ± 0.01	0.50 ± 0.10	0.60 ± 0.20	0.20 ± 0.10

-; not detected.

Table 3. Recovery obtained from the determination of polycyclic aromatic hydrocarbons in the studied grilled pork samples (n = 3).

Analytes	Spiked (mg·kg^{-1})	Grilled pork I		Grilled pork II		Grilled pork III		Grilled pork IV	
		RR (%)	RSD (%)	RR (%)	RSD (%)	RR (%)	RSD (%)	RR (%)	RSD (%)
Fluorene	0.01	83.45	1.52	87.64	4.73	95.73	6.63	91.57	8.76
	0.05	87.93	3.30	90.62	5.67	90.67	2.54	89.72	6.78
	0.10	84.63	6.35	96.87	3.79	85.33	2.63	98.74	6.87
Phenanthrene	0.01	82.93	5.36	92.39	4.33	87.73	3.87	89.30	7.45
	0.05	91.58	4.74	83.76	5.76	93.48	4.57	95.78	6.87
	0.10	98.74	6.79	90.78	7.63	90.63	7.86	90.87	7.86
Anthracene	0.01	82.97	4.53	87.33	3.63	92.78	6.75	90.63	3.67
	0.05	87.62	5.73	88.63	4.76	90.78	7.67	89.78	4.38
	0.10	90.73	8.63	93.74	6.78	91.87	1.56	89.90	3.78
Fluoranthrene	0.01	89.97	3.44	93.35	4.32	94.67	3.39	89.93	3.45
	0.05	91.56	4.67	89.90	6.32	91.75	4.13	83.63	6.78
	0.10	95.73	7.78	91.72	4.36	89.73	5.62	90.57	8.98
Pyrene	0.01	89.93	3.65	89.97	8.78	87.78	7.89	93.65	8.35
	0.05	90.67	4.78	91.63	5.67	88.98	6.78	91.32	5.56
	0.10	89.92	7.87	89.73	6.72	90.56	5.76	90.01	4.65

RR: Relative recovery; RSD: Relative standard deviation.

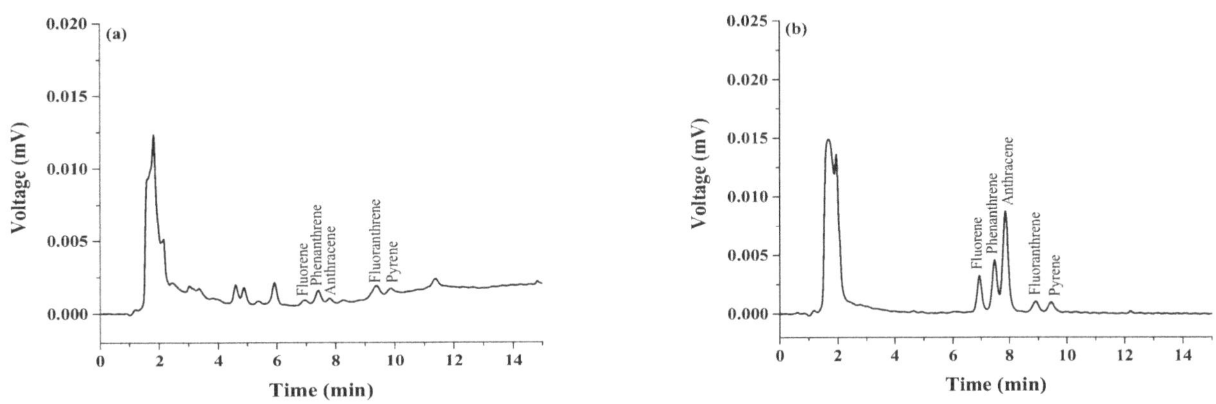

Figure 7. *Cont.*

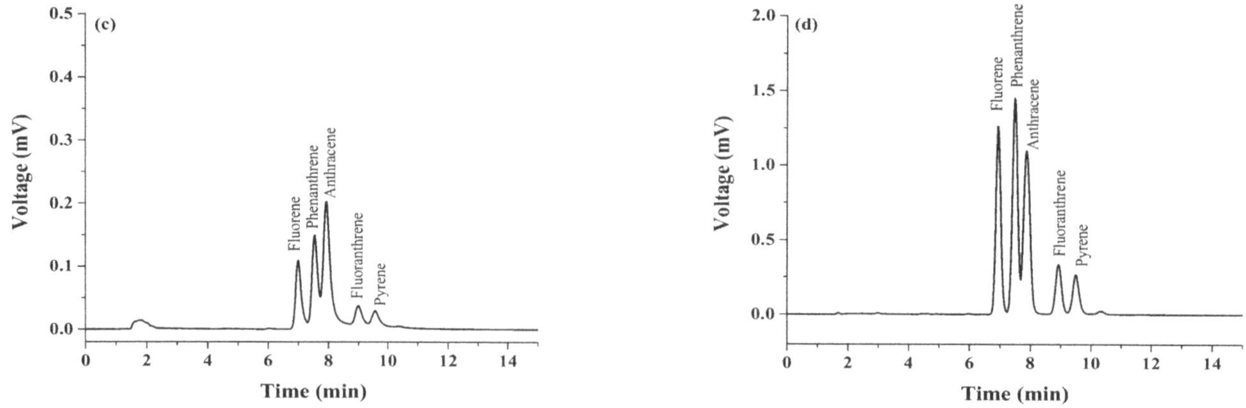

Figure 7. Chromatograms of (**a**) grilled pork sample, (**b**) grilled pork sample spiked with 0.01 mg·kg^{-1} of each polycyclic aromatic hydrocarbon, (**c**) grilled pork sample spiked with 0.05 mg·kg^{-1} of each polycyclic aromatic hydrocarbon, and (**d**) grilled pork sample spiked with 0.10 mg·kg^{-1} of each polycyclic aromatic hydrocarbon.

4. Conclusions

A fast and simple combining apparatus magnetic stirring assisted demulsified dispersive liquid–liquid microextraction has been investigated for extraction and preconcentration of polycyclic aromatic hydrocarbon residues coupled to high-performance liquid chromatographic analysis. In this procedure, a magnetic stirring-assisted process was used to accelerate the formation of fine droplets, which increased the extraction efficacy and decreased the extraction time. No centrifugation step was required. High enrichment factors, low LODs and good repeatability were obtained. The proposed method is both easy and rapid. As it uses less toxic organic solvent, it is also an environmentally friendly technique for the determination of polycyclic aromatic hydrocarbons in grilled pork samples.

Author Contributions: J.V., Y.S., and R.B. participated in designing the study. J J.V. and Y.S. conducted the study. Data was collected and analyzed by J.V. Manuscript was written by J.V., Y.S., R.B. and S.S.

Acknowledgments: This article is written in honor of the 60th birthday of Professor Supalax Srijaranai. The authors gratefully acknowledge financial supports for this research from the Center of Excellence for Innovation in Chemistry (PERCH-CIC), and the National Research Council of Thailand (NRCT). J. Vichapong would like to acknowledge the Thailand Research Fund (TRF) and the Commission on Higher Education (CHE), through the grant number MRG6180024).

References

1. Dost, K.; İdeli, C. Determination of polycyclic aromatic hydrocarbons in edible oils and barbecued food by HPLC/UV–Vis detection. *Food Chem.* **2012**, *133*, 193–199. [CrossRef]
2. Mohammadi, A.; Ghasemzadeh-Mohammadi, V.; Haratian, P.; Khaksar, R.; Chaichi, M. Determination of polycyclic aromatic hydrocarbons in smoked fish samples by a new microextraction technique and method optimisation using response surface methodology. *Food Chem.* **2013**, *141*, 2459–2465. [CrossRef]
3. Mohd Hassan, F.W.; Raoov, M.; Kamaruzaman, S.; Sanagi, M.M.; Yoshida, N.; Hirota, Y.; Nishiyama, N.; Yahaya, N. Dispersive liquid–liquid microextraction combined with dispersive solid-phase extraction for gas chromatography with mass spectrometry determination of polycyclic aromatic hydrocarbons in aqueous matrices. *J. Sep. Sci.* **2018**, *41*, 3751–3763. [CrossRef]
4. Shi, Y.; Wu, H.; Wang, C.; Guo, X.; Du, J.; Du, L. Determination of polycyclic aromatic hydrocarbons in coffee and tea samples by magnetic solid-phase extraction coupled with HPLC–FLD. *Food Chem.* **2016**, *199*, 75–80. [CrossRef]
5. Plaza-Bolanos, P.; Frenich, A.G.; Vidal, J.L.M. Polycyclic aromatic hydrocarbons in food and beverages. Analytical methods and trends. *J. Chromatogr. A* **2010**, *1217*, 6303–6326. [CrossRef]

6. Wang, W.; Ma, R.; Wu, Q.; Wang, C.; Wang, Z. Magnetic microsphere-confined graphene for the extraction of polycyclic aromatic hydrocarbons from environmental water samples coupled with high performance liquid chromatography–fluorescence analysis. *J. Chromatogr. A* **2013**, *1293*, 20–27. [CrossRef]
7. Guo, J.; Jiang, R.; Pawliszyn, J. Determination of polycyclic aromatic hydrocarbons in solid matrices using automated cold fiber headspace solid phase microextraction technique. *J. Chromatogr. A* **2013**, *1307*, 66–72. [CrossRef]
8. Luo, Y.-B.; Chen, X.-J.; Zhang, H.-F.; Jiang, X.-Y.; Li, X.; Li, X.-Y.; Zhu, F.-P.; Pang, Y.-Q.; Hou, H.-W. Simultaneous determination of polycyclic aromatic hydrocarbons and tobacco-specific N-nitrosamines in mainstream cigarette smoke using in-pipette-tip solid-phase extraction and on-line gel permeation chromatography-gas chromatography–tandem mass spectrometry. *J. Chromatogr. A* **2016**, *1460*, 16–23.
9. Lian, W.; Ren, F.; Tang, L.; Dong, D. Analysis of polycyclic aromatic hydrocarbons in cigarette samples using gel permeation chromatography clean-up by gas chromatography–tandem mass spectrometry. *Microchem. J.* **2016**, *129*, 194–199. [CrossRef]
10. Sun, M.; Feng, J.; Bu, Y.; Luo, C. Highly sensitive copper fiber-in-tube solid-phase microextraction for online selective analysis of polycyclic aromatic hydrocarbons coupled with high performance liquid chromatography. *J. Chromatogr. A* **2015**, *1408*, 41–48. [CrossRef]
11. Fernández, M.; Clavijo, S.; Forteza, R.; Cerdà, V. Determination of polycyclic aromatic hydrocarbons using lab on valve dispersive liquid–liquid microextraction coupled to high performance chromatography. *Talanta* **2015**, *138*, 190–195. [CrossRef] [PubMed]
12. Moazzen, M.; Ahmadkhaniha, R.; Gorji, M.E.; Yunesian, M.; Rastkari, N. Magnetic solid-phase extraction based on magnetic multi-walled carbon nanotubes for the determination of polycyclic aromatic hydrocarbons in grilled meat samples. *Talanta* **2013**, *115*, 957–965. [CrossRef]
13. Xue, S.-W.; Tang, M.-Q.; Xu, L.; Shi, Z. Magnetic nanoparticles with hydrophobicity and hydrophilicity for solid-phase extraction of polycyclic aromatic hydrocarbons from environmental water samples. *J. Chromatogr. A* **2015**, *1411*, 9–16. [CrossRef]
14. Payanan, T.; Leepipatpiboon, N.; Varanusupakul, P. Low-temperature cleanup with solid-phase extraction for the determination of polycyclic aromatic hydrocarbons in edible oils by reversed phase liquid chromatography with fluorescence detection. *Food Chem.* **2013**, *141*, 2720–2726. [CrossRef] [PubMed]
15. Wu, S.; Yu, W. Liquid–liquid extraction of polycyclic aromatic hydrocarbons in four different edible oils from China. *Food Chem.* **2012**, *134*, 597–601. [CrossRef]
16. Krüger, O.; Christoph, G.; Kalbe, U.; Berger, W. Comparison of stir bar sorptive extraction (SBSE) and liquid–liquid extraction (LLE) for the analysis of polycyclic aromatic hydrocarbons (PAH) in complex aqueous matrices. *Talanta* **2011**, *85*, 1428–1434. [CrossRef] [PubMed]
17. Rezaee, M.; Assadi, Y.; Hosseini, M.R.M.; Aghaee, E.; Ahmadi, F.; Berijani, S. Determination of organic compounds in water using dispersive liquid–liquid microextraction. *J. Chromatogr. A* **2006**, *1116*, 1–9. [CrossRef] [PubMed]
18. Chen, H.; Chen, R.; Li, Sh. Low-density extraction solvent-based solvent terminated dispersive liquid–liquid microextraction combined with gas chromatography-tandem mass spectrometry for the determination of carbamate pesticides in water samples. *J. Chromatogr. A* **2010**, *1217*, 1244–1248. [CrossRef]
19. Zacharis, C.K.; Tzanavaras, P.D.; Roubos, K.; Dhima, K. Solvent-based de-emulsification dispersive liquid–liquid microextraction combined with gas chromatography–mass spectrometry for determination of trace organochlorine pesticides in environmental water samples. *J. Chromatogr. A* **2010**, *1217*, 5896–5900. [CrossRef]
20. Vichapong, J.; Burakham, R.; Srijaranai, S. Vortex-assisted surfactant-enhanced-emulsification liquid–liquid microextraction with solidification of floating organic droplet combined with HPLC for the determination of neonicotinoid pesticides. *Talanta* **2013**, *117*, 221–228. [CrossRef]
21. Vichapong, J.; Burakham, R. Novel ultrasound-assisted mixed anionic–cationic surfactant-enhanced emulsification microextraction combined with HPLC for the determination of carbamate pesticides. *Anal. Methods* **2012**, *4*, 2101–2108. [CrossRef]
22. Horstkotte, B.; Suárez, R.; Solich, P.; Cerdà, V. In-syringe-stirring: A novel approach for magnetic stirring-assisted dispersive liquid–liquid microextraction. *Anal. Chim. Acta* **2013**, *788*, 52–60. [CrossRef] [PubMed]

23. Zhang, P.-P.; Shi, Z.-G.; Yu, Q.-W.; Feng, Y.-Q. A new device for magnetic stirring-assisted dispersive liquid–liquid microextraction of UV filters in environmental water samples. *Talanta* **2011**, *83*, 1711–1715. [CrossRef] [PubMed]
24. Seebunrueng, K.; Santaladchaiyakit, Y.; Srijaranai, S. Vortex-assisted low density solvent liquid–liquid microextraction and salt-induced demulsification coupled to high performance liquid chromatography for the determination of five organophosphorus pesticide residues in fruits. *Talanta* **2015**, *132*, 769–774. [CrossRef] [PubMed]
25. Regueiro, J.; Llompart, M.; Garcia-Jares, C.; Garcia-Monteagudo, J.C.; Cela, R. Ultrasound-assisted emulsification–microextraction of emergent contaminants and pesticides in environmental waters. *J. Chromatogr. A* **2008**, *1190*, 27–38. [CrossRef] [PubMed]
26. Seebunrueng, K.; Santaladchaiyakit, Y.; Srijaranai, S. Vortex-assisted low density solvent based demulsified dispersive liquid–liquid microextraction and high-performance liquid chromatography for the determination of organophosphorus pesticides in water samples. *Chemosphere* **2014**, *103*, 51–58. [CrossRef]
27. Xia, Y.; Cheng, M.; Guo, F.; Wang, X.; Cheng, J. In-syringe demulsified dispersive liquid–liquid microextraction and high performance liquid chromatography–mass spectrometry for the determination of trace fungicides in environmental water samples. *Anal. Chim. Acta* **2012**, *724*, 47–53. [CrossRef]
28. Wang, X.; Cheng, J.; Zhou, H.; Wang, X.; Cheng, M. Development of a simple combining apparatus to perform a magnetic stirring-assisted dispersive liquid–liquid microextraction and its application for the analysis of carbamate and organophosphorus pesticides in tea drinks. *Anal. Chim. Acta* **2013**, *787*, 71–77. [CrossRef]
29. Vichapong, J.; Santaladchaiyakit, Y.; Burakham, R.; Srijaranai, S. Determination of Benzimidazole Anthelminthics in Eggs by Advanced Microextraction with High-Performance Liquid Chromatography. *Anal. Lett.* **2015**, *48*, 617–631. [CrossRef]

4

Multi-(myco)toxins in Malting and Brewing By-Products

Kristina Mastanjević [1,*], Jasmina Lukinac [1], Marko Jukić [1], Bojan Šarkanj [2], Vinko Krstanović [1] and Krešimir Mastanjević [1]

1 Faculty of Food Technology Osijek, Josip Juraj Strossmayer University of Osijek, F. Kuhača 20, 31000 Osijek, Croatia; ptfosptfos2@gmail.com (J.L.); ptfosptfos@gmail.com (M.J.); vkrstano@ptfos.hr (V.K.); kmastanj@gmail.com (K.M.)

2 Department of Food Technology, University North, University Center Koprivnica, Trg dr. Žarka Dolinara 1, 48000 Koprivnica, Croatia; bsarkanj@unin.hr

* Correspondence: kristinahabschied@gmail.com

Abstract: Fungi, yeasts, and bacteria are common microorganisms on cereals used in malting and brewing industries. These microorganisms are mostly associated with the safety and quality of malt and beer, but also with the health safety of by-products used in animal nutrition. The real problem is their harmful metabolites—toxins that, due to their thermostable properties, can easily be transferred to malting and brewing by-products. Besides fungal metabolites, other toxins originating from plants can be harmful to animal health. Precise and accurate analytical techniques broadened the spectrum of known toxins originating from microorganisms and plants that can pose a threat to animal health. Multi-(myco)toxin analyses are advanced and useful tools for the assessment of product safety, and legislation should follow up and make some important changes to regulate yet unregulated, but highly occurring, microbial and plant toxins in malting and brewing by-products used for animal feed.

Keywords: Multi-toxins; mycotoxins; malting and brewing by-products; animal feed

Key Contribution: The emerging multi-toxins can represent a serious threat to human and animal health. The unknown consequences of these compounds are yet to be pinpointed, but as a precaution measure, the presence of such toxins in malting and brewing by-products should be actively monitored and, with time, included into legislation regarding animal feed.

---

1. Introduction

Beer can be produced not only from barley malt, but other cereals (malted or unmalted wheat, corn, rice, sorghum) can be added to the brew to cut production costs or improve the colloidal stability of beer. During malting and brewing processes, by-products are formed and adjusted to be reused as feed. Nowadays, modern processes can be applied and transform these by-products into functional foods or additives for the food industry. Many studies on myco- and other toxins are available in professional and scientific literature, but only a few investigations have focused on by-products formed during the malting and brewing process, such as germ and rootlets, spent grains, and spent yeast. These by-products represent a nutritious and valuable, low-cost source of feed for livestock. In terms of using it as functional food, malting and brewing (M/B) by-products can be added to different cereal products to enhance their nutritional properties. However, the above-mentioned by-products can also be contaminated with mycotoxins and other toxins originating from plants or other microorganisms. Most mycotoxins are thermo-stable and can survive the unit operations applied during the malting and brewing processes. In addition, M/B by-products are susceptible to microbial contamination and have a short shelf life (due to the original high water content). To ensure the food safety for the final consumers, whether animals or humans, it is necessary to investigate, determine, and regularly monitor the existing toxins in these by-products.

Encouraged by the recent findings of our research on multi-toxins in malting and brewing by-products [1,2], we decided to round up similar papers and point out how important and urgent this matter of emerging myco- and other toxins in animal feed is.

2. The Production of Malting and Brewing By-Products

Malting is a process of forced germination of cereal grains to develop enzymes needed to break down starch and proteins to more simple molecules (glucose, maltose, maltotriose, maltodextrines, and aminoacids), so that yeast can be employed in alcohol production. When germ and rootlets are shown, germination is halted by hot air to dry out the excess water. After that, the germinated grain is subjected to degermination, germ and rootlets get separated, and are adapted to be sold as animal feed. Since they are low in bulk density, they are mixed with barley and malt dust and small barley grains and pelletized [3]. Fungi originating from field and storage facilities thrive under malting conditions.

The brewing process consists of several continuous phases. Firstly, milled malt is subjected to mashing (slow heating with water). The residual precipitate from this process is called spent grains and the liquid part, wort, undergoes further processing to become beer.

After precipitation, wort is boiled, and, at this point, hops are added. When boiling is done, hopped wort is cooled and separated from hop residues/spent hops. Spent hops get discarded, but some of it ends up in the trub/cold break (the protein precipitate leftover after cooling of wort).

After cooling, hopped wort is inoculated with yeast and left for main fermentation in a fermentation vessel. During this time, the yeast biomass increases three- to six fold, and at the end of this phase, the yeast biomass is called spent yeast [4].

All these by-products can be utilized for human or animal consumption. However, the native microflora and accompanying produced or bio-transformed (myco)toxins can be a cause of serious economy- and health-related problems for humans and animals.

3. Common Fungal Species in Malting and Brewing

The most common fungal species in malting and brewing by-products are different Fusarium species. *F. acuminatum*, *F. anthophilium*, *F. avanceum*, *F. cerealis* (crookwellense), *F. chlamydosporum*, *F. culmorum*, *F. equiseti*, *F. graminearum*, *F. heterosporum*, *F. nygamai*, *F. oxysporum*, *F. poae*, *F. proliferatum*, *F. sambucinum*, *F. semitectum*, *F. sporotrichoides*, *F. subglutaminans*, *F. tricintum*, and *F. verticilioides* all belong to genus *Fusarium*, but only some of them show inclination to invade grains in our region. According to Krstanović et al. [5], the most prevalent species in Eastern Croatia is *F. graminearum*, being present in both wheat and barley samples (Figure 1). The highest contaminations (20%) were detected in samples from Nova Gradiška. *F. culmorum* was found in significantly lower percentages, with 3.1% being the highest contamination, and statistically did not represent any threat to human health. Generally, colder Northern European regions support the growth of *F. culmorum*, and warmer parts [6] of Europe fight battles with *F. graminearum*. However, the global climate changes caused a shift in Fusarium species' distribution. Namely, according to Parikka et al. [7], *F. culmorum* lost its place in being the prevalent specie in Nordic countries. *F. graminearum* is expected to populate the Northern European areas very soon. Ward et al. [8] reported on toxicity of *F. graminearum* in North America and similar research was conducted in China, where the results indicated that more aggressive isolates (producing a wider variety of mycotoxins) are taking over [9].

(a) (b)

Figure 1. Heavily infected malt (*F. graminearum*) before (**a**) and after (**b**) kilning.

Other fungal species invading cereals and making problems in malting and brewing industries include: *Alternaria* spp., *Epicoccum* spp., *Penicillium* spp., *Aspergillus* spp., etc. In research conducted on Slovakian barley, many fungal species were detected: *Acremonium strictum*, *Alternaria* spp., *Aspergillus niger*, *Aspergillus flavus*, *Botrytis cinerea*, *Cladosporium herbarum*, *Cladosporium cladosporioides*, *Cochliobolus sativus* (*Helminthosporium sativum*), *Epicoccum nigrum*, *Fusarium* spp., *Chrysonilia sitophila*, *Microdochium nivale*, *Mucor* sp., *Nigrospora sphaerica*, *Penicillium* sp., *Phaeosphaeria nodorum* (*Septoria nodorum*), *Pyrenophora teres* (*Helminthosporium teres*), *Rhizoctonia* sp., *Rhizopus nigricans*, *Stemphylium* sp., *Tanatephorus cucumeris* (*Rhizoctonia solani*), *Trichoderma* sp., and *Verticillium albo-atrum* [10].

In a study conducted by Gonzalez Pereyra et al. [11], the toxigenic species belonging to the genera, *Fusarium*, *Aspergillus*, *Penicillium*, *Cladosporium*, *Geotrichum*, and *Alternaria*, were identified in malted barley and brewer's spent grains. The detected *Fusarium* spp. made 30% and *Aspergillus* spp. made 27.3% of contamination, and 30% of brewer's grain samples were contaminated with *Mucorales*. Most prevalent *Fusarium* species in brewer's spent grain were *Fusarium verticillioides* (50%) and *Fusarium proliferatum* (25%) while *Fusarium equiseti* (12.5%) and *Fusarium oxysporum* (12.5%).

Piacentini et al. [12] published the results from their research involving the detection of fungal species and mycotoxins (deoxynivalenol (DON) and fumonisins (FB_1 and FB_2)) on malting barley. They also reported that the most prevalent genera were *Fusarium*. *F. graminearum* and *F. verticillioides* were the dominant species, with an incidence of occurrence of 26% and 12%. Other fungal genera determined in their research included *Penicilium*, *Alternaria*, and *Rhizophus*. Mycotoxin analysis revealed that DON contaminated 18% of samples while FB_1 and FB_2 were detected at a lesser extent, 10% for FB_1 and 1% for FB_2.

Fungi on malting barley are problematic because of the favourable conditions [13], which support fungal mycelium growth and mycotoxins' production. Fungi-thriving process conditions are assisted by the easily available nutrients from the grain, emergence in water, exposure to lower temperatures (10–14 °C) and aeration, high air humidity (>90%), and overall high grain moisture during germination and the first phase of kilning (cca. 45% grain moisture). Subsequent mycotoxin biosynthesis occurs at almost every stage of malting and brewing. Colder temperatures during steeping and germination enable fungi to produce mycotoxins. Mycotoxins synthesis occurs [2,14] during the whole malting process, especially after kilning, and continues during brewing. The fermenting stage also allows mycotoxins' content increase. In case there are some leftover spores, mycotoxins can be found even during aging. Some toxins stem from yeast (tryptophol) and are produced after the inoculation of wort. According to our findings [2], mycotoxin concentrations vary during the brewing stage, and after inoculation, mycotoxins (mainly DON) can show a significant increase in comparison to wort.

4. General Toxicity and Occurrence of Multi- and Myco-Toxins

Fungal infections are not unusual and they often contaminate cereals, diminishing their quality and lowering their marketability. However, the most important problem related to fungi is the occurrence of different myco- and other toxins that, directly or indirectly, affect human and animal health.

The sources of human exposure to mycotoxins can be direct (from plant-derived foods contaminated with toxins) or indirect (through a carryover of mycotoxins and their metabolites in animal products, such as meat and eggs) [15]. The exposure to air and dust containing toxins can also cause serious consequences to human health [16]. Therefore, it is important to wear suitable protection (masks-preferably with HEPA (High Efficiency Particulate Air) filter, goggles) during handling of infected grain [17].

As is well-known, mycotoxins influence vital organs and tissues, having carcinogenic properties, inducing immunosuppression and reproductive problems in animals and humans [15]. Most mycotoxins affect the health of humans and animals by causing nephropathy, infertility, cancer, or death [18]. Intense research on mycotoxins reveals that cca. 400 mycotoxins are known till today [19] and about 200 of them belong to the group of trichothecenes [18]. Especially dangerous are mycotoxins that accumulate in animal tissues, meat, and fat, because they can cause problems when ingested by humans.

Some studies report aflatoxins, fumonisins, trichothecenes, ochratoxin A, and zearalenone as the most significant mycotoxin in the barley-beer chain [20,21], but for animal feed, only seven mycotoxins: Aflatoxin B_1 (AFB_1), DON, zearalenone (ZEA), fumonisins (FUM) and ochratoxin A (OTA), T-2, and HT-2 are regulated by the EU (European Union) legislation [22–24] and are under strict governmental control.

Aflatoxins (AFs) are produced by different species of *Aspergillus* fungi (*A. flavus*, *A. parasitous*, and *A. nomius*). They can occur in different foods (groundnuts, tree nuts, maize, corn, rice, figs, and other dried foods, spices, crude vegetable oils, and cocoa beans) as a result of fungal contamination before and after harvest. Aflatoxins are potentially toxic, carcinogenic, mutagenic, and immunosuppressive compounds [25]. There are 18 different types of AFs, with the major members being AFB_1, AFB_2, AFG_1, AFG_2, AFM_1, and AFM_2. AFB_1 is the only mycotoxin whose maximum content in mg/kg (relative to a feedstuff moisture content of 12%) is determined by the Commission Directive 2003/100/EC [22]. Even though AF's can end up in beer, the European Union has not set a maximum allowable limit for AFB_1 in beer, although there is a limit for barley and malt [26]. AFB_1 has been described as a high carcinogenic natural toxin and has been classified as a Group 1 (carcinogenic to humans) [27]. AFB_1 can be found in beer [28–30] and raw materials, such as barley [31,32] and corn [33,34], used for the production of beer. Thus, AFB_1 can be found in brewing by-products that can be used as animal feed and cause health problems for animals, as confirmed by Gerbaldo et al. [35] and Gonzalez Pereyra et al. [11].

Most *Fusarium* spp. produce trichothecenes, and amongst them, *F. graminearum*, *F. culmorum*, *F. proliferatum*, and *F. equiseti* are characterized as main producers of trichotecenes [36]. However, many other fungal species can produce trichotecenes: *Myrotecium*, *Trichoderma*, *Cephalosporium*, *Verticimonosporium*, and *Stachybotrys* [37]. According to Krstanović et al. [5], the most prevalent species on cereals in Eastern Croatia are *F. graminearum* Schw. (teleomorph *Gibberella zeae*) and *F. culmorum* (Wm. G. Sm.) Sacc. (teleomorph unknown) can be found at a lesser extent. DON, being the most prevalent mycotoxin in cereals, is identified as one of the indicators of the quality and safety of barley and wheat used for malting [38]. In the research by Velić et al. [39] conducted with *F. graminearum*, a correlation between fungal contamination and DON appearance was shown. DON is identified as a protein synthesis inhibitor in eukaryotic cells due to its ability to bind to ribosomes. Low to moderate concentrations of DON can affect the health of animals, causing anorexia (lower economical profit), and higher doses can cause vomiting [40,41]. Besides DON, other trichotecenes can be found in malting and brewing by-products, and many of them occur in coherence with their derives. Many trichotecenes are well described in scientific literature and are well known in popular circles. They are nivalenol (NIV),

T-2 toxin, diacetoxyscirpenol (DAS), etc. However, they have not been yet included into legislative regarding barley, malt, or beer for that matter. This can represent a serious problem since many of them can be found in malting and brewing by-products used as animal feed [1,2].

ZEA is one of the infamous mycotoxins that acts similarly to estrogen and is related with clinical manifestations of various estrogenic effects in humans and farm animals. The exposure to this mycotoxin can been seen in pubertal changes in young children in Puerto Rico [42] and gynecomastia with testicular atrophy in southern Africa [43]. Although ZEA is mostly determined in maize, it can easily be found in other cereals, such as wheat and barley. Maize is one of the unmalted cereals that can be added to a brew to cut costs and improve the colloidal stability of beer, and thus can cause damage to the brewing industry if ZEA is present in the raw material. ZEA has different modified forms that derive from two different metabolic phases:

1. Phase I includes: α-zearalenol, β-zearalenol (α-ZEL and β-ZEL), zearalenone-4-O-β-glucoside (ZEA-14Glc), zearalenone-16-O-β-glucoside (ZEA-16Glc), and zearalenone sulphate (ZEA-14-sulphate, ZEA-14S), and many of them have the ability to transform into their basic form during digestion in the mammalian intestinal tract [44]. ZEA-14Glc can generate complex compounds, such as zearalenone-malonyl-glucoside (ZEA-MalGlc), zearalenone-di-hexoside (ZEA-di-hexoside), and zearalenone-pentosylhexoside [45].
2. Phase II derivate, β-zearalenol-4-glucoside (β-ZEL-4Glc), may be formed by maize itself during xenobiotic detoxification [46] and can easily be transferred back into the original form-zearalenone [47].

Binder et al. [48] conducted research in which they monitored the biotransformation of two plants and one fungal metabolite: ZEN-14-sulfate, ZEN-14-O-β-glucoside, and ZEN-16-O-β-glucoside in pigs. Although the total biological recoveries in urine and faeces were below 50%, the authors attributed this to the extensive metabolization by intestinal bacteria to yet unknown metabolites and offered in conclusion that the complete hydrolyzation of all monitored metabolites (ZEN-14-sulfate, ZEN-14-O-β-glucoside, and ZEN-16-O-β-glucoside) occurred, contributing to the total ZEN toxicity.

Fumonisins are mycotoxins produced by fungi belonging to the genus, *Fusarium*. There are many detected fumonisins, but type B fumonisins are most prevalent in feed and food chains [49]. Fumonisins B_1, B_2, B_3, and B_4 all comprise this group, with FB_1 and FB_2 being the most common members [50,51]. Although primarily found in corn, fumonisins can be found in other cereals. Symptoms of fumonisin intoxication are various [52] for various animals. Pigs, for example, can suffer from porcine pulmonary edema [53] and horses can manifest hemorrhagic-liquefactive brain lesions. In humans, neural tube defects and growth retardation in children can occur [54]. According to Voss and Riley [52], FB_1 is characterized as a potent carcinogen as some reports link exposure to high doses of FB_1 to human esophageal cancers [55]. However, no strong empirical relationship has been established between the manifestation of cancer and FB_1, so FB_1 is still being considered as possibly carcinogenic to humans and is categorized as a group 2B carcinogen [56]. Fumonisins can also exist in modified forms, which Braun and Wink [56] call "cryptic" fumonisins. They form when basic fumonisins undergo hydrolyzation, usually during digestion in the gastrointestinal tract. In 2018, EFSA (European Food Safety Authority) [57] issued a scientific opinion on fumonisins and their modified and hidden forms in feed. According to EFSA, modified forms are extractable from the matrix and can be formed during biotransformations in fungal, plant, or animal organisms or during thermal and chemical processes to which feed and food can be subjected to. The term 'hidden forms' refers to the fraction of fumonisins strongly bonded with non-covalent bonds to the matrix and they are basically non-extractable.

Ochratoxins are secondary metabolites primarily produced by *Penicillium* and *Aspergillus* species. This group involves ochratoxin (OTA), its methyl ester, its ethyl ester also known as ochratoxin C (OTC), 4-hydroxyochratoxin A (4-OH OTA), ochratoxin B (OTB) and its methyl and ethyl esters, and ochratoxin α (OTα) [58]. OTA is nephrotoxic, carcinogenic, teratogenic, genotoxic, and immunotoxic [59]. OTA is one of the highest occurring ochratoxins and is incorporated into legislative recommendations and regulations. As reported by Huff and Doerr [60], OTA and aflatoxin can act synergistically and significantly retard the growth of broiler chickens.

T-2 and HT-2 are *Fusarium* mycotoxins incorporated into EU legislative. Like other mycotoxins, their occurrence is climate dependent and they usually occur together. They are toxic to most animals and humans as they interfere with protein synthesis and haematopoiesis inhibition, lymphoid depletion, and necrotic lesions [61]. Plants and yeasts can modify these mycotoxins into different glucosides: T-2 3-O-glucoside (T-2Glc) and HT-2-O-3-glucoside (HT-2Glc), HT-2-di-glucoside (HT-2di-Glc) [62], di-glucosides (T-2di-Glc, and HT-2di-Glc) [63]. Yeasts of the genus, *Trichomonascus* and *Blastobotrandy*, have three biotransformation levels for the T-2 toxin: Acetylation to 3-acetyl-T-2; glycosylation; and the removal of an isovaleryl group from the molecule, leading to the formation of neosolaniol [64]. Other modified forms of T-2 and HT-2 mycotoxins include: HT-2-malonyl-glucoside, hydroxy-HT-2-glucoside, dehydro-HT-2-glucoside, T2-triol-glucoside, and T-2-feruloyl-T-2 toxin [65].

Emerging mycotoxins in food and feed, such as enniatins (ENNs) and beauvericin (BEA), have been under the scientific light for the past few years. The toxicity of enniantin B (ENNB) and BEA has been recently described by a group of authors in EFSA's [57] external scientific report, where it was reported that ENNB is genotoxic to mice and BEA targeted thyroid, kidneys, and reproductive systems in both sexes. Repeated oral exposure to BEA seems to affect male mice more, but further studies should be conducted.

Besides these mycotoxins and their derivates stemming from plants itself, there are many others that can be found in malting and brewing by-products. Some of them are reported in a recent publication by Mastanjević et al. [1]. Emerging toxins are currently the focal point of the scientific community since these toxins are not included into legislation and some of them are very poorly studied; in a sense, there is little or no information of how they affect human or animal health.

5. Multi- and Myco-Toxins in Malting and Brewing By-Products

Since M/B by-products are rich in proteins, digestible fibre material, vitamins, and minerals, their utilization as feed supplements has become an alternative for animal feed [11,66,67]. These by-products represent low cost and nutritious animal feed, but also promote rumen fermentation and digestion and overall act beneficially on animal health [11].

Cavaglieri et al. [67] confirmed the correlation between fungal contamination and natural incidence of selected mycotoxins in barley rootlets. Namely, Cavaglieri et al. [67] found FB_1 contamination in 100% of barley rootlet samples. However, FB_1 levels found in this substrate were subjectively high (254–2043 μg/kg); while Gerbaldo et al. [35] carried out another study that confirmed AFB_1 contamination on brewer's grain.

In a research conducted by Gonzalez Pereyra et al. [11], the detected mycotoxins in malting barley and BSG (brewer's spent grain) were fumonisin B_1 and AFB_1. Although FB_1 toxin levels were below the recommended values (104–145 μg/kg), the occurrence of the fumonisins' contamination in these samples was 100%. Aflatoxin B_1 was detected in 18% of brewer's spent grain samples (19–45 μg/kg) and was presumably produced during storage [68]. Other mycotoxins (AFB_2, AFG_1, AFG_2, or ZEA) were not found in the analysed samples. This indicates that, in the case of brewer's grain and other barley by-products intended for animals, mycotoxin contamination increases with the storage time if storage conditions are not closely controlled. Similar research was conducted by Habschied et al. [69], in which the ZEA concentration in barley germ and rootlets after degermination increased with prolonged storage time and higher water activity (a_w).

Barley germ and rootlets are the result of the ending phase of the malting process [67]. After the final phase of drying, degermination is carried out. Germ and rootlets have a lower water content, but because of their high nutritive value in amino acids and fat content, they are susceptible to microbial degradation. Inappropriate storage and management of these materials can lead to the loss of nutritive substances and mycotoxins concentration increase [67,69]. Figure 2 shows a significant reduction of germ/rootlets in malt severely contaminated with *Fusarium graminearum* (sample b). The higher the initial *Fusarium* contamination, the poorer the germ development will be.

Figure 2. Germ/rootlets of (**a**) healthy malt (**b**) severely infected malt with *F. graminearum*.

As mentioned before, Mastanjević et al., Habschied et al., and Krstanović et al. investigated several cases in which mycotoxins and multi-toxins were determined in M/B by-products [1,2,69–71]. They reported that not only mycotoxins can be found in such by-products, but also in many other toxins and originating from plants or other sources.

Brewer's spent grain is obtained after the filtration of mash and are a moist brewing by-product. Spent grain can be used as animal feed in various forms: fresh (wet) (Figure 3), ensiled, or dehydrated (dried) grain [72]. Spent grain is often used as a supplement for lactating cows [73]. Since spent grains and spent yeast are very nutritious, and contain high amounts of available water [11], they are susceptible to microbial growth and are very perishable. Before storage, any excess water must be removed to ensure microbial safety and to prolong the shelf life. Spent yeast also must be treated in additional processes to inactivate the living yeast cells because ruminants have a complex mixture of micro flora to help breakdown the cellulose into simpler carbohydrates they can use for energy. Although small amounts of live yeast can be safe for ruminants, excessive amounts of live yeast cells can disrupt the gut microbiology. This ends up as a gas build up that can block their airways, causing suffocation and even death. Drying also improves the storability of these by-products and enables easier transportation and handling [73].

Figure 3. Spent grains and cold break (protein residue) after filtration.

A research conducted by Mastanjević et al. [2] followed the transition of eight *Fusarium culmorum* mycotoxins (fusarenone-X (FUS-X), 3-ADON, DAS, HT-2, T-2, DON, NIV, ZEA) from starting wheat to beer and associated by-products (steeping water, germ/rootlets, spent grains, and spent yeast). The results showed that steeping water can withhold extremely high DON concentrations amounting up to 20,326 μg/kg. Germ/rootlets also retain high amounts of DON (5636 μg/kg) and NIV (1691 μg/kg). This is in accordance with an earlier research by Krstanović et al. [71] conducted on wheat samples with different initial *F. culmorum* infection levels (0–50%). Namely, germ/rootlets here too retained significantly high DON concentrations (5000–12,600 μg/kg), depending on the infection level. The germ is rich in fats and made of soft tissue, hence it is a place of the most intense fungal growth, which explains why DON levels in germ are so high. Similar research was conducted by Habschied et al. [69]

on barley malt where ZEA was the followed mycotoxin. This research showed that ZEA levels in malt bran and germ samples were significantly higher than in malt flour (endosperm) in almost all growing conditions.

This affirms that it is necessary to monitor and determine the exact types and quantity of mycotoxins in cereals, final products, and by-products to ensure the food and feed safety [70]. In addition, according to Habschied et al. [69], storage conditions for these by-products are of immense importance. Thus, attention should be paid to ensure the suitable storage conditions.

Except mycotoxins, many malting and brewing by-products contain other toxins stemming from the plant itself, residual fungicides, or herbicides applied in the field. Namely, plants possess two detoxifying mechanisms (chemical modification and compartmentalization) that help them transfigure (myco)toxins to a less toxic form [47]. To explain this, Berthiller et al. [47] and Rychlik et al. [74] described deoxynivalenol-3-glucoside (D-3-G) as a product of a detoxifying plant mechanism, where glycosylation of DON occurs via a conjugation reaction and the result is D3G. Modified mycotoxins usually co-occur with free mycotoxins [47], and can even exceed the concentration of free form in processed foods and feed [75,76].

In research by Mastanjević et al. [1], a wide array of multi-toxins was detected in malting and brewing by-products. Most abundantly found mycotoxins in spent grains and spent yeast were lotaustralin and tryptophol. Other detected myco- and multi-toxins were: Aurofusarin, beauvericin, brevianamid F, chrysogin, culmorin, 5-hydroxyculmorin, 15-hydroxyculmorin, deoxynivalenol, deoxynivalenol-3-glucoside, linamarin, tentoxin, and zearalenon. Some of these toxins can be found in Table 1 and Figure 4.

Table 1. Known and emerging detectable toxins that can be found in M/B by-products.

		Mycotoxins	Source
Fungal	*Fusarium* mycotoxins	Deoxynivalenol, Nivalenol, 3-Acetyldeoxynivalenol, 15-Acetyldeoxynivalenol, Zearalenon, HT-2 toxin, T2-toxin, Moniliformin, Chrysogin, Culmorin,, Aurofusarin, Fumonisin B_1, Fumonisin B_2, Fumonisin B_3, Beauvericin, Enniantins (A, A_1, B, B_1)	[1,2,12,15,77]
	Other species mycotoxins (*Aspergillus, Alternaria*)	Aflatoxin B_1, Tentoxin, Brevianamid F	[1,11,77]
Yeast metabolite		Tryptophol	[1]
Plant metabolites		DON-3-glucoside, Lotaustralin, Linamarin	[1,77]

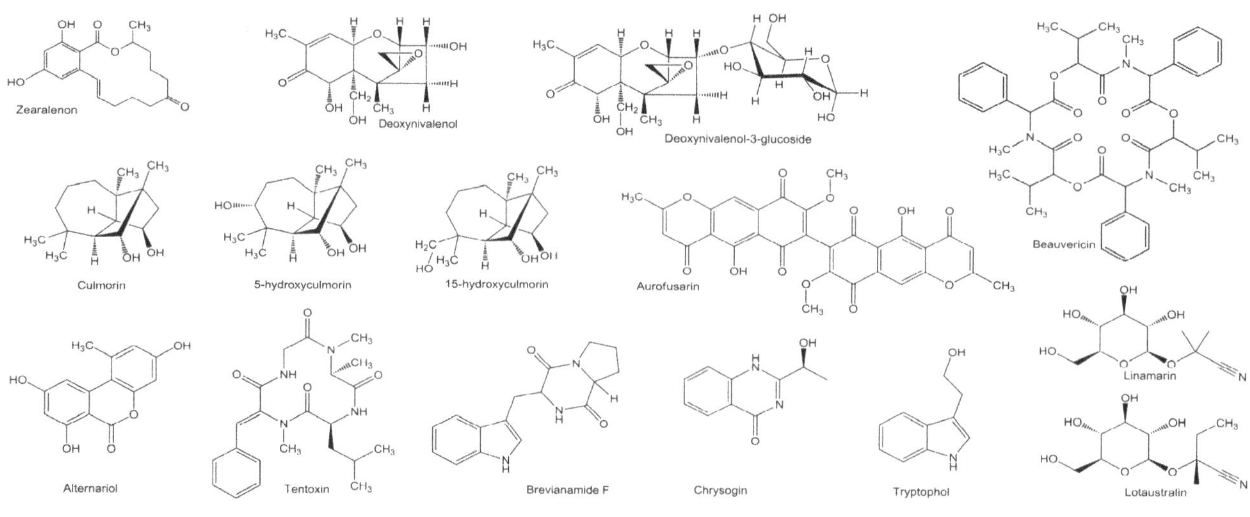

Figure 4. Chemical structure of some of known and emerging toxins.

Since the spectrum of found mycotoxins is very broad, an update regarding the detection and determination of emerging multitoxins in legislation should be made to ensure the health of animals and, subsequently, humans. Surveillance of mycotoxin concentrations should be regular and stricter and analysis spectrum should be expanded, especially on masked mycotoxins (3-acetyl-DON (3-ADON) and 15-acetyl-DON (15-ADON), cyanogenic glucosides (D-3-G), and yeast secondary metabolites (tryptophol).

6. Malting and Brewing By-Products Adding Value to Different Food Products

Besides the use in animal feed, malting and brewing by-products can be used to enrich and improve the functional properties of different food products [4,78,79]. Considering the increasing, almost pandemic, occurrence of obesity, diabetes, cancer, and cardiovascular diseases in today's world, a general idea of functional food derived from food industry by-products has been diligently implemented in contemporary research investigations.

Being produced in large quantities and available all year long, brewing by-products, such as spent grains, spent hops, and spent yeast, can be utilized in food production by adding nutritive value to different products. The main constituents -minerals, nitrogen, and carbon—are the most important compounds from a biotechnological point of view and they have a tremendous potential for use in biotechnological processes. As mentioned before, being rich in carbon and nitrogen makes them suitable for the extraction of sugars, proteins, acids, and antioxidants [4].

Brewer's spent grain and spent yeast can also be applied in the baking industry as a source of dietary fibers (β-glucans). BSG contains around 17% of cellulose and 28% of non-cellulose polysaccharides (arabinoxylans and lignine). According to some research, BSG retains significant amounts of polyphenols [80–82]. Ktenioudaki et al. [83] published an article where brewer's spent grain can be used as functional food in baked snacks, specifically in breadsticks in this case. However, as reported by the authors, some baking properties were significantly affected and this can be correlated with the high fiber content in BSG. Namely, breadsticks containing BSG were darker, less crispy, and of a lower baking volume. Nevertheless, the quality of breadsticks after 50 days of storage remained unchanged.

From an environmental point of view, the elimination of industrial by-products represents a sane solution to pollution problems and deserves attention.

Much research has been conducted and is still ongoing to reveal and determine the chemistry and biology of mycotoxin production in fungi on many cereals. However, the literature data on mycotoxins in wheat malt and beer are modest and even less data are found on mycotoxins in malting and brewing by-products. Therefore, it is relevant to research and monitor the occurrence of these mycotoxins in wheat food and feed, including malting and brewing products and by-products.

While the FDA proposed a rule on Good Manufacturing Practice, which suggested that separating and selling by-products derived from malting and brewing industries should obligate the manufacturers to honor food safety regulations, the Beer Institute and American Malting Barley Association Inc. strongly disagreed because this would mean additional costs for maltsters and brewers on which they could not agree upon [84].

7. The Disposal/Use of Infected Grains

Considering all the risks infected grains can do to the economy in malting and brewing industries, the end question is what to do with infected and contaminated grains. This is rarely mentioned in the scientific papers, and is of great concern for growers, farmers, and M/B industries. However, there are some operations that can reduce the damage caused by fungal infection. Namely, diluting the contaminated grains (only for feed) with healthy ones, removing the outer layer of contaminated grain, and applying chemical and physical decontamination can lessen the detrimental consequences of fungal contamination. Alkaline treatments, binders that can reduce the bioavailability of mycotoxins in the intestinal tract, detoxifiers, and enzymes can be beneficial to toxins' reduction. Biocontrol is also a possible method. However, the question remains: What to do with heavily infected *Fusarium* grains? Handling the infected grains demands appropriate equipment since grain dust can contain DON and other toxins or fungal spores. During storage and handling, such grains should be excluded from healthy ones and one should be careful not to disperse the spores in the local fields [85]. Badly infected grains could be burnt or possibly used in digestion for biogas or bioethanol production. Composting is also an option. Agnew and Kirsch [86] reported the results from their project, where they discovered that composting seemed to be the safest and most efficient way to eliminate DON from the grains. However, it should be taken into an account that other microorganisms from the composting pile probably modified DON into a less toxic form. Namely, modified forms appear to be less toxic, but when degraded in the gut, they reform into their basic toxic form.

In a nutshell, so far, the most acceptable methods for mycotoxins' decontamination are [87]:

a. Clean and blend—with the uncontaminated grain; exclusively for animal feed;
b. Dumping—grains with high toxin levels can be disposed in a bush, slough, or in a hole;
c. Gravity sorter—grains infected with *Fusarium* have a lower weight and are easy to sort with this technique;
d. Burning—the most common method of disposal. This requires an investment in a grain-burning stove, but the produced heat can be used for heating. However, according to Agnew and Kirsch [86], ash can still contain DON and fungal spores and thus represents a danger to human health;
e. Composting;
f. Anaerobic digestion; and
g. Gamma irradiation [88].

8. Conclusions

Literature data indicate that malting and brewing by-products withhold significant amounts of various mycotoxins and for that reason, attention should be paid to ensure the health of animal feed. It is necessary to monitor multi-(myco)toxin concentrations and also to expand the analysis spectrum on masked toxins, since masked mycotoxins can be degraded during digestion by lactic acid bacteria in the intestinal track and recover the toxicity of unmasked mycotoxin [47,89,90]. New masked forms of DON (DON sulphates) that can occur in grains naturally have been confirmed recently [89], but no occurrence or metabolism studies were made. However, in the several research papers Mastanjević et al. published [1,2], the problem of mycotoxins in malting and brewing by-products was referred to.

Mycotoxins are retained in germ/rootlets, spent grains, and spent yeast depending on the genotype susceptibility, starting mycotoxin concentrations, and fungicide treatment [91]. Since the spectrum of found mycotoxins is so broad, as confirmed by Mastanjević et al. [1], this calls for attention and updates on legislation to ensure the health of animals and, subsequently, humans.

According to the literature [68,69], suitable storage conditions for these by-products are indisputably important and should be provided (low a_w and temperature and shorter storage time).

Author Contributions: Writing review & Editing, K.M. (Kristina Mastanjević); Conceptualization, J.L.; Investigation, M.J. and B.Š.; Visualization, V.K.; Data Acquisition, K.M. (Kristina Mastanjević); Supervising, K.M. (Krešimir Mastanjević), all contributed equally.

Acknowledgments: Many thanks to all the authors who took part in writing this manuscript and showed a professional and scientific approach to the subject.

References

1. Mastanjević, K.; Šarkanj, B.; Warth, B.; Krska, R.; Sulyok, M.; Mastanjević, K.; Šantek, B.; Krstanović, V. *Fusarium culmorum* multi-toxin screening in malting and brewing byproducts. *LWT* **2018**, *98*, 642–645. [CrossRef]
2. Mastanjević, K.; Šarkanj, B.; Šantek, B.; Mastanjević, K.; Krstanović, V. *Fusarium culmorum* mycotoxin transfer from wheat to malting and brewing products and by-products. *World Mycotoxin J.* **2018**, in press. [CrossRef]
3. MAGB. Available online: http://www.ukmalt.com/malting-co-products (accessed on 16 March 2018).
4. Mussato, S.I. Biotechnological potential of brewing industry by-products. In *Biotechnology for Agro-Industrial Residues Utilization*; Singh nee' Nigam, P., Pandey, A., Eds.; Springer: Dordrecht, The Netherlands, 2009; pp. 313–326.
5. Krstanović, V.; Klapec, T.; Velić, N.; Milaković, Z. Contamination of malt barley and wheat by *Fusarium graminearum* and *Fusarium culmorum* from the crop years 2001–2003 in Eastern Croatia. *Microbiol. Res.* **2005**, *160*, 353–359. [CrossRef]
6. Snijders, C.H.A. Systemic fungal growth of *Fusarium culmorum* in stems of winter wheat. *J. Phytopathol.* **1990**, *129*, 133–140. [CrossRef]
7. Parikka, P.; Hakala, K.; Tiilikkala, K. Expected shifts in *Fusarium* species' composition on cereal grain in Northern Europe due to climatic change. *Food Addit. Contam. A* **2012**, *29*, 1543–1555. [CrossRef]
8. Ward, T.J.; Clear, R.M.; Rooney, A.P.; O'Donnell, K.; Gaba, D.; Patrick, S.; Starkey, D.E.; Gilbert, J.; Geiser, D.M.; Nowicki, T.W. An adaptive evolutionary shift in Fusarium head blight pathogen populations is driving the rapid spread of more toxigenic *Fusarium graminearum* in North America. *Fungal Genet. Biol.* **2008**, *45*, 473–484. [CrossRef]
9. Zhang, H.; Van der Lee, T.; Waalwijk, C.; Chen, W.; Xu, J.; Xu, J.; Zhang, Y.; Feng, L. Population analysis of the *Fusarium graminearum* species complex from wheat in china show a shift to more aggressive isolates. *PLoS ONE* **2012**, *7*, e31722. [CrossRef]
10. Roháčik, T.; Hudec, K. Fungal infection of malt barley kernels in Slovak Republic. *Plant Prot.Sci.* **2007**, *43*, 86–93. [CrossRef]
11. Gonzalez Pereyra, M.L.; Rosa, C.A.R.; Dalcero, A.M.; Cavaglieri, L.R. Mycobiota and mycotoxins in malted barley and brewer's spent grain from Argentinean breweries. *Lett. Appl. Microbiol.* **2011**, *53*, 649–655. [CrossRef]
12. Piacentini, K.C.; Savi, G.D.; Pereira, M.E.V.; Scussel, V.M. Fungi and the natural occurrence of deoxynivalenol and fumonisins in malting barley (*Hordeum vulgare* L.). *Food Chem.* **2015**, *187*, 204–209. [CrossRef]
13. Mastanjević, K.; Krstanović, V.; Mastanjević, K.; Šarkanj, B. Malting and Brewing Industries Encounter *Fusarium* spp. Related Problems. *Fermentation* **2018**, *4*, 3. [CrossRef]
14. Habler, K.; Hofer, K.; Geissinger, C.; Schuler, J.; Huckelhoven, R.; Hess, M.; Gastl, M.; Rychlik, M. Fate of *Fusarium* toxins during the malting process. *J. Agric. Food Chem.* **2016**, *64*, 1377–1384. [CrossRef]
15. CAST (Councile for Agricultural Science and Technology). *Mycotoxins: Risk in Plant, Animal and Human Systems*; Task Force Report No. 139; CAST: Ames, Iowa, 2003.
16. Jarvis, B.B. Chemistry and toxicology of molds isolated from water-damaged buildings, Mycotoxins and Food Safety. *Adv. Exp. Med. Biol.* **2002**, *504*, 43–52.
17. Pennsylvania Mycotoxin Management Guidance Document. Deoxynivalenol (DON) in Corn. Pennsylvania Department of Agriculture, Pennsylvania Department of Environmental Protection, 2012. Available online: https://www.agriculture.pa.gov/Food/pet_food/Documents/ManagementofDONincorn.pdf (accessed on 8 November 2018).
18. Varga, E.; Wiesenberger, G.; Hametner, C.; Ward, T.J.; Dong, Y.; Schöfbeck, D.; McCormick, S.; Broz, K.; Stückler, R.; Schuhmacher, R.; et al. New tricks of an old enemy: Isolates of *Fusarium graminearum* produce a type A trichothecene mycotoxin. *Environ. Microbiol.* **2015**, *17*, 2588–2600. [CrossRef]
19. Berthiller, F.; Sulyok, M.; Krska, R.; Schuhmacher, R. Chromatographic methods for the simultaneous determination of mycotoxins and their conjugates in cereals. *Int. J. Food Microbiol.* **2007**, *119*, 33–37. [CrossRef]
20. Dupire, S. Mycotoxins and other contaminants in the malting and brewing industries. In Proceedings of the European Brewery Convention Congress, Dublin, Ireland, 17–22 May 2003.

21. Melotte, L. Survey on the analysis of mycotoxins. Analysis Committee of the European Brewery Convention, European Brewery Association. *J. Inst. Brew.* **2004**, *110*, 235–239.
22. Directive 2002/32/EC. OJEC, L140/10. Available online: https://eur-lex.europa.eu/resource.html?uri=cellar:aca28b8c-bf9d-444f-b470-268f71df28fb.0004.02/DOC_1&format=PDF (accessed on 8 January 2019).
23. Recommendation 2006/576/EC. OJEC, L229/7. Available online: https://eur-lex.europa.eu/LexUriServ/LexUriServ.do?uri=OJ:L:2006:229:0007:0009:EN:PDF (accessed on 8 January 2019).
24. Recommendation 2013/165/EC. OJEC L 91, 12–15. Available online: https://eur-lex.europa.eu/LexUriServ/LexUriServ.do?uri=OJ:L:2013:091:0012:0015:EN:PDF (accessed on 8 January 2019).
25. Ruiz-Medina, A.; Fernández-de Córdova, M.L. Aflatoxin B_1 in beer at different Stages of Production. In *Processing and Impact on Active Components in Food*; Preedy, V., Ed.; Elsevier/Academic Press: San Diego, CA, USA, 2015; pp. 517–523.
26. Commission Regulation 2174/2003 EC. OJEC, L326, 12–15. Available online: https://publications.europa.eu/en/publication-detail/-/publication/c849b56b-42ec-4e34-8c90-3a3668737420 (accessed on 8 January 2019).
27. World Health Organization International Agency for Research on Cancer IARC. Monographs on the Evaluation of Carcinogenic Risks to Humans Volume 82. 2002. Available online: http://monographs.iarc.fr/ENG/Monographs/vol82/mono82.pdf (accessed on 8 July 2018).
28. Mably, M.; Mankotia, M.; Cavlovic, P.; Tam, J.; Wong, L.; Pantazopoulos, P.; Calway, P.; Scott, P.M. Survey of aflatoxins in beer sold in Canada. *Food Addit. Contam.* **2005**, *22*, 1252–1257. [CrossRef]
29. Nakajima, M.; Tsubouchi, H.; Miyabe, M. A survey of ochratoxin A and aflatoxins in domestic and imported beers in Japan by immunoaffinity and liquid chromatography. *J. AOAC Int.* **1999**, *82*, 897–902.
30. Inoue, T.; Nagatomi, Y.; Uyama, A.; Mochizuki, N. Fate of mycotoxins during beer brewing and fermentation. *Biosci. Biotechnol. Biochem.* **2013**, *77*, 1410–1415. [CrossRef]
31. Reddy, K.R.N.; Salleh, B. A preliminary study on the occurrence of *Aspergillus* spp. and aflatoxin B_1 in imported wheat and barley in Penang, Malaysia. *Mycotoxin Res.* **2010**, *26*, 267–271. [CrossRef]
32. Park, J.W.; Kim, E.K.; Shon, D.H.; Kim, Y.B. Natural co-occurrence of aflatoxin B_1, fumonisin B_1 and ochratoxin A in barley and corn foods from Korea. *Food Addit. Contam.* **2002**, *19*, 1073–1080. [CrossRef]
33. Scudamore, K.A.; Patel, S. Survey for aflatoxins, ochratoxin A, zearalenone and fumonisins in maize imported into United Kingdom. *Food Addit. Contam.* **2000**, *17*, 407–416. [CrossRef]
34. Lewis, L.; Onsongo, M.; Njapau, H.; Shurz-Rogers, H.; Luber, G.; Kieszak, S.; Nyamongo, J.; Backer, L.; Dahiye, A.M.; Misore, A.; et al. Aflatoxin contamination of commercial maize products during an outbreak of acute aflatoxicosis in eastern and central Kenya. *Environ. Health Perspect.* **2005**, *113*, 1763–1767. [CrossRef]
35. Gerbaldo, G.A.; Pereyra, C.M.; Cavaglieri, L.R.; Ruiz, F.; Pascual, L.; Dalcero, A.M.; Barberis, A.M. Surveillance of aflatoxin and microbiota related to brewers grain destined for swine feed in Argentina. *Vet. Med. Int.* **2011**, 912480. [CrossRef]
36. Papadopoulou-Bouraoui, A.; Vrabcheva, T.; Valzacchi, S.; Stroka, J.; Anklam, E. Screening survey of deoxynivalenol in beer from the European market by an enzyme-linked immunosorbent assay. *Food Addit. Contam.* **2004**, *21*, 607–617. [CrossRef]
37. Milićević, D.R.; Škrinjar, M.; Baltić, T. Real and perceived risks for mycotoxin contamination in foods and feeds: Challenges for food safety control. *Toxins* **2010**, *2*, 527–592. [CrossRef]
38. Salas, B.; Steffenson, B.J.; Casper, H.H.; Tacke, B.; Prom, L.K.; Fetch, T.G., Jr.; Schwarz, P.B. *Fusarium* species pathogenic to barley and their associated mycotoxins. *Plant Dis.* **1999**, *83*, 667–674. [CrossRef]
39. Velić, N.; Pavlović, H.; Ćosić, J.; Kanižai, G.; Krstanović, V. A survey of *Fusarium graminearum* and deoxynivalenol contamination of malt barley from the crop year 2004 in eastern Croatia. *Cereal Res. Commun.* **2007**, *35*, 1293–1296. [CrossRef]
40. Legzdina, l.; Buerstmayr, H. Comparison of infection with Fusarium head blight and accumulation of mycotoxins in grain of hulles and covered barley. *J. Cereal Sci.* **2004**, *40*, 61–67. [CrossRef]
41. Rotter, B.A.; Prelusky, D.B. Toxicology of deoxynivalenol (vomitoxin). *J. Toxicol. Environ. Health* **1996**, *48*, 1–34. [CrossRef]
42. Saenz de Rodrigues, C.A.; Bongiovanni, A.M.; Conde de Borrego, L. An epidemic of precocious development in Puerto Rican children. *J. Pediatr.* **1985**, *107*, 393–396. [CrossRef]
43. Campbell, G.D. Trichothecene mycotoxicosis B a new entity? *S. Afr. Med. J.* **1991**, *80*, 361–362.

44. Dall'Erta, A.; Cirlini, M.; Dall'Asta, M.; Del Rio, D.; Galaverna, G.; Dall'Asta, C. Masked mycotoxins are efficiently hydrolyzed by human colonic microbiota releasing their aglycones. *Chem. Res. Toxicol.* **2013**, *26*, 305–312. [CrossRef]
45. Berthiller, F.; Werner, U.; Sulyok, M.; Krska, R.; Hauser, M.-T.; Schuhmacher, R. Liquid chromatography coupled to tandem mass spectrometry (LC-MS/MS) determination of phase II metabolites of the mycotoxin zearalenone in the model plant *Arabidopsis thaliana*. *Food Addit. Contam.* **2006**, *23*, 1194–1200. [CrossRef]
46. González Pereyra, M.L.; Sulyok, M.; Baralla, V.; Dalcero, A.M.; Krska, R.; Chulze, S.; Cavaglieri, L.R. Evaluation of zearalenone, -zearalenol, -zearalenol, zearalenone 4-sulfate and -zearalenol 4-glucoside levels during the ensiling process. *World Mycotoxin J.* **2014**, *7*, 291–295. [CrossRef]
47. Berthiller, F.; Crews, C.; Dall'Asta, C.; Saeger, S.D.; Haesaert, G.; Karlovsky, P.; Oswald, I.P.; Seefelder, W.; Speijers, G.; Stroka, J. Masked mycotoxins: A review. *Mol. Nutr. Food Res.* **2013**, *57*, 165–186. [CrossRef]
48. Binder, S.B.; Schwartz-Zimmermann, H.E.; Varga, E.; Bichl, G.; Michlmayr, H.; Adam, G.; Berthiller, F. Metabolism of zearalenone and its major modified forms in pigs. *Toxins* **2017**, *9*, 56. [CrossRef]
49. Dall'Asta, C.; Battilani, P. Fumonisins and their modified forms, a matter of concern in future scenario? *World Mycotoxin J.* **2016**, *9*, 727–739. [CrossRef]
50. Scott, P.M.; Lawrence, G.A. Analysis of beer for fumonisins. *J. Food Prot.* **1995**, *58*, 1379–1382. [CrossRef]
51. Daško, L'.; Rauová, D.; Belajová, E.; Kováč, M. Determination of fumonisins B_1 and B_2 in beer. *Czech. J. Food Sci.* **2005**, *23*, 20–26. [CrossRef]
52. Voss, K.A.; Riley, R.T. Fumonisin toxicity and mechanism of action: Overview and current perspectives. *Food Saf.* **2013**, *1*, 49–69. [CrossRef]
53. Scott, P.M. Recent research on fumonisins: A review. *Food Addit. Contam. A* **2012**, *29*, 242–248. [CrossRef]
54. Missmer, S.A.; Suarez, L.; Felkner, M.; Wang, E.; Merrill, A.H., Jr.; Rothman, K.J.; Hendricks, K.A. Exposure to fumonisins and the occurrence of neural tube defects along the Texas-Mexico border. *Environ. Health Perspect.* **2006**, *114*, 237–241. [CrossRef]
55. Rheeder, J.; Marasas, W.; Theil, P.; Sydenham, E.; Shephard, G.; Van Schalkwyk, D. *Fusarium moniliforme* and fumonisins in corn in relation to human esophageal cancer in Transkei. *Phytopathology* **1992**, *82*, 353–357. [CrossRef]
56. Braun, M.S.; Wink, M. Exposure, occurrence, and chemistry of fumonisins and their cryptic derivatives. *Compr. Rev. Food Sci. Food Saf.* **2018**, *17*, 769–791. [CrossRef]
57. Maranghi, F.; Tassinari, R.; Narciso, L.; Tait, S.; Rocca, C.L.; Felice, G.D.; Butteroni, C.; Corinti, S.; Barletta, B.; Cordelli, E.; et al. In Vivo Toxicity and Genotoxicity of Beauvericin and Enniatins. Combined Approach to Study In Vivo Toxicity and Genotoxicity of Mycotoxins Beauvericin (BEA) and Enniatin B (ENNB). *EFSA Supporting Publication* **2018**, *15*, 1406E. Available online: https://efsa.onlinelibrary.wiley.com/doi/epdf/10.2903/sp.efsa.2018.EN-1406 (accessed on 8 January 2019). [CrossRef]
58. Anli, E.; Mert Alkis, I. Ochratoxin A and brewing technology: A Review. *J. Inst. Brew.* **2010**, *116*, 23–32. [CrossRef]
59. Creppy, E.E. Human ochratoxicosis. *J. Toxicol.* **1999**, *18*, 277–293. [CrossRef]
60. Huff, W.E.; Doerr, J.A. Synergism between aflatoxin and ochratoxin A in broiler chickens. *Poult. Sci.* **1981**, *60*, 550–555. [CrossRef]
61. Peraica, M.; Rašić, D. The impact of mycotoxicoses on human history. *Arch. Ind. Hyg. Toxicol.* **2012**, *63*, 513–518. [CrossRef]
62. Veprikova, Z.; Vaclavikova, M.; Lacina, O.; Dzuman, Z.; Zachariasova, M.; Hajslova, J. Occurrence of mono and di-glycosylated conjugates of T-2 and HT-2 toxins in naturally contaminated cereals. *World Mycotoxin J.* **2012**, *5*, 231–240. [CrossRef]
63. Nakagawa, H.; Sakamoto, S.; Sago, Y.; Nagashima, H. Detection of type A trichothecene di-glucosides produced in corn by high-resolution liquid chromatography-orbitrap mass spectrometry. *Toxins* **2013**, *5*, 590–604. [CrossRef]
64. McCormick, S.P.; Price, N.P.J.; Kurtzman, C.P. Glucosylation and other biotransformations of T-2 toxin by yeasts of the *Trichomonascus* clade. *Appl. Environ. Microbiol.* **2012**, *78*, 8694–8702. [CrossRef]
65. Nathanail, A.V.; Varga, E.; Meng-Reiterer, J.; Bueschl, C.; Michlmayr, H.; Malachova, A.; Fruhmann, P.; Jestoi, M.; Peltonen, K.; Adam, G.; et al. Metabolism of the *Fusarium* mycotoxins T-2 toxin and HT-2 toxin in wheat. *J. Agric. Food Chem.* **2015**, *63*, 7862–7872. [CrossRef]

66. Wen-Shyg Chiou, P.; Chen, C.R.; Chen, K.J.; Yu, B. Wet brewers' grains or bean curd pomance as partial replacement of soybean meal for lactating cows. *Anim. Feed Sci. Technol.* **1998**, *74*, 123–134. [CrossRef]
67. Cavaglieri, L.R.; Keller, K.M.; Pereya, C.M.; Gonzalez Pereya, M.L.; Alonso, V.A.; Rojo, F.G.; Dalcero, A.M.; Rosa, C.A.R. Fungi and natural incidence of selected mycotoxins in barley rootlets. *J. Stored Prod. Res.* **2009**, *45*, 147–150. [CrossRef]
68. Kensler, T.W.; Roebuck, B.D.; Wogan, G.N.; Groopman, J.D. Aflatoxin: A 50-year odyssey of mechanistic and translational toxicology. *Toxicol. Sci.* **2011**, *120*, S28–S48. [CrossRef]
69. Habschied, K.; Šarkanj, B.; Klapec, T.; Krstanović, V. Distribution of zearalenone in malted barley fractions dependent on *Fusarium graminearum* growing conditions. *Food Chem.* **2011**, *129*, 329–332. [CrossRef]
70. Habschied, K.; Šarkanj, B.; Krstanović, V.; Velić, N.; Novak, M.; Mastanjević, K. *Fusarium* mycotoxins in malting and brewing by-products. In Proceedings of the World Mycotoxin Forum 8th Conference, Mycotoxin Control: The System Approach, Vienna, Austria, 10–12 November 2014; p. 85.
71. Krstanović, V.; Mastanjević, K.; Velić, N.; Pleadin, J.; Perši, N.; Španić, V. The influence of *Fusarium culmorum* contamination level on deoxynivalenol content in wheat, malt and beer. *Roman. Biotechnol. Lett.* **2015**, *20*, 10901–10910.
72. Blezinger, S.B. Feed Supplements Come in Several Different Forms: Part 4. Cattle Today Online. Available online: http://www.cattletoday.com/archive/2003/February/CT251.shtml (accessed on 16 March 2018).
73. Ha, J.-K.; Kim, S.W.; Kim, W.Y. *Use of Agro-Industrial by-Products as Animal Feeds in Korea*; ASPAC Food & Fertilizer Technology Center: Taipei, Taiwan, 1996; pp. 1–15.
74. Rychlik, M.; Humpf, H.U.; Marko, D.; Dänicke, S.; Mally, A.; Berthiller, F.; Klaffke, H.; Lorenz, N. Proposal of a comprehensive definition of modified and other forms of mycotoxins including "masked" mycotoxins. *Mycotoxin Res.* **2014**, *30*, 197–205. [CrossRef]
75. Maul, R.; Pielhau, R.; Koch, M. Evaluation of an extraction method and spin column cleanup procedure for *Fusarium* mycotoxins and their masked derivatives from grain matrix. *Food Control* **2014**, *40*, 151–156. [CrossRef]
76. De Boevre, M.; Di Mavungu, J.D.; Landschoot, S.; Audenaert, K.; Eeckhout, M.; Maene, P.; Haesaert, G.; De Saeger, S. Natural occurrence of mycotoxins and their masked forms in food and feed products. *World Mycotoxin J.* **2012**, *5*, 207–219. [CrossRef]
77. Zachariasova, M.; Dzuman, Z.; Veprikova, Z.; Hajkova, K.; Jiru, M.; Vaclavikova, M.; Pospichalova, M.; Florian, M.; Hajslova, J. Occurrence of multiple mycotoxins in European feedingstuffs, assessment of dietary intake by farm animals. *Anim. Feed Sci. Technol.* **2014**, *193*, 124–140. [CrossRef]
78. Jozinović, A.; Šubarić, D.; Ačkar, Đ.; Miličević, B.; Babić, J.; Jašić, M.; Lendić, V.K. Food industry by-products as raw materials in functional food production. *Hrana u Zdravlju i Bolesti* **2014**, *3*, 22–30.
79. Obradović, V.; Babić, J.; Šubarić, D.; Ačkar, Đ.; Jozinović, A. Improvement of nutritional and functional properties of extruded food products. *J. Food Nutr. Res.* **2014**, *53*, 189–206.
80. Meneses, N.G.T.; Martins, S.; Teixeira, J.A.; Mussatto, S.I. Influence of extraction solvents on the recovery of antioxidant phenolic compounds from brewer's spent grains. *Sep. Purif. Technol.* **2013**, *108*, 152–158. [CrossRef]
81. Moreira, M.M.; Morais, S.; Carvalho, D.O.; Barros, A.A.; Delerue-Matos, C.; Guido, L.F. Brewer's spent grain from different types of malt: Evaluation of the antioxidant activity and identification of the major phenolic compounds. *Food Res. Int.* **2013**, *54*, 382–388. [CrossRef]
82. McCarthy, A.L.; O'Callaghan, C.; Connolly, A.; Piggott, C.O.; FitzGerald, R.J.; O'Brien, N.M. Phenolic extracts of brewers' spent grain (BSG) as functional ingredients-Assessment of their DNA protective effect against oxidant-induced DNA single strand breaks in U937 cells. *Food Chem.* **2012**, *134*, 641–646. [CrossRef]
83. Ktenioudaki, A.; Chaurin, V.; Reis, S.; Gallagher, E. Brewer's spent grain as a functional ingredient for breadsticks. *Int. J. Food Sci. Technol.* **2012**, *47*, 1765–1771. [CrossRef]
84. Corwell, C. Spent Grain Regulation: Commentary from the Beer Institute. 2014. Available online: https://www.craftbrewingbusiness.com/business-marketing/spent-grain-regulation-commentary-beer-institute/ (accessed on 16 March 2018).
85. Scott, T. Feed and Processing Options for Heavily Downgraded Wheat. 2014. Available online: https://cigi.ca/wp-content/uploads/2014/11/Tom-Scott-Feed-and-Processing-Options-for-Heavily-Downgraded-Wheat.pdf (accessed on 12 December 2017).

86. Agnew, J.; Kirsch, J. Evaluation of Disposal Options for *Fusarium* Damaged Grain and Screenings. 2017. Available online: http://pami.ca/wp-content/uploads/2017/04/R8415-Research-Report.pdf (accessed on 16 March 2018).
87. King, C. Extracting Value from Fusariem-Damage Grain and Screenings. 2017. Available online: https://www.topcropmanager.com/diseases/from-waste-to-worth-20793 (accessed on 4 August 2018).
88. Snizhko, O.; Lomova, N.; Narizhnyy, S.; Mingaleeva, Z. Enhancing food safety of pollen by means of irradiation. *Ukr. Food J.* **2015**, *4*, 32–39. Available online: http://ufj.ho.ua/Archiv/UKRAINIAN%20FOOD%20JOURNAL%202015%20V.4%20Is.1.pdf (accessed on 15 November 2018).
89. Berthiller, F.; Krska, R.; Domig, K.J.; Kneifel, W.; Juge, N.; Schuhmacher, R.; Adam, G. Hydrolytic fate of deoxynivalenol-3-glycoside during digestion. *Toxicol. Lett.* **2011**, *206*, 264–267. [CrossRef] [PubMed]
90. Warth, B.; Fruhmann, P.; Wiesenberger, G.; Kluger, B.; Sarkanj, B.; Lemmens, M.; Hametner, C.; Fröhlich, J.; Adam, G.; Krska, R.; et al. Deoxynivalenol-sulfates: Identification and quantification of novel conjugated (masked) mycotoxins in wheat. *Anal. Bioanal. Chem.* **2015**, *407*, 1033–1039. [CrossRef]
91. Mastanjević, K.; Španić, V.; Horvat, D.; Mastanjević, K.; Šarkanj, B.; Krstanović, V.; Šantek, B. Establishing the impact of *Fusarium culmorum* infection and fungicide treatment on wheat malt quality. *J. Food Process. Preserv.* **2018**, *42*, e13714. [CrossRef]

5

Aromatic Characterization of Mangoes (*Mangifera indica* L.) using Solid Phase Extraction Coupled with Gas Chromatography–Mass Spectrometry and Olfactometry and Sensory Analyses

Haocheng Liu, Kejing An, Siqi Su, Yuanshan Yu, Jijun Wu, Gengsheng Xiao and Yujuan Xu *

Sericulture & Agri-Food Research Institute Guangdong Academy of Agricultural Sciences/Key Laboratory of Functional Foods, Ministry of Agriculture and Rural Affairs/Guangdong Key Laboratory of Agricultural Products Processing, Guangzhou 510610, China; AnsisHC@163.com (H.L.); ankejing@gdaas.cn (K.A.); lijun@gdaas.cn (S.S.); yuyuanshan@gdaas.cn (Y.Y.); wujijun@gdaas.cn (J.W.); gshxiao@yahoo.com.cn (G.X.)

* Correspondence: xuyujuan@gdaas.cn

Abstract: Mangoes (*Mangifera indica* L.) are wildly cultivated in China with different commercial varieties; however, characterization of their aromatic profiles is limited. To better understand the aromatic compounds in different mango fruits, the characteristic aromatic components of five Chinese mango varieties were investigated using headspace solid-phase microextraction (HS-SPME) coupled with gas chromatography-mass spectrometry-gas chromatography-olfactometry (GC-MS-O) techniques. Five major types of substances, including alcohols, terpenes, esters, aldehydes, and ketones were detected. GC-O (frequency detection (FD)/order-specific magnitude estimation (OSME)) analysis identified 23, 20, 20, 24, and 24 kinds of aromatic components in Jinmang, Qingmang, Guifei, Hongyu, and Tainong, respectively. Moreover, 11, 9, 9, 8, and 17 substances with odor activity values (OAVs) ≥1 were observed in Jinmang, Qingmang, Guifei, Hongyu, and Tainong, respectively. Further sensory analysis revealed that the OAV and GC-O (FD/OSME) methods were coincided with the main sensory aromatic profiles (fruit, sweet, flower, and rosin aromas) of the five mango pulps. Approximately 29 (FD ≥ 6, OSME ≥ 2, OAV ≥ 1) aroma-active compounds were identified in the pulps of five mango varieties, namely, γ-terpinene, 1-hexanol, hexanal, terpinolene trans-2-heptenal, and *p*-cymene, which were responsible for their special flavor. Aldehydes and terpenes play a vital role in the special flavor of mango, and those in Tainong were significantly higher than in the other four varieties.

Keywords: mango; volatile compounds; frequency detection (FD); order-specific magnitude estimation (OSME); odor activity value; sensory analysis

---

1. Introduction

Mango (*Mangifera indica* L.), a native crop of South Asia, is a member of the *Anacardiaceae* family [1] and has been historically grown for more than 4000 years [2], thereby earning the title "king of fruits". It is recognized as one of the most popular fruits around the world, with the highest rates of production, marketing, and consumption [3,4]. Among tropical fruits, mango is the second most common crop involved in international trade, following banana. Global mango production has been estimated at 50.65 million tons, with China being the second largest mango-growing country, 2017 production reaching close to 4.94 million tons [5]. Common mango cultivars in China featuring specific regional characteristics include Guifei, Hongyu Jinmang, Qingmang, and Tainong.

Aroma is a major factor that influences the quality and consumer acceptance of mango products. Investigating various aromatic components would improve our understanding and facilitate controlling critical quality parameters that could influence mango processing. Hundreds of compounds have been characterized in various mango cultivars, which mainly include aldehydes, alcohols, esters, ketones, and terpenes [6]. However, previous studies have mainly focused on the volatiles in various mangoes in China [7,8], and a few volatiles were detected because of their odor threshold. Thus, scientific information relating to aromatic constituents as well as sensory characteristics of various mango cultivars is limited. Therefore, an in-depth investigation is required to identify the volatile or aromatic components of various mango cultivars in China.

To determine the aromatic components of mango, simultaneous solvent-assisted flavor evaporation (SAFE), solid phase microextraction (SPME), and distillation and extraction (SDE) have been employed in food aroma extraction [9–12]. The procedures of SDE and SAFE isolate aromatic compounds from food matrices using organic solvents [13]; however, these methods are highly laborious, time-consuming, and entail preconcentration of extracts. Unlike well-established protocols, SPME has been extensively used in the preparation of volatile and semi-volatile compounds from various types of samples [14]. This technique was developed more than two decades ago and is a rapid, simple, sensitive, and solvent-free technique for the analysis of volatile organic compounds (VOCs) [15]. Coupling of the methods of aroma extraction with the GC-MS/O technique, in particular, headspace solid-phase microextraction (HS-SPME) for extraction, together with detection frequencies (FD) and order-specific magnitude estimation (OSME) for GC-MS/O, creates a highly reliable method of identifying potent odorants.

Thus, to fully understand the aromatic compounds present in typical mango varieties, this study conducted the following studies: (1) identification and quantification of volatiles of different types of mangoes by HS-SPME-GC-MS; (2) discrimination of the major aroma-active compounds in various types of mangoes using combined GC-O detection (FD, OSME) and odor activity value (OAV); and (3) validation of the sensory differences using quantitative descriptive analysis (QDA). In summary, objectives of this investigation were to reveal the major aromatic compounds of mangoes in China.

2. Materials and Methods

2.1. Samples Preparation

This study used seven mature samples of the cultivars Jinmang (JM), Qingmang (QM), Hongyu (HY), Guifei (GF), and Tainong (TN), which were purchased from the regional market in Guangzhou (133.35° N, 23.12° E), Guangdong Province (April 2019). The samples were shipped to the laboratory and kept at 25 °C until complete maturity (2 days). Full maturity and maturity of the mangoes were based on fruit color (green to yellow-orange or red, except for the green lawn, Table S1), odor (sweet scent), and hardness (pulp hardness index changed from 5.28 to 4.32 N). Finally, the mango samples with the same maturity were washed, and the peeled pulp was immediately frozen in liquid nitrogen, and then stored at −80 °C for further studies.

2.2. Chemicals

Humulene, 2-penten-1-ol, 2-hexen-1-ol, (E)-3-hexen-1-ol, 1-hexanol, p-cymen-8-ol, 2-vinyloxy)-ethanol, 3-methyl-1-butanol, 1,3,8-*p*-menthatriene, allo-ocimene, 2-carene, α-phellandrene, 3-carene, terpinolene, 1,3-cyclohexadiene, 1-methyl-4-(1-methylethyl), isovaleraldehyde, 3-hexenal, 2,4-dimethyl-benzaldehyde, heptanal, *trans*-2-heptenal, trans-2-pentenal, 1-nonanal, decanal, citral, 1-penten-3-one, 2-cyclohepten-1-one, 3-methylcyclohex-3-en-1-one, 6-methyl-5-hepten-2-one, ethyl-propionate, ethyl cyclopropanecarboxylate, ethyl crotonate, isoamyl acetate, tetraethyl orthosilicate, ethyl butyrate, and γ-octanoic were purchased from TCI (Tokyo, Japan); and linalool, α-pinene, β-pinene, β-myrcene, D-limonene, *p*-cymene, β-ocimene, hexanal, γ-terpinene, and *trans*-2-hexenal were purchased from Sigma (St. Louis, MO, USA). All of the chemical standards were of GC quality.

2.3. HS-SPME-GC-MS

Based on previous studies, the optimized SPME experimental conditions were established [16–18]. Approximately 5.0 g of the juice with 1.5 g of NaCl were blended in a 15 mL vial tightly capped with a PTFE-silicon septum at a stirring speed of 80 rpm. The flavor compounds in mango pulp are formed during equilibrium and during the extraction process. Therefore, the extraction temperature was set at 40 °C. After the vial containing the sample was equilibrated at 40 °C for 10 min on a heating platform agitation, the pretreated (conditioned at 270 °C for 30 min) SPME fiber (50/30 μm DVB/Carboxen/PDMS, Supelco, Bellefonte, PA, USA) was then inserted into the headspace, and extraction was performed for 30 min with continued heating and agitation. Afterward, the fiber was withdrawn and instantly introduced to the GC for desorption and analysis.

GC-MS analysis was performed on an Agilent 7890 (Agilent Technologies, Palo Alto, CA, USA) gas chromatography and Agilent 5977 mass selective detector. Samples were separated using both HP-5 and DB-WAX (both 30 m × 0.25 mm i.d., 0.25 mm film thickness, Agilent Technologies, Palo Alto, CA, USA). Helium was used as carrier gas at a flow rate of 1.7 mL/min, and the GC inlet was set in the split-less mode. The injector temperature was 250 °C. The temperature program was from 40 °C (2 min hold) to 160 °C at 4 °C/min and finally raised to 280 °C at 50 °C/min. Then, electron ionization mode (EI) was used with a 70 eV ionization energy. The ion source temperature was 230 °C, and the mass range was from *m/z* 35 to 450. The volatile compounds were determined by authentic standards, retention indices (RI), and NIST 14.0 library. The retention indices (RIs) of compounds were determined via sample injection with a homologous series of straight-chain alkanes (C6–C30) (Sigma Aldrich, St. Louis, MO, USA).

2.4. Identification of Aroma-Active Compounds by GC-O

The odorant compounds were analyzed using a sniffing port (ODP3, Gerstel, Germany) coupled with a GC-MS (7890B–5977B, Agilent Technologies, Inc.). Upon exiting the capillary column, the effluents were divided to a ratio of 3:1 (by volume) into a sniffing port as well as an MS detector using an Agilent capillary flow technique. The transfer line directed to the GC-O sniffing port was set at a temperature of 270 °C. The GC-MS settings were similar to those described earlier. Aroma extraction was conducted by four highly skilled personnel (in an alternate order of 50 min intervals) using reference compounds. All personnel were extensively trained on the GC-O technique for at least 90 h.

Frequency analysis was conducted by four trained sensory panelists (i.e., two males and two females). Retention time and odor quality, together with substance detection, were recorded. Frequency analysis was performed in duplicate by every panelist. Odorants with an FD ≥2 (determined by at least two analysts) were considered to have potential aroma activity [19].

The OSME reflected the aromatic intensity of the stimulus that was based on a five-point scale that ranged from 0 to 5, where 0 = none, 3 = moderate, and 5 = extreme. Every sample was sniffed thrice by each panelist, and then the average aromatic intensity values were calculated. If the panelists did not utilize a similar attribute for an aroma that was eluted by GC, the analysis was repeated, and only the descriptors used for the same aroma were included in the analysis [20].

2.5. Quantitative Analysis of Aromatic Compounds

Quantitative data on the identified compounds were gathered by calculating their relative quantitative correction factors (RQCFs) with the "single-point correction method", which is similar to the standard addition method. Similar GC conditions as described above were used in GC-MS analysis, with solvent delay time set at 3 min to prevent the solvent in the standard solutions to reach the filament. The method of obtaining RQCFs consisted of the following: 5.0 g of mango pulp was analyzed using SPME-GC-MS, resulting in an ion peak area for each identified compound. A similar volume of mango pulp with defined amounts of various authentic and internal standards (to avoid run-to-run variations) was then analyzed to generate a new quantifying ion peak for each detected compound and quantitative correction factor for the internal standard. The RQCF of each volatile was generated using the following equation:

$$f_i' = \frac{f_{wi}}{f_{ws}} = \frac{m_i/A_i}{m_s/A_S} = \frac{A_S m_i}{A_i m_s}, \tag{1}$$

where f_i' was the RQCF of a detected compound (i); m_s and m_i were the respective known contents of authentic (i) and internal standard (s); A_s was the peak area of the quantifying ions of (s); and A_i was the peak area of the quantifying ions of (i) before and after the addition of the standard solution to the juice sample.

To determine the amounts of the identified volatiles in mango pulp, approximately 5.0 g of juice per volume containing a similar amount of internal standard as that of the calculated RQCF was prepared and used in GC-MS analysis. The concentration was computed using the equation:

$$m_i = f_i' \times A_i \times \frac{m_s}{A_s} \tag{2}$$

where m_i was the amount of compound (i); m_s was the known amount of (s), A_i and A_s were the respective peak areas of the quantifying ions of detected compound (i) and internal (s) standards; and f_i' was the RQCF of (i). The peak area of the quantifying ion of every component in selected ion chromatograms was assessed in triplicate, and the average value was computed. Then, the concentration of every identified volatile in mango pulp is described in nanograms per milliliter of juice [19].

2.6. Odor Activity Value (OAV)

OAV pertains to the concentration of the odor divided by its threshold in water. Compounds with an OAV ≥1 were considered as major contributors to the aromatic profile of each sample [21].

2.7. Sensory Panel and Aroma Profile Analysis

Aromatic profiling was performed using descriptive sensory analysis, as earlier described [22]. The mango pulps were analyzed by a highly skilled panel of 10 members consisting of five males and five females. Prior to quantitative descriptive analysis, 50 mL of mango pulp was placed in a 100 mL cubage of a plastic cup with a Teflon lid, which was handed over to a panelist in the laboratory without peculiar smell at a temperature of 25 °C. Then, the panelists assessed the aromatic profile of the mango pulp using three preliminary sessions (each spent approximately 2 h), until all of them attained a consensus as to the degree of aromatic flavor. Then, the organoleptic characteristic descriptors were assessed using eight sensory features (i.e., "overall aroma", "tropical fruit", "citrus", "floral", "fresh", "rosin", "honey/sweet", and "green") to evaluate aroma negative and positive mango pulp features. The descriptors were described as the following odors: linalool for the "floral" descriptor, β-phellandrene for "citrus", butyl acetate for the "fruity" descriptor, β-damascenone for the "honey/sweet" descriptor, phenylacetaldehyde for "fresh", (E)-2-hexenal for "green and grassy", and terpinolene for "rosin". The complete profile of each sample was randomly assessed in triplicate for each treatment. The assessors then rated the odor intensities a seven-point scale ranging from 0 to 3, with 0 as not perceivable, 1 as weak, 2 as significant, and 3 as strong. The results were then averaged for every odor note and plotted using a spider web diagram.

3. Results

3.1. Sensory Analysis of Five Mango Cultivars

Figure 1 shows that the five mango pulps had similar aromatic intensities, and these can be divided into six aroma attributes, including tropical fruit, flower aroma, sweet aroma, green grass aroma, green melon aroma, wood aroma, and rosin aroma, although to different extents. SPSS was used to distinguish differences between the mango samples through sensory evaluation scores (Table S2). In most cases, the five mango pulps exhibited significant differences in intensity of fruit aroma ($p < 0.05$), with some exceptions for "tropical fruit" and sweetness between JM and TN, tropical fruit between GF and QM, green between JM and HY, melon between GF and TN, and wood between JM and GF, where no significant difference was observed ($p > 0.05$). In general, innate differences among the mango varieties significantly influenced the intensities of most of the key sensory attributes.

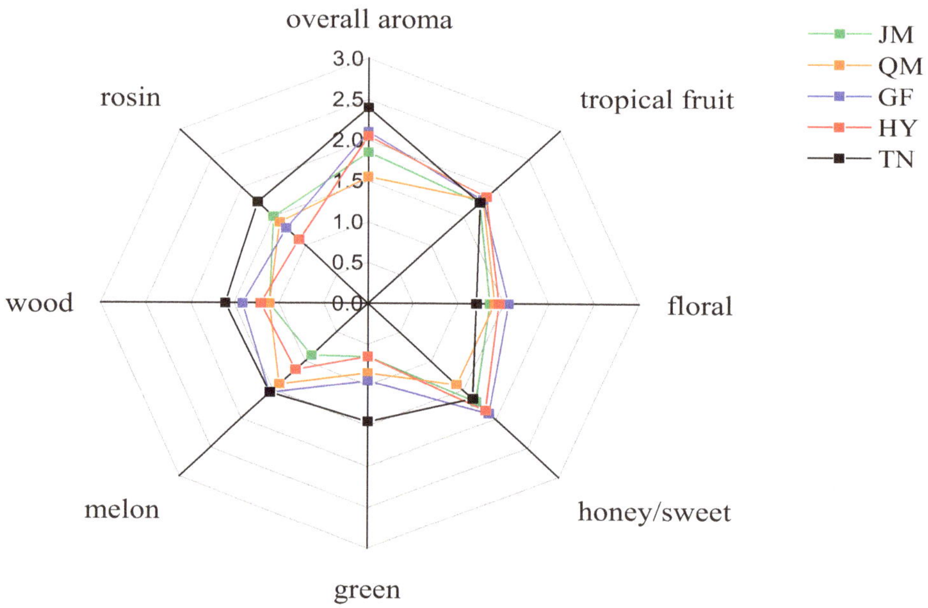

Figure 1. Spider plot for flavor attributes of five varieties mango samples. Jinmang—JM, Qingmang—QM, Hongyu—HY, Guifei—GF, and Tainong—TN.

The overall (decreasing) order of flavor scores was: TN > GF > HY > JM > QM. The intensity of fruit aroma was highest in both GF and QM, followed by TN and JM. The intensity of flower aroma was highest in both GF and HY, followed by QM and JM. TN had the lowest. GF and HY had the highest intensity of sweet aroma, followed by TN, QM, and JM. TN had the highest intensity of green grass aroma, followed by GF and QM. The intensity of green grass aroma in JM and HY was almost identical. The intensities of green melon aroma in QM, GF, and TN were markedly higher than those of JM and HY. TN had the highest intensity of green melon aroma. In addition, the intensity of wood aroma in GF and TN was higher than in JM, QM, and HY. TN had the highest intensity of wood aroma.

In general, GF had the strongest flower and sweet aromas, HY had the strongest fruit aroma, and TN had the strongest green grass, green melon, wood, and Rosin aromas. According to Bonneau [9], different mango varieties significantly influenced the intensities of key sensory attributes of mango, which was due to the different amounts of aroma-active components in mango.

3.2. Comparasion of the GC-MS Results of Five Mango Cultivars

The MS and RI results preliminarily identified 47 volatile compounds (Figure 2). There were 25, 24, 24, 29, and 23 volatile compounds in JM, QM, GF, HY, and TN, respectively. These were generally composed of alcohols, alkenes, aldehydes, esters, ketones, and ethers. Most of the volatiles found in this study were similar to those in the findings of previous studies [1,6,23]. Six compounds, namely, *p*-cymen-8-ol, 2-(vinyloxy)ethanol, 1-methyl-4-(1-methylethyl)-1,3-cyclohexadiene, 3-methyl butanal, 2-cyclohepten-1-one, and tetraethyl orthosilicate, were first observed in the volatile composition of mango, although those were not the major contributors to the mango aroma. This difference might be caused by habitat, maturity conditions, and aromatic extraction method.

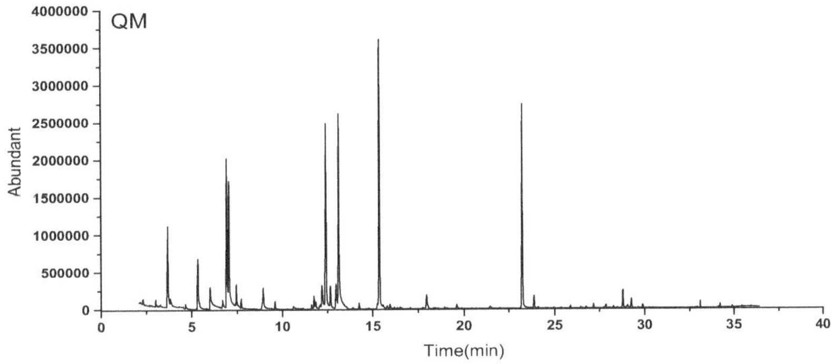

(a)

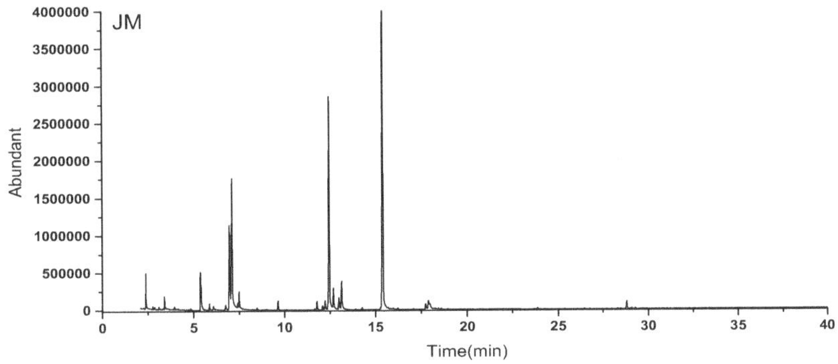

(b)

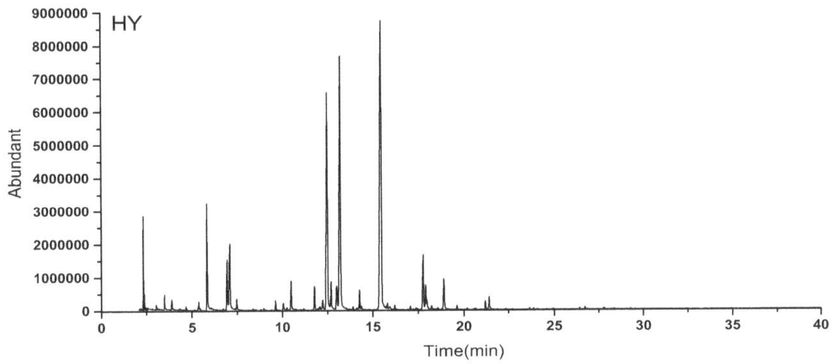

(c)

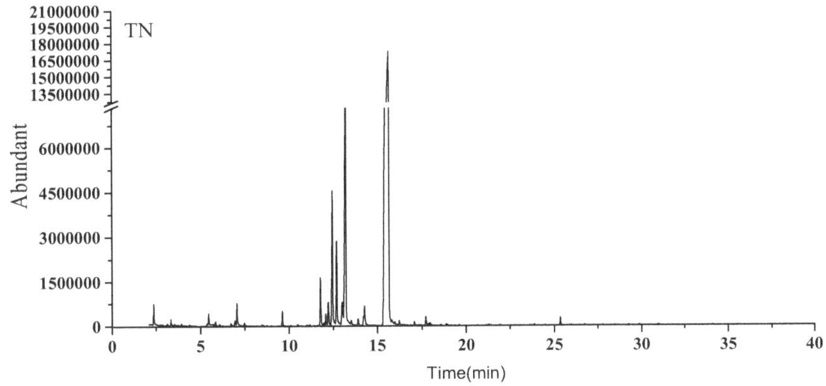

(d)

Figure 2. *Cont.*

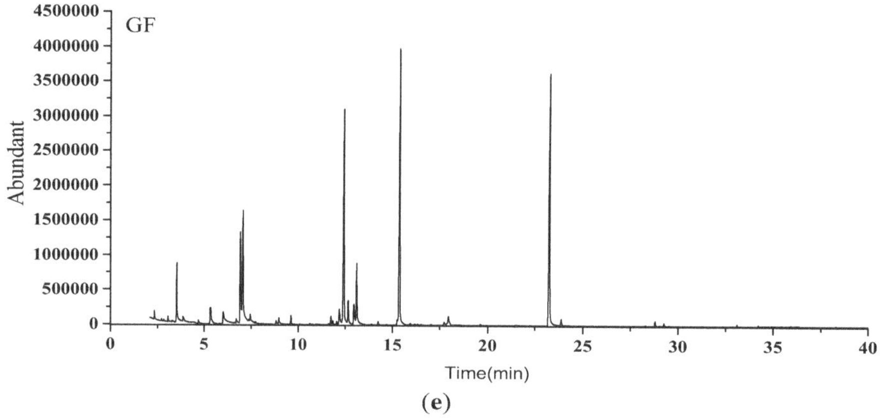

Figure 2. The representative Total ion chromatograms of five varieties mango samples. **a**: QM, **b**: JM, **c**: HY, **d**: TN, **e**: GF.

Figure 3 shows that the number of alkenes in the five mangoes was much higher than that of the other kinds of volatile compounds, indicating that alkenes were the most important volatile substances. In addition, the highest number of alcohols, terpenes, and aldehydes were found in JM, TN, and HY, respectively. The number of ketones in QM and HY was higher than in other mango varieties, whereas the number of esters in GF was higher than in other mangoes. Moreover, only nine volatile components, including one alcohol, one aldehyde, and seven terpenes, were identified in the five kinds of mango pulp. This showed that the volatiles of the five mango cultivars varied.

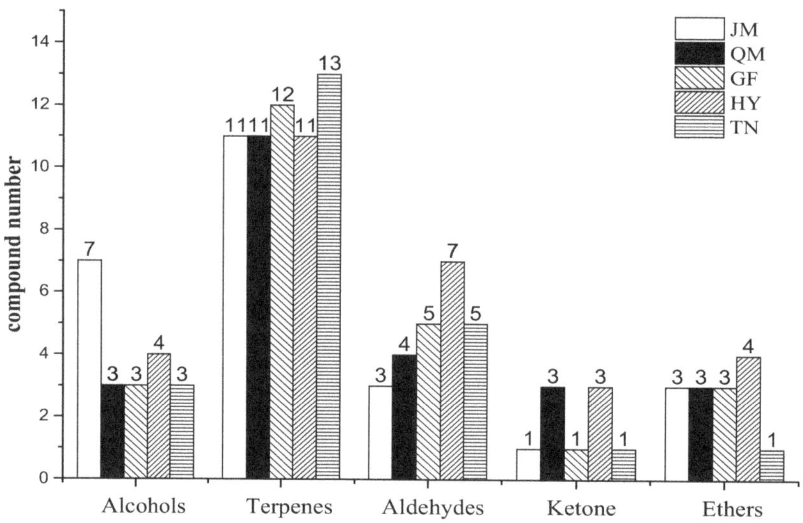

Figure 3. Comparisons of the numbers of volatile compounds detected in different mango samples

Full quantifications using standards were conducted for the volatile substances in the five mango varieties (Table 1 and Figure 4). In JM, the total amount of volatiles was 539.29 μg/kg; the content of terpinolene was the highest (147.07 μg/kg); then came 3-carene (103.63 μg/kg), 1-hexanol (55.83 μg/kg), e-3-hexen-1-ol (41.50 μg/kg), isoamyl alcohol (38.87 μg/kg), and *p*-cymene (27.43 μg/kg). Isoamyl alcohol (38.87 μg/kg) and 2-(vinyloxy) ethanol (3.49 μg/kg) were the unique volatile components. In QM, the total amount of volatiles was 662.92 μg/kg; the content of (e)-2-hexenal was the highest (180.60 μg/kg); then came e-3-hexen-1-alcohol (132.56 μg/kg), 3-carene (60.36 μg/kg), and terpinolene (59.42 μg/kg). The unique volatile components in QM were ethyl crotonate (9.83 μg/kg), e-2-nonenal (7.60 μg/kg), and 3-methyl-3-cyclohexene-1-one (0.62 μg/kg). In GF, the total amount of volatiles was 381.12 μg/kg; (e)-2-hexenal (84.03 μg/kg), trans-3-hexen-1-ol (76.82 μg/kg), and terpinolene (43.53 μg/kg) were the main volatile substances; the unique volatile component was isovaleraldehyde (1.77 μg/kg). In HY, the total amount of volatiles was 400.50 μg/kg; (e)-2-hexenal (85.64 μg/kg), hexanal (75.54 μg/kg), (e)-2-heptenal (65.65 μg/kg), 3-carene (27.67 μg/kg), and terpinolene (24.90 μg/kg) were the main volatile substances; HY is the most special kind of mango. Its unique volatile components were β-caryophyllene (2.18 μg/kg), (e)-2-heptenal (65.65 μg/kg), *p*-cymen-8-ol (12.23 μg/kg), ethyl propionate (6.39 μg/kg), 3-hexen-1-ol (6.59 μg/kg), humulene (2.93 μg/kg), β-caryophyllene (2.18 μg/kg), 2-cyclohepten-1-one (0.66 μg/kg), (e,z)-2,6-nonadienal (0.66 μg/kg), and ethyl cyclopropanecarboxylate (0.12 μg/kg). In the TN pulp, the total amount of volatiles was 1279.68 μg/kg; the content of terpinolene was the highest (811.61 μg/kg), followed by *p*-cymene (132.96 μg/kg), 3-hexenal (44.13 μg/kg), 3-carene (70.92 μg/kg), and phellandrene (30.17 μg/kg), with 3-hexenal being the unique volatile component. Based on the above analysis, the contents of volatile substances in the five cultivars of mangoes were different, and the total content of volatile substances in TN was the highest. And all five kinds of mangoes have unique volatile substances with individual aromatic components; especially for ruby, further analyses by GC-O are needed.

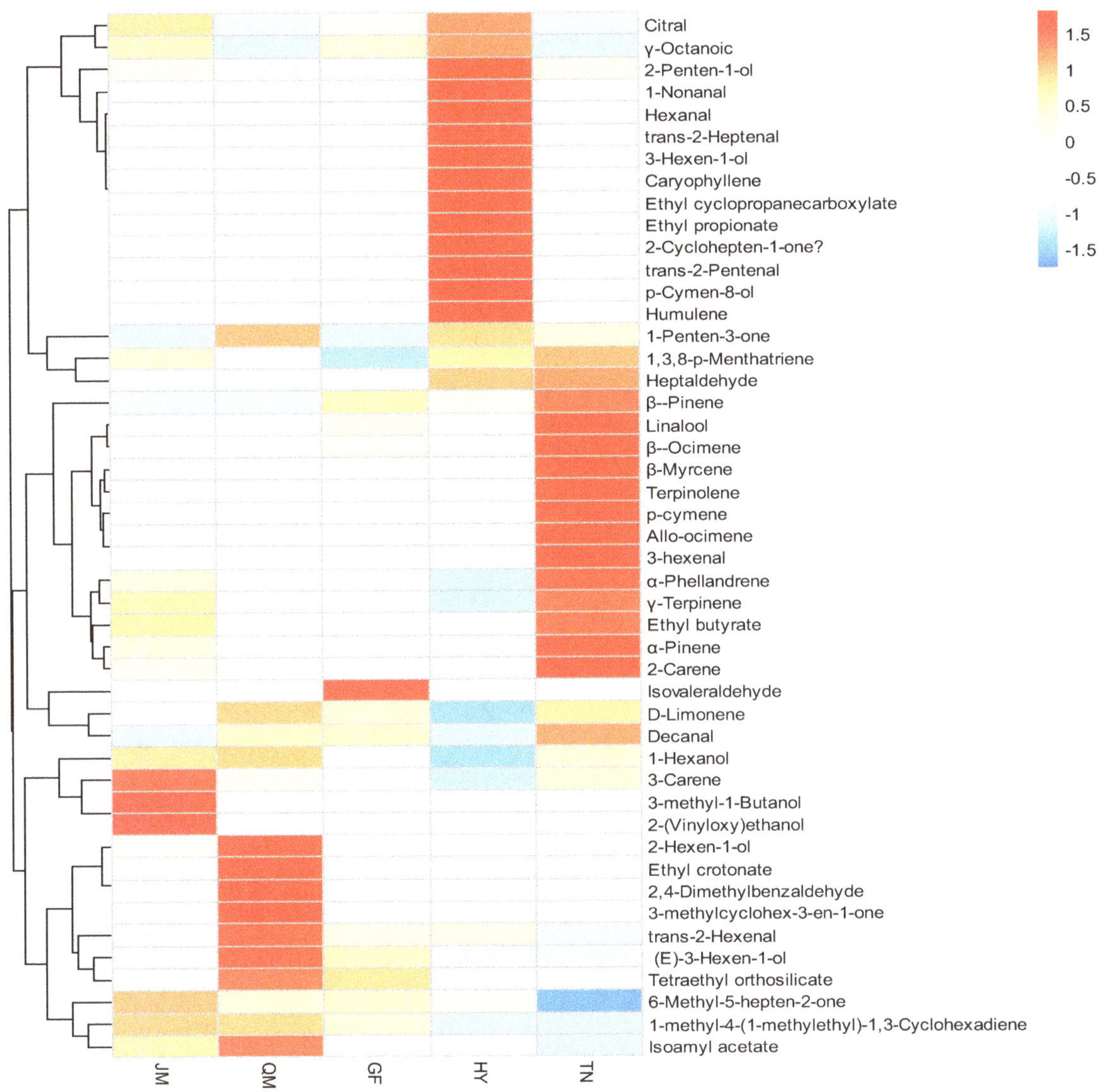

Figure 4. Hierarchical clustering of the all compounds in different mango samples.

3.3. Identification of Key Aromatic Compounds in the Pulps of Five Mango Cultivars

Not all volatile substances in mango contribute to its aroma, and thus aromatic intensity identification was conducted to determine whether the high content or unique volatile components in mango contribute to its overall aroma. The frequency-of-detection (FD) method requires the evaluator to smell the same aromatic substance and record the peak time and aromatic properties of the aromatic substance. The more times the aromatic substance was detected, the greater the contribution to the overall aroma. The intensity (OSME) method was a GC-O detection method that was used to evaluate flavor contribution based on the odor intensity of the aroma substance. After the aromatic substances are separated by GC-MS, the method directly describes the change in odor intensity (measured) and the frequency of aroma attributes. It is considered to be the simplest and most effective GC-O analysis method because it is less time-consuming and less demanding on assessors. Thirty-three characteristic aromatic components were identified in the five mango pulps by GC-O combined with FD and OSME methods, including three alcohols, 10 aldehydes, 11 terpenes, three esters, and four ketones. The differences in aromatic components in the pulp of five mango cultivars were not significant.

Further analysis, the FD analysis identified 23 aromatic substances (FD $\geq$ 2) in JM. According to the FD statistical data of each substance in the Table 2, one of the substances (FD = 8) was identified in all of the tests: γ-terpinene with citrus aroma. The substances with FD = 7 were phellandrene (citrus-like aroma) and γ-octanoic (floral, violet aroma). The substances with FD = 6 were ethyl cyclopropanecar-boxylate (fruit aroma) and terpinolene (rosin aroma). In QM, 18 aromatic substances with FD $\geq$2 were identified. According to the FD statistical data of each substance in the table, three kinds of substances (FD = 8) were identified in all of the tests, including 1-penten-3-one, 3-hexenal, and terpinolene. 1-penten-3-one had mushroom-like aroma, 3-hexenal had a grassy aroma, and terpinolene had a rosin-like aroma. The substance with FD = 7 was β-pinene, which has a green-grass-type aroma. The substances with FD = 6 were *p*-cymene (citrus and green aroma) and γ-terpinene (citrus and lemon-like aroma). In GF, 18 aromatic substances with FD $\geq$2 were identified. According to the FD statistical data of each substance in Table, 2 substances with FD = 8 were identified in all of the tests: ethyl cyclopropanecarboxylate (fruit aroma) and γ-terpinene (citrus and lemon aroma). The substances with FD = 7 were 3-hexenal (green grass aroma) and terpinolene (rosin-like aroma). The substance with FD = 6 was *p*-cymene (citrus and green aroma). In HY, 23 aromatic substances with FD $\geq$2 were identified. The substances with FD = 8 were ethyl cyclopropanecarboxylate (fruit aroma). The substances with FD = 7 were cis-2-penten-1-ol (grass and tea aroma), 3-hexenal (green grass aroma), and terpinolene (rosin-like). The substance with FD = 6 was *p*-cymene (citrus and green aroma). In TN, there were 24 aromatic substances with FD $\geq$ 2. The substances with FD = 8 were terpinolene and γ-terpinene, which mainly had a rosin citrus-like aroma. The substances with FD = 6 were ethyl cyclopropanecarboxylate, β-myrcene, 2-carene, phellandrene, and *p*-cymene. They had a fruity rosin and sweet aroma.

The OSME analysis showed that the aromatic intensities of the five mango pulps had significant differences (Figure 5). There were 23 kinds of aromatic components in JM, and the substances with the aromatic intensity greater than 2 were phellandrene (4.1), ocimene (2.33), γ-terpinene (2.3), and terpinolene (2.3). This suggests an overall aromatic profile of citrus, green grass, and flowers. There were 19 kinds of aromatic components in QM, and the substances with an aromatic intensity greater than 2 were 1-hexanol (2.12), methyl-3-methylcyclohex-3-en-1-one (2.67), β-phellandrene (2.1), and decanal (3) showing an overall aromatic profile of lemon, sweet, and green grass. Seventeen aromatic components were identified in GF, and the substances with aroma intensities greater than 2 were 1-hexanol (2.13), phellandrene (2.5), terpinolene (2.33), and p-menthol (2.25) showing an overall aroma profile of citrus and lemon, fresh green leaves, and green grass. There were 22 aromatic components in HY, and the substances with aromatic intensity greater than 2 were phellandrene (3), ocimene (2.5), α-pinene (2.2), γ-terpinene (2.5), terpinolene (2.2), and 1,3,8-*p*-menthatriene (2.2) showing an overall aroma profile of flower, citrus, green grass, and pine wood. TN had 22 aromatic components, including hexanal (2.2), 2-carene (2.2), phellandrene (3.53), *p*-cymene (8.7), γ-terpinene (2.5), terpinolene (2.8), and decanal (3) with aromatic intensity values greater than 2. This has an overall aroma profile of sweet, fruit, green grass, and pine wood.

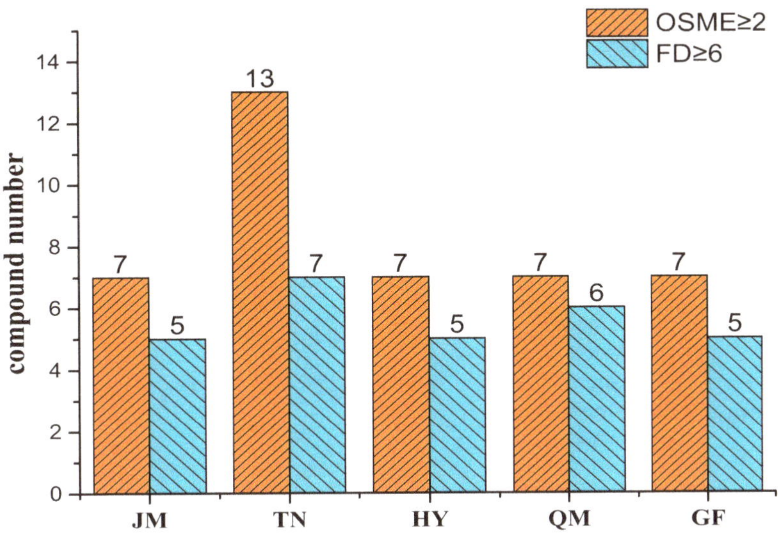

Figure 5. Identification of key aromatic compounds by frequency detection (FD) ≥6 and order-specific magnitude estimation (OSME) ≥2 in five mango pulps.

There were no significant differences in the numbers of overall aromatic substances identified using the two methods (FD and OSME). The number of substances with an FD ≥6 identified in TN by the FD method was the highest, followed by QM, JM, HY, and GF. The highest number of aromatic components with intensity values >2 was also found in TN, followed by HY and JM. QM and GF had the same number of aromatic substances. Therefore, the identification results obtained using FD and OSME were similar. These were highly consistent in identifying key aromatic substances. 2-Cyclohepten-1-one, ethyl cyclopropanecarboxylate, 1,3,8-*p*-menthatriene, and citral were identified for the first time and regarded as a characteristic aroma substance in mangoes, regardless of analytical method used (FD or OSME), although 1,3,8-*p*-menthatriene and citral were previously detected in lychee fruits [16]. In contrast to previous findings of aromatic substances in mangoes, some potent odorants, such as linalool and isoamyl acetate, were not detected, which might be due to differences in varieties, storage conditions, and extraction techniques. According to sensory analysis, the main aromatic profiles (fruit, sweet, flower, and rosin aromas) of pulp from the five mango cultivars were similar to those identified using the FD and intensity methods, indicating that the intensity method combined with the FD method can accurately illustrate the characteristic aromatic components with high or low intensity.

3.4. Odor Activity Values (OAVs) for the Pulps of Five Mango Cultivars

Aromatic analysis techniques such as the FD and intensity methods can effectively analyze the major aromatic compounds in mango pulp, but these cannot accurately reflect the contributions of individual components to the overall aroma characteristics [24]. Therefore, OAVs may be a more accurate scale to evaluate the contribution of volatile substances to fully consider the interactions between the food matrix and aromatic substances. The main component of mango pulp is water, and the calculation of the OAVs of each substance was based on the results of accurate quantitative analysis performed in this study, and the aroma threshold of each compound in water was previously reported [25,26].

According to the literature [19], substances with OAVs >1 contribute to the overall aroma of the sample. Table 3 and Figure 6 show the results of OAV analysis of pulp from five mango cultivars. There were 25 characteristic aromatic components with an OAV ≥ 1 in pulps of five mango cultivars, including two alcohols, seven aldehydes, three esters, 11 terpenes, and two ketones. Terpenes (44%) and aldehydes (28%) were the main aromatic components of mango, of which γ-terpinene had the highest OAV (3.04–10.04), followed by β-phellandrene (2.41–3.41), hexanal (1.10–16.97), and 1-nonanal (5.37–56.2), which were also considered as major aroma-active compounds in Australian mango cultivars. In contrast, although alcohols were the predominant component of all substances (Table 1), these showed minimal contribution (8%), due to their relatively high odor threshold. For instance, the OAV of the highest concentration of (e)-3-hexen-1-ol was only within the range of 0.38–1.21. In addition, 2-cyclohepten-1-one, ethyl cyclopropanecarboxylate, 1,3,8-*p*-menthatriene, and citral were first identified to be useful in aroma activity in mango based on OVA, which coincides with the results of GC-O.

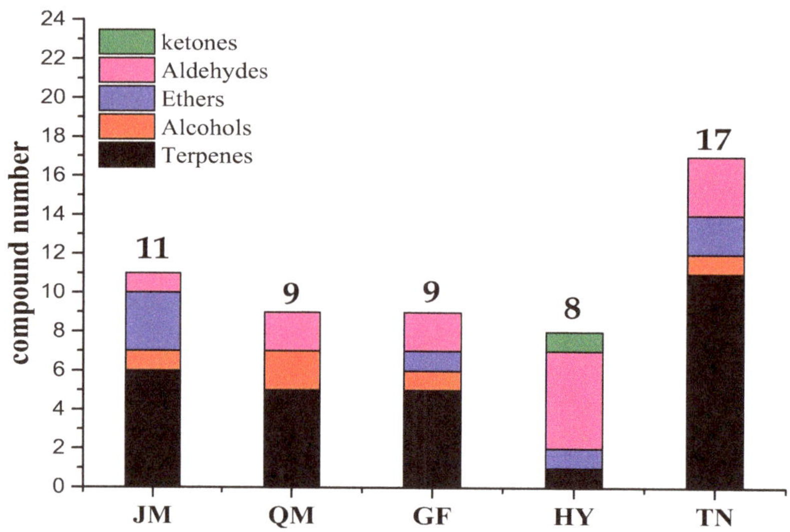

Figure 6. Identification of key aromatic compounds by OVA ≥ 1 in five mango pulps.

OAV values of characteristic aromatic substances were also different in different varieties of mango. Table 3 shows 11 substances with OAV ≥1 in JM. The OAV of γ-terpinene was the largest (7.70), followed by 1-hexanol (6.20), γ-octanoic (2.66), phellandrene (2.41), and ethyl cyclopropanecar-boxylate (2.17). In QM, nine aromatic substances with OAV ≥1 were identified, of which 1-nonanal (9.73) had the largest OAV, followed by 1-hexanol (6.58), γ-terpinene (3.37), *p*-cymene (1.83), and β-phellandrene (1.55). GF has nine substances with OAV ≥1, among which 1-nonanal (6.07) has the largest OAV, followed by γ-terpinene (3.04), 1-hexanol (2.55), γ-octanoic (2.36), decanal (1.41), and *p*-cymene (1.05). HY has eight substances with OAV ≥1, among which 1-nonanal (56.2) has the largest OAV value, followed by 1-hexanal (16.79), trans-2-heptenal (5.05), 2-cyclohepten-1-one (4.71), and γ-octanoic (3.76). HY has 17 substances with OAV ≥1, among which γ-terpinene (10.44) has the largest OAV, followed by *p*-cymene (7.19), terpinolene (5.80), 1-nonanal (5.73), 1-hexanol (5.13), and β-phellandrene (4.31).

3.5. Comparison of GC-O(FD/OSME) and OAV Aroma-Active Compounds

The joint analysis revealed 29 components (FD ≥ 6, OSME ≥ 2, OAV ≥ 1) as aroma-active compounds in the pulps of five mango cultivars (Figure 7). A total of 28 components were detected by GC-O (FD/OSME), whereas 25 substances were detected only by OVA. Compounds with high OAVs, such as 1-nonanal (5.73–56.20), ethyl butyrate (1.56–5.40), and heptanal (1.65–1.83), were not detected using GC-O (FD/OSME). Among the components discriminated by all the panelists in GC-O (FD/OSME), the contributions of 2-penten-1-ol, β-pinene, 3-methylcyclohex-3-en-1-one, and 6-methyl-5-hepten-2-one to the overall mango aroma were limited, as their OAVs were ≤0.1. The discrepancies between the two assessments mainly resulted from differences in application principles [19]. The calculations of the OAVs were based on the odor threshold in water instead of the food matrix. For the actual food matrix, the release of aroma is promoted or inhibited by interactions of volatiles with food components [27–29]. Therefore, OAV identification may not precisely match the actual results generated using GC-O (FD/OSME). However, biological variations, including respiratory rate and receptor state, may lead to errors in aromas based on GC-O (FD/OSME). This explains why the synergetic use of the two methods is strongly recommended for the identification of aroma-active compounds. In this study, the results of OAV coincided with those of GC-O (FD/OSME) to a certain extent, and the 29 key contributors to the five Chinese mango pulps were thus identified. These included 1-penten-3-one, 2-vyclohepten-1-one, 2-penten-1-ol, hexanal, ethyl cyclopropanecarboxylate, *trans*-2-hexenal, 3-hexenal, 1-hexanol, heptanal, α-pinene, β-pinene, e-2-heptenal, 3-methylcyclohex-3-en-1-one, 6-methyl-5-hepten-2-one, β-myrcene, ethyl butyrate, 2-carene, β-phellandrene, 3-carene, *p*-cymene, d-limonene, (E)-beta-ocimene, γ-terpinene, terpinolene, 1-nonanal, 1,3,8-*p*-menthatriene, citral, decanal, and γ-octanoic. These were all recognized as potent and major aroma contributors to mango pulp flavor. Further investigation showed that 1-hexanol, γ-terpinene, β-phellandrene, terpinolene, ethyl cyclopropanecarboxylate, and γ-octanoic were aroma-active compounds in JM; β-phellandrene, *p*-cymene, d-limonene, decanal, 1-hexanol, 1-nonanal were the most important aromatic substances in QM; 1-nonanal, 2-cyclohepten-1-one, 1,3,8-*p*-menthatriene, hexanal, 2-cyclohepten-1-one, and γ-octanoic were the most important aromatic substances in HY; and 1-hexanol, γ-terpinene, γ-octanoic, β-phellandrene, and 1-nonanal were the most important aromatic substances in GF. TN was significantly higher than in the other four aromatic substances. β-myrcene, 2-carene, β-phellandrene, 3-carene, *p*-cymene, d-limonene, γ-terpinene, terpinolene, 1,3,8-*p*-menthatriene, and decanal were the most important aromatic substances in HY.

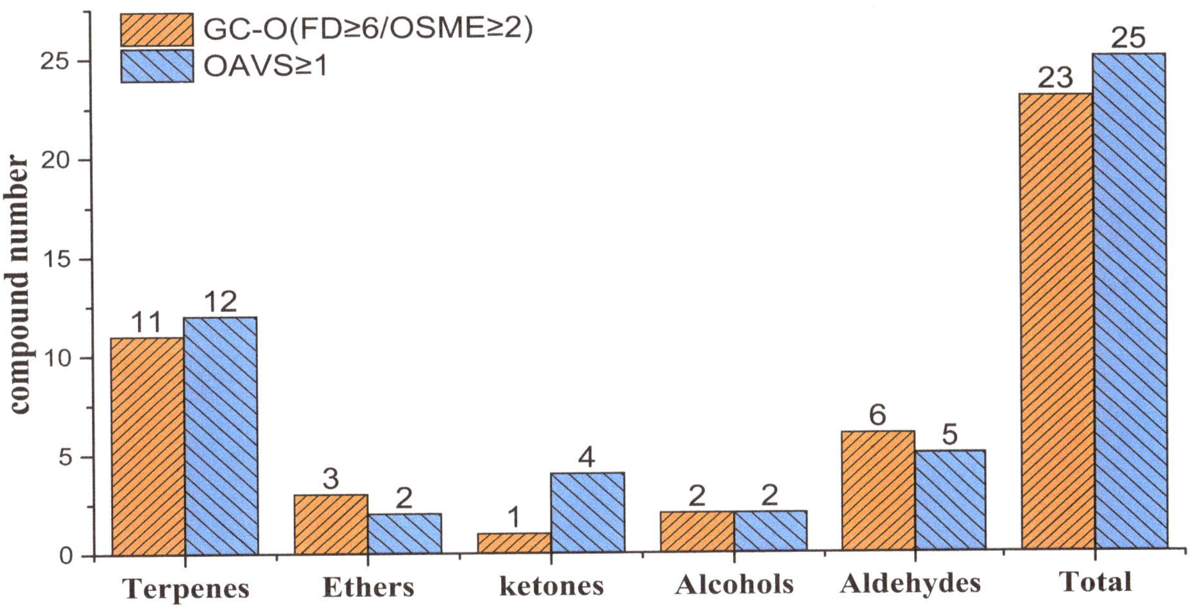

Figure 7. Comparison of GC-O (FD/OSME) and OAVs for aroma-active compounds.

Table 1. Volatile compounds identified in five different cultivars of Chinese mango samples using SPME-GC/MS.

Code	Compounds	RI [1]		Identification [2]	RQCF (fi′) [3]	Concentration (μg/kg Mango Must) [4]									
		HP-5	Reference			JM	RSD (%)	QM	RSD (%)	GF	RSD (%)	HY	RSD (%)	TN	RSD (%)
						Alcohols									
A1	2-Penten-1-ol	761	769	RI, Std, MS,	1.25	1.92 ^{b5}	0.02	- [8]	-	-	-	5.45 ^{a}	0.22	1.87 ^{c}	0.07
A2	(E)-3-Hexen-1-ol	844	852	RI, Std, MS,	0.52	41.50 ^{c}	0.03	132.56 ^{a}	0.12	76.82 ^{b}	0.12	-	-	-	-
A3	2-Hexen-1-ol	852	852	RI, Std, MS,	1.09	7.08 ^{b}	0.23	35.53 ^{a}	0.11	-	-	-	-	-	-
A4	3-Hexen-1-ol	854	857	RI, Std, MS,	4.3	-	-	-	-	-	-	6.59	0.11	-	-
A5	1-Hexanol	856	865	RI, Std, MS,	2.01	55.83 ^{b}	0.08	59.24 ^{a}	0.13	22.91 ^{d}	0.10	7.71 ^{e}	0.12	46.14 ^{c}	0.17
A6	isoamyl alcohol	921	910	RI, Std, MS,	0.74	38.87	0.03	-	-	-	-	-	-	-	-
A7	p-Cymen-8-ol	945	ND	RI, Std, MS,	11.08	-	-	-	-	-	-	12.23	0.04	-	-
A8	2-(Vinyloxy) ethanol	1012	ND	RI, Std, MS,	3.34	3.49	0.06	-	-	-	-	-	-	-	-
A9	Linalool	1103	1098	RI, Std, MS,	0.80	-	-	-	-	1.92 ^{b}	0.07	-	-	6.90 ^{a}	0.2
						Terpenes									
B1	α-Pinene	918	916	RI, Std, MS,	0.36	4.05 ^{b}	0.24	1.76 ^{d}	0.17	1.93 ^{c}	0.12	1.01 ^{e}	0.17	7.38 ^{a}	0.04
B2	Allo-ocimene	947	950	RI, Std, MS,	2.56	-	-	-	-	-	-	-	-	0.08	0.12
B3	β–Pinene	958	946	RI, Std, MS,	2.10	-	-	-	-	2.33 ^{b}	0.08	1.34_{c}	0.17	3.54 ^{a}	0.01
B4	β-Myrcene	973	990	RI, Std, MS,	0.43	6.28 ^{b}	0.24	3.83 ^{d}	0.34	4.5 ^{c}	0.07	1.41 ^{e}	0.17	22.94 ^{a}	0.05
B5	2-Carene	981	995	RI, Std, MS,	0.22	1.60 ^{b}	0.31	0.67 ^{c}	0.14	0.54 ^{d}	0.22	-	-	4.44 ^{a}	0.05
B6	Humulene	984	990	RI, Std, MS,	0.81	-	-	-	-	-	-	2.93	0.03	-	-
B7	α-Phellandrene	985	989	RI, Std, MS,	0.50	16.84 ^{b}	0.24	10.84 ^{c}	0.01	6.90 ^{d}	0.07	-	-	30.17 ^{a}	0.04
B8	3-Carene	990	1107	RI, Std, MS,	0.25	103.63 ^{a}	0.24	60.36 ^{c}	0.23	43.32 ^{d}	0.07	27.67 ^{e}	0.17	70.92 ^{b}	0.01
B9	D-Limonene	1007	1026	RI, Std, MS,	0.10	12.4 ^{d}	0.24	32.54 ^{a}	0.23	26.2 ^{c}	0.07	2.56 ^{e}	0.17	28.87 ^{b}	0.04
B10	*p*-cymene	1005	1016	RI, Std, MS,	1.11	27.43 ^{c}	0.08	33.85 ^{b}	0.14	19.44 ^{d}	0.06	15.57 ^{e}	0.09	132.96 ^{a}	0.04
B11	β-Caryophyllene	1018	ND	RI, Std, MS,	0.64	-	-	-	-	-	-	2.18	0.17	-	-
B12	1,3-Cyclohexadiene, 1-methyl-4-(1-methylethyl)	1021	1030	RI, Std, MS,	0.18	4.96 ^{a}	0.07	4.84 ^{b}	0.02	3.42 ^{c}	0.06	-	-	-	-
B13	γ-Terpinene	1036	1045	RI, Std, MS,	0.68	15.44 ^{b}	0.16	6.73 ^{c}	0.19	6.08 ^{d}	0.26	1.10 ^{e}	0.07	20.88 ^{a}	0.25
B14	β-Ocimene	1050	1060	RI, Std, MS,	1.52	-	-	-	-	6.97 ^{b}	0.08	-	-	9.11 ^{a}	0.04
B15	Terpinolene	1066	1065	RI, Std, MS,	0.19	147.07 ^{b}	0.24	59.42 ^{c}	0.23	43.5 ^{d}	0.07	24.90 ^{e}	0.17	811.60 ^{a}	0.04
B16	1,3,8-*p*-Menthatriene	1086	1110	RI, Std, MS,	1.37	11.92 ^{c}	0.18	3.79 ^{d}	0.19	-	-	13.58 ^{b}	0.16	16.38 ^{a}	0.14
						Aldehydes									
C1	Hexanal	792	800	RI, Std, MS,	1.30	4.94 ^{b}	0.07	-	-	-	-	75.54 ^{a}	1.50	-	-
C2	Isovaleraldehyde	837	ND	RI, Std, MS,	2.47	-	-	-	-	1.77	0.17	-	-	-	-
C3	trans-2-Hexenal	840	844	RI, Std, MS,	2.20	8.38 ^{c}	0.02	180.60 ^{a}	0.26	84.03 ^{b}	0.15	85.64 ^{b}	1.50	4.11 ^{d}	0.26
C4	3-hexenal	857	847	RI, Std, MS,	11.55	-	-	-	-	-	-	-	-	44.13	0.12
C5	Heptanal	888	899	RI, Std, MS,	0.07	-	-	-	-	-	-	1.65 ^{b}	1.50	1.83 ^{a}	0.44
C6	trans-2-Heptenal	940	957	RI, Std, MS,	1.35	-	-	-	-	-	-	65.65	1.50	-	-

Table 1. *Cont.*

Code	Compounds	RI [1]		Identification [2]	RQCF (fi′) [3]	Concentration (μg/kg Mango Must) [4]									
		HP-5	Reference			JM	RSD (%)	QM	RSD (%)	GF	RSD (%)	HY	RSD (%)	TN	RSD (%)
C7	1-Nonanal	1080	1140	RI, Std, MS,	0.45	-	-	1.46 ^{b}	0.16	0.91 ^{c}	0.28	8.43 ^{a}	1.50	0.86 ^{c}	0.26
C8	(E,Z)-2,6-Nonadienal	1152	1152	RI, Std, MS,	0.64	-	-	-	-	-	-	0.66	0.10	-	-
C9	E-2-Nonenal	1162	1160	RI, Std, MS,	1.87	-	-	7.60	0.09	-	-	-	-	-	-
C10	Decanal	1175	1205	RI, Std, MS,	6.05	-	-	8.48 ^{c}	0.06	8.45 ^{c}	0.15	-	-	12.46 ^{a}	0.04
C11	Citral	1236	1242	RI, Std, MS,	0.02	0.18 ^{b}	0.28	-	-	0.09 ^{c}	0.05	1.23 ^{a}	0.36	-	-
						ketones									
D1	1-Penten-3-one	652	687	RI, Std, MS,	0.05	-	-	0.52 ^{a}	0.56	-	-	0.47 ^{b}	0.28	0.31 ^{c}	0.22
D2	2-Cyclohepten-1-one	673	ND	RI, Std, MS,	1.50	-	-	-	-	-	-	0.66	0.14	-	-
D3	3-methylcyclohex-3-en-1-one	967	986	RI, Std, MS,	3.15	-	-	0.62	0.17	-	-	-	-	-	-
D4	6-Methyl-5-hepten-2-one	969	985	RI, Std, MS,	0.42	0.39 ^{a}	0.05	0.28 ^{c}	0.22	0.30 ^{b}	0.25	0.23 ^{d}	0.32	-	-
						Ethers									
E1	Ethyl propionate	765		RI, Std, MS,	1.10	-	-	-	-	-	-	6.39	0.11	-	-
E2	Ethyl cyclopropanecarboxylate	808	755	RI, Std, MS,	0.26	-	-	-	-	-	-	0.12	0.55	-	-
E3	Ethyl crotonate	876	802	RI, Std, MS,	1.60	-	-	9.83	0.05	-	-	-	-	-	-
E4	Isoamyl acetate	879	876	RI, Std, MS,	1.37	5.31 ^{b}	0.09	7.45 ^{a}	0.22	2.21 ^{c}	0.19	1.28 ^{d}	0.14	-	-
E5	Tetraethyl orthosilicate	923	880	RI, Std, MS,	0.06	-	-	0.12 ^{a}	0.54	0.09 ^{b}	0.93	-	-	-	-
E6	Ethyl butyrate	978	ND	RI, Std, MS,	0.84	1.17 ^{b}	0.06	-	-	-	-	-	-	1.80 ^{a}	0.23
E7	γ-Octanoic	1222	994	RI, Std, MS,	2.48	18.60 ^{b}	0.21	-	-	16.49 ^{c}	0.20	26.32 ^{a}	0.11	-	-

[1] Retention index of volatile compounds on HP-5 columns according to equation proposed [30]; reference: comparing linear retention indices (LRI) on columns (HP-5) in the literature. "ND" not detected in literature. [2] Method of identification: RI, retention index (HP-5) in agreement with literature value; Std, confirmed by authentic standards; MS, mass spectrum comparisons using NIST14 library; [3] RQCF (fi′) equals the ratio of quantitative factor of identified components standards to that of internal standard (ethyl hexanoate). [4] Concentrations are expressed in nanograms per milliliter of mango must, with ethyl hexanoate as the internal standard, and data listed are the means of three assays ± RSDs (%); all RSDs were <15%. [5] Values in total data with different letters are significantly different ($p < 0.05$). [8] "-" not detected in samples.

Table 2. GC-O identified aroma-active compounds in mango samples with the method of aromatic intensity and frequency.

No.	Compounds	RI (Calculate) [A]		RI (Reference) [B]		Aroma description [C]	Identification [D]	Aromatic Intensity [E]										Frequency [F]				
		HP-5	Wax	HP-5	Wax			JM	RSD (%)	TN	RSD (%)	HY	RSD (%)	QM	RSD (%)	GF	RSD (%)	JM	TN	HY	QM	GF
1	1-Penten-3-one	652	ND	687 [6]	ND	mushroom	R, S, M, O	0	0	2.25	10.00	1.67	5.99	1.33	11.28	0	0	0	5	4	8	0
2	2-Cyclohepten-1-one	673	ND	ND	ND	coffee-like	R, S, M, O	0	0	0	0	2.00	0.50	0	0	0	0	0	0	5	0	0
3	2-Penten-1-ol	761	1316	769 [31]	1305 [6]	grassy, tea	R, S, M, O	2.00	0.01	1.25	0.80	1.00	1.00	0	0	0	0	4	2	7	0	0
4	Hexanal	792	1061	800 [32]	1088 [32]	grass, tallow, fat	R, S, M, O	1.83	0.49	2.2	0.91	2.00	1.00	0	0	1.80	0.56	5	2	2	0	2
5	Ethyl cyclopropanecarboxylate	808	ND	ND	ND	fruity	R, S, M, O	1.50	0.23	1.23	0.20	1.30	0.90	0	2.11	0.56	0	6	6	8	0	8
6	trans-2-Hexenal	840	1202	844 [32]	1192 [32]	green, leaf	R, S, M, O	1.50	0.23	1.67	6.95	1.67	3.78	2.00	0.50	2.00	0.50	3	5	4	5	4
7	3-hexenal	857	ND	857 [6]	ND	grass	R, S, M, O	1.27	0.67	0	0	1.36	3.78	1.60	8.13	1.43	0.45	3	1	7	8	7
8	3-Hexen-1-ol, (E)-	844	1356	852 [31]	1343 [31]	grassy, tea	R, S, M, O	1.00	0.01	0	0	0	0	1.33	0.75	1.75	5.71	4	2	0	2	3
9	1-Hexanol	856	1360	865 [32]	1362 [32]	resin, flower, green	R, S, M, O	1.60	0.28	1.83	2.73	1.23	0.56	2.12	0.34	2.13	0.67	5	2	4	4	3
10	Heptanal	888	1174	899 [6]	1184 [32]	fruity	R, S, M, O	0	0	1.5	0.67	1.67	2.34	1.00	1.00	0	0	0	5	4	0	0
11	α-Pinene	918	1032	916 [31]	1056 [23]	pine-like	R, S, M, O	1.10	0.34	1.80	10.00	2.20	20.00	1.00	1.00	1.5	0.67	2	4	2	2	2
12	trans-2-Heptenal	940	1300	957 [31]	1307 [31]	grassy, irritant	R, S, M, O	0	0	0	0	0	0	1.25	1.60	1.56	0.9	0	2	1	3	3
13	Allo-ocimene	947	1125	ND	1135 [1]	floral	R, S, M, O	0	0	0	0	1.00	1.00	0	0	0	0	0	0	3	0	0
14	β-Pinene	958	1113	946 [23]	1108 [23]	grass	R, S, M, O	0	0	1.56	0.80	1.23	1.50	1.67	0.60	0	0	0	5	4	7	0
15	3-methylcyclohex-3-en-1-one	967	ND	986 [1]	ND	grass	R, S, M, O	0	0	0	0	0	0	2.67	2.25	0	0	0	1	0	3	0
16	6-Methyl-5-hepten-2-one	969	ND	985 [23]	ND	fruit, grass	R, S, M, O	1.67	0.32	2.00	0.50	1.33	4.51	1.56	0.78	1.75	3.43	2	4	2	2	3
17	β-Myrcene	973	1137	990 [32]	1158 [32]	light balsam, wood	R, S, M, O	1.80	1.20	2.00	0.50	0	0	0	0	0	0	4	6	0	0	0
18	Ethyl butyrate	978	1070	994 [31]	1076 [31]	fruity, apple	R, S, M, O	1.23	5.79	1.67	6.95	0	0	0	0	0	0	3	5	0	0	0
19	2-Carene	981	ND	995 [23]	ND	sweet, rosin	R, S, M, O	1.89	0.98	2.20	2.27	0	0	1.54	0.56	1.23	0.44	5	6	0	4	2
20	β-Phellandrene	985	1166	989 [31]	1171 [31]	citrus like	R, S, M, O	4.10	1.34	3.53	6.67	3.00	0.33	2.10	0.90	2.50	0.67	7	6	2	5	4
21	3-Carene	990	1148	1007 [31]	1153 [31]	sweet, rosin	R, S, M, O	1.50	0.87	2.00	0.50	1.14	0.34	1.45	0.76	1.23	1.56	3	4	2	2	2
22	*p*-Cymene	1005	1275	1026 [32]	1274 [32]	citrus, green	R, S, M, O	1.00	0.29	8.70	0.45	1.67	6.95	2.00	15.0	1.80	6.67	5	6	6	6	6
23	D-Limonene	1028	1205	1030 [32]	1208 [32]	citrus-like, sweet	R, S, M, O	0	0	2	0.5	0	0	2.00	0.50	0	0	0	5	0	4	0
24	(E)-beta-ocimene	1050	1241	1048 [33]	1250 [33]	floral and green	R, S, M, O	2.33	0.01	0	0	2.50	0.40	0	0	0	0	3	0	3	0	0
25	γ-Terpinene	1057	1249	1060 [33]	1245 [33]	citrus, lemon	R, S, M, O	2.33	0.02	2.50	1.60	2.33	1.72	1.67	6.95	2.00	2.50	8	8	5	6	8
26	Terpinolene	1066	1275	1065 [31]	1276 [31]	rosin-like	R, S, M, O	2.30	0.02	2.28	2.19	2.20	3.18	1.33	0.75	2.33	9.31	6	8	7	8	7

Table 2. *Cont.*

No.	Compounds	RI (Calculate) [A]		RI (Reference) [B]		Aroma description [C]	Identification [D]	Aromatic Intensity [E]										Frequency [F]				
		HP-5	Wax	HP-5	Wax			JM	RSD (%)	TN	RSD (%)	HY	RSD (%)	QM	RSD (%)	GF	RSD (%)	JM	TN	HY	QM	GF
27	1-Nonanal	1080	1328	1104 [31]	1349 [31]	cucumber-like	R, S, M, O	1.67	0.26	1.50	3.33	0	0	0	0	1.40	7.14	5	4	0	0	3
28	1,3,8-*p*-Menthatriene	1086	ND	1110 [34]	ND	minty-like	R, S, M, O	1.00	0.01	2.00	0.50	2.20	0.45	0	0	2.25	4.44	5	4	4	0	4
29	(E,Z)-2,6-Nonadienal	1152	1460	1152 [31]	1469 [6]	fresh, green, cucumber	R, S, M, O	0	0	0	0	1.67	0.60	0	0	0	0	0	0	2	0	0
30	E-2-Nonenal	1162	1416	1160 [31]	1436 [31]	cucumber-like	R, S, M, O	1.20	0.01	0	0	0	0	0	0	0	0	2	0	0	0	0
31	Citral	1236	1700	1242 [34]	1679 [33]	lemon-like	R, S, M, O	2.00	0.01	0	0	0	0	0	0	0	0	2	0	0	0	0
32	Decanal	1175	1835	1205 [34]	1846 [34]	fruity, citrus, orange	R, S, M, O	0	0	3	0.33	0	0	3.00	0.33	0	0	0	2	0	2	0
33	γ-Octanoic	1222	1721	1236 [31]	1733 [31]	floral, violet	R, S, M, O	2.00	0.01	0	0	1.50	2.00	0	0	2.00	0.50	7	0	4	0	3

[A] Retention index of volatile compounds on columns (HP-5 and WAX) according to equation proposed [30]; "ND" not detected in samples. [B] RI (reference): comparing linear retention indices (LRI) on columns (HP-5 and WAX) in the literature. [C] Odor note perceived at the sniffing port. [D] Method of identification: RI, retention index (HP-5) in agreement with literature value; Std, confirmed by authentic standards; MS, mass spectrum comparisons using NIST14 library. [E] Aromatic intensity, the data listed are the means of three assays ± RSDs (%); all RSDs were <15%. [F] Aroma frequency.

Table 3. Concentrations and calculations of odor activity values (OAVs) of the important aroma-active compounds in mango samples.

No	Compounds	Threshold [a] (μg kg^{-1})	Source [b]	OAV [c]				
				JM	QM	GF	HY	TN
1	1-Penten-3-one	1.00 [35]	mango	0	0.52	0	0.47	0.31
2	2-Cyclohepten-1-one	0.14 [35]	new	0	0	0	4.71	0
3	2-Penten-1-ol	720.00 [23]	mango	<0.01	0	0	<0.01	<0.01
4	Hexanal	4.50 [23]	mango	1.10	0	0.29	16.79	0.40
5	Ethyl cyclopropanecarboxylate	0.12 [23]	new	2.17	0	0	0	1.00
6	trans-2-Hexenal	400.00 [23]	mango	0.02	0.45	0.21	0.21	0.01
7	3-hexenal	550.00 [23]	mango	0	0	0	0	0.08
8	3-Hexen-1-ol, (E)-	110.00 [23]	mango	0.38	1.21	0.70	0	0
9	1-Hexanol	9.00 [26]	mango	6.20	6.58	2.55	0.86	5.13
10	Heptanal	1.00 [23]	mango	0	0	0	1.65	1.83
11	α-Pinene	6.00 [35]	mango	0.68	0.29	0.32	0.17	1.23
12	trans-2-Heptenal	13.00 [35]	mango	0	0	0	5.05	0
13	Allo-ocimene	140.00 [35]	mango	0	0	0.07	0.04	0.10
14	β-Pinene	100.00 [23]	mango	0	0	0	<0.01	0
15	3-methylcyclohex-3-en-1-one	7.00 [36]	mango	0	0.09	0	0	0
16	6-Methyl-5-hepten-2-one	50.00 [23]	mango	<0.01	<0.01	<0.01	<0.01	0
17	β-Myrcene	20.00 [23]	mango	0.31	0.19	0.23	0.07	1.15
18	Ethyl butyrate	0.75 [23]	mango	1.56	0	0	0	2.40
19	2-Carene	4.00 [23]	mango	0.40	0.17	0.14	0	1.11
20	β-Phellandrene	7.00 [23]	mango	2.41	1.55	1.00	0	4.31
21	3-Carene	50.00 [23]	mango	2.07	1.21	0.87	0.55	1.42
22	*p*-Cymene	18.50 [23]	mango	1.48	1.83	1.05	0.84	7.19
23	D-Limonene	26.00 [23]	mango	0.48	1.25	1.01	0.10	1.11
24	(E)-beta-ocimene	6.70 [23]	mango	0	0	1.04	0	1.36
25	γ-Terpinene	2.00 [23]	mango	7.72	3.37	3.04	0.55	10.44
26	Terpinolene	140.00 [23]	mango	1.05	0.42	0.31	0.18	5.80
27	1-Nonanal	0.15 [23]	mango	0	9.73	6.07	56.20	5.73
28	1,3,8-*p*-Menthatriene	6.80 [23]	Lychee	1.75	0.56	0	2.00	2.41
29	(E,Z)-2,6-Nonadienal	4.50 [23]	mango	0.57	0	0	0	0.02
30	E-2-Nonenal	50.00 [36]	mango	0	0.15	0	0	0
31	Citral	1.00 [23]	Lychee	0.18	0	0.09	1.23	0
32	Decanal	6.00 [23]	mango	0	1.41	1.41	0	2.08
33	γ-Octanoic	7.00 [35]	mango	2.66	0	2.36	3.76	0

[a] OT odor threshold in water (ppb) found in the newly determined and taken from the literature. [b] Source: It indicates substances found in the related literature for mangoes and litchis; New: first identified to be useful in aroma activity in mango. [c] An OAV was calculated by dividing the concentration of an odorant by its orthonasal odor threshold.

4. Conclusions

Mango has a pleasing sensory quality and rich nutritional components, and thus it is essential to study the composition of flavor components in mango. A total of 47 volatile compounds were preliminarily identified by GC-MS, which were subsequently classified into alcohols, alkenes, aldehydes, esters, ketones, and ethers. The results of GC-O (FD/OSME) analysis showed that there were 23, 20, 20, 24, and 24 kinds of aromatic components in JM, QM, GF, HY, and TN, respectively. Sensory analysis indicated that the main sensory aroma profiles (fruit, sweet, flower, and rosin aromas) of the pulps of five mango cultivars were consistent with those of identified using the FD and OSME methods, indicating that the intensity method combined with the FD method could accurately reflect the characteristic aromatic components with high or low intensities. Moreover, OAV calculations indicated that there were 11 substances with OAVs ≥1 in JM, nine in QM, nine in GF, eight in HY, and 17 in TN. Analysis of OAV and GC-O(FD/OSME) identified 29 predominant aroma-active compounds (FD ≥ 6, OSME ≥ 2, OAV ≥ 1) in the pulps of five mango cultivars, which included citrus, lemon-like γ-terpinene and β-phellandrene, rosin-like terpinolene, floral, green-like 1-hexanol and γ-octanoic, and fruit-like ethyl cyclopropanecarboxylate in JM. The predominant aroma-active compounds of cucumber, fruity, floral, citrus, green-like cucumber-like β-phellandrene, *p*-cymene, d-limonene, decanal, 1-hexanol, and 1-nonanal were observed in QM. The predominant aroma-active compounds of minty, citrus, green, floral, violet, coffee, cucumber-like 1-nonanal, 2-cyclohepten-1-one, 1,3,8-*p*-menthatriene, hexanal, 2-cyclohepten-1-one, and γ-octanoic were detected in HY. The predominant aroma-active compounds of resin, flower, green, citrus, lemon, cucumber-like 1-hexanol, γ-terpinene, γ-octanoic, β-phellandrene, and 1-nonanal were observed in GF. Light balsam, wood, sweet, rosin, citrus, minty, fruity, citrus, orange-like β-myrcene, 2-carene, β-phellandrene, 3-carene, *p*-cymene, d-limonene, γ-terpinene, terpinolene, 1,3,8-*p*-menthatriene, and decanal were the most important aromatic substances in HY. TN was significantly higher than HY in four other aromatic substances. In addition, 2-cyclohepten-1-one, ethyl cyclopropanecarboxylate, 1,3,8-*p*-menthatriene, and citral were identified to be associated for the first time with aroma activity in mango based on OVA and GC-O(FD/OSME). Hence, this research not only revealed the aroma-active compounds in different mangoes, but also improved our understanding and control of critical aroma parameters in different mango cultivars in China.

Author Contributions: Conceptualization, H.L. and K.A.; Data curation and S.S., H.L.; Formal analysis, G.X. and Y.X.; Funding acquisition, Y.X.; Methodology, H.L.; Project administration, Y.Y., J.W. and Y.X.; Resources, Y.Y. and J.W.; Writing—original draft, H.L.; Writing—review & editing, Y.X. and K.A. All authors have read and agreed to the published version of the manuscript.

References

1. Pandit, S.S.; Chidley, H.G.; Kulkarni, R.S.; Pujari, K.H.; Giri, A.P.; Gupta, V.S. Cultivar relationships in mango based on fruit volatile profiles. *Food Chem.* **2009**, *114*, 363–372. [CrossRef]
2. Munafo, J.P., Jr.; Didzbalis, J.; Schnell, R.J.; Schieberle, P.; Steinhaus, M. Characterization of the major aroma-active compounds in mango (*Mangifera indica* L.) cultivars Haden, White Alfonso, Praya Sowoy, Royal Special, and Malindi by application of a comparative aroma extract dilution analysis. *J. Agric. Food Chem.* **2014**, *62*, 4544–4551. [CrossRef] [PubMed]
3. Dar, M.S.; Oak, P.; Chidley, H.; Deshpande, A.; Giri, A.; Gupta, V. Nutrient and flavor content of mango (*Mangifera indica* L.) cultivars: An appurtenance to the list of staple foods. In *Nutritional Composition of Fruit Cultivars*; Elsevier: Amsterdam, The Netherlands, 2016; pp. 445–467.
4. Musharraf, S.G.; Uddin, J.; Siddiqui, A.J.; Akram, M.I. Quantification of aroma constituents of mango sap from different Pakistan mango cultivars using gas chromatography triple quadrupole mass spectrometry. *Food Chem.* **2016**, *196*, 1355–1360. [CrossRef] [PubMed]
5. FAOSTAT. Available online: http://www.fao.org/faostat/en/#data/QC/visualize (accessed on 1 October 2019).
6. Pino, J.A.; Mesa, J.; MUNoz, Y.; Martí, M.P.; Marbot, R. Volatile components from mango (*Mangifera indica* L.) cultivars. *J. Agric. Food Chem.* **2005**, *53*, 2213–2223. [CrossRef] [PubMed]
7. Li, L.; Ma, X.W.; Zhan, R.L.; Wu, H.X.; Yao, Q.S.; Xu, W.T.; Luo, C.; Zhou, Y.G.; Liang, Q.Z.; Wang, S.B. Profiling of volatile fragrant components in a mini-core collection of mango germplasms from seven countries. *PLoS ONE* **2017**, *12*, e0187487. [CrossRef]
8. Ma, X.W.; Su, M.Q.; Wu, H.X.; Zhou, Y.G.; Wang, S.B. Analysis of the Volatile Profile of Core Chinese Mango Germplasm by Headspace Solid-Phase Microextraction Coupled with Gas Chromatography-Mass Spectrometry. *Molecules* **2018**, *23*, 1480. [CrossRef]
9. Bonneau, A.; Boulanger, R.; Lebrun, M.; Maraval, I.; Gunata, Z. Aroma compounds in fresh and dried mango fruit (*Mangifera indica* L. cv. Kent): Impact of drying on volatile composition. *Int. J. Food Sci. Technol.* **2016**, *51*, 789–800. [CrossRef]
10. Kung, T.L.; Chen, Y.J.; Chao, L.K.; Wu, C.S.; Lin, L.Y.; Chen, H.C. Analysis of Volatile Constituents in Platostoma palustre (Blume) Using Headspace Solid-Phase Microextraction and Simultaneous Distillation-Extraction. *Foods* **2019**, *8*, 415. [CrossRef] [PubMed]
11. Lau, H.; Liu, S.Q.; Xu, Y.Q.; Lassabliere, B.; Sun, J.; Yu, B. Characterising volatiles in tea (*Camellia sinensis*). Part I: Comparison of headspace-solid phase microextraction and solvent assisted flavor evaporation. *LWT Food Sci. Technol.* **2018**, *94*, 178–189. [CrossRef]

12. Usami, A.; Nakahashi, H.; Marumoto, S.; Miyazawa, M. Aroma evaluation of setonojigiku (*Chrysanthemum japonense var. debile*) by hydrodistillation and solvent-assisted flavor evaporation. *Phytochem. Anal. PCA* **2014**, *25*, 561–566. [CrossRef]
13. Schuh, C.; Schieberle, P. Characterization of the key aroma compounds in the beverage prepared from Darjeeling black tea: Quantitative differences between tea leaves and infusion. *J. Agric. Food Chem.* **2006**, *54*, 916–924. [CrossRef] [PubMed]
14. He, C.; Guo, X.; Yang, Y.; Xie, Y.; Ju, F.; Guo, W. Characterization of the aromatic profile in "zijuan" and "pu-erh" green teas by headspace solid-phase microextraction coupled with GC-O and GC-MS. *Anal. Methods* **2016**, *8*, 4727–4735. [CrossRef]
15. Bianchin, J.N.; Nardini, G.; Merib, J.; Dias, A.N.; Martendal, E.; Carasek, E. Screening of volatile compounds in honey using a new sampling strategy combining multiple extraction temperatures in a single assay by HS-SPME–GC–MS. *Food Chem.* **2014**, *145*, 1061–1065. [CrossRef] [PubMed]
16. Feng, S.; Huang, M.; Crane, J.H.; Wang, Y. Characterization of key aroma-active compounds in lychee (*Litchi chinensis* Sonn.). *JFDA* **2018**, *26*, 497–503. [CrossRef] [PubMed]
17. San, A.T.; Joyce, D.C.; Hofman, P.J.; Macnish, A.J.; Webb, R.I.; Matovic, N.J.; Williams, C.M.; De Voss, J.J.; Wong, S.H.; Smyth, H.E. Stable isotope dilution assay (SIDA) and HS-SPME-GCMS quantification of key aroma volatiles for fruit and sap of Australian mango cultivars. *Food Chem.* **2017**, *221*, 613–619. [CrossRef]
18. Zhao, J.-H.; Liu, F.; Pang, X.-L.; Xiao, H.-W.; Wen, X.; Ni, Y.-Y. Effects of different osmo-dehydrofreezing treatments on the volatile compounds, phenolic compounds and physicochemical properties in mango (*Mangifera indica* L.). *Int. J. Food Sci. Technol.* **2016**, *51*, 1441–1448. [CrossRef]
19. Pang, X.; Guo, X.; Qin, Z.; Yao, Y.; Hu, X.; Wu, J. Identification of aroma-active compounds in Jiashi muskmelon juice by GC-O-MS and OAV calculation. *J. Agric. Food Chem.* **2012**, *60*, 4179–4185. [CrossRef]
20. Mastello, R.B.; Janzantti, N.S.; Monteiro, M. Volatile and odoriferous compounds changes during frozen concentrated orange juice processing. *Food Res. Int.* **2015**, *77*, 591–598. [CrossRef]
21. Munafo, J.P., Jr.; Didzbalis, J.; Schnell, R.J.; Steinhaus, M. Insights into the Key Aroma Compounds in Mango (*Mangifera indica* L. Haden) Fruits by Stable Isotope Dilution Quantitation and Aroma Simulation Experiments. *J. Agric. Food Chem.* **2016**, *64*, 4312–4318. [CrossRef]
22. Matheis, K.; Granvogl, M. Characterization of Key Odorants Causing a Fusty/Musty Off-Flavor in Native Cold-Pressed Rapeseed Oil by Means of the Sensomics Approach. *J. Agric. Food Chem.* **2016**, *64*, 8168–8178. [CrossRef]
23. Bonneau, A.; Boulanger, R.; Lebrun, M.; Maraval, I.; Valette, J.; Guichard, E.; Gunata, Z. Impact of fruit texture on the release and perception of aroma compounds during in vivo consumption using fresh and processed mango fruits. *Food Chem.* **2018**, *239*, 806–815. [CrossRef] [PubMed]
24. Song, H.; Liu, J. GC-O-MS technique and its applications in food flavor analysis. *Food Res. Int.* **2018**, *114*, 187–198. [CrossRef] [PubMed]
25. Zhao, D.; Shi, D.; Sun, J.; Li, A.; Sun, B.; Zhao, M.; Chen, F.; Sun, X.; Li, H.; Huang, M.; et al. Characterization of key aroma compounds in Gujinggong Chinese Baijiu by gas chromatography-olfactometry, quantitative measurements, and sensory evaluation. *Food Res. Int.* **2018**, *105*, 616–627. [CrossRef] [PubMed]
26. Zhu, J.; Wang, L.; Xiao, Z.; Niu, Y. Characterization of the key aroma compounds in mulberry fruits by application of gas chromatography-olfactometry (GC-O), odor activity value (OAV), gas chromatography-mass spectrometry (GC-MS) and flame photometric detection (FPD). *Food Chem.* **2018**, *245*, 775–785. [CrossRef]
27. Chalier, P.; Angot, B.; Delteil, D.; Doco, T.; Gunata, Z. Interactions between aroma compounds and whole mannoprotein isolated from Saccharomyces cerevisiae strains. *Food Chem.* **2007**, *100*, 22–30. [CrossRef]
28. Kühn, J.; Considine, T.; Singh, H. Interactions of milk proteins and volatile flavor compounds: Implications in the development of protein foods. *J. Food Sci.* **2006**, *71*, R72–R82. [CrossRef]
29. Xu, J.; He, Z.; Zeng, M.; Li, B.; Qin, F.; Wang, L.; Wu, S.; Chen, J. Effect of xanthan gum on the release of strawberry flavor in formulated soy beverage. *Food Chem.* **2017**, *228*, 595–601. [CrossRef]
30. Van den Dool, H.; Kratz, P.D. A Generalization of the Retention Index System including Linear Temperature Programmed Gas-Liquid Partition Chromatography. *J. Chromatogr.* **1963**, *11*, 463–471. [CrossRef]

31. Zhang, W.; Dong, P.; Lao, F.; Liu, J.; Liao, X.; Wu, J. Characterization of the major aroma-active compounds in Keitt mango juice: Comparison among fresh, pasteurization and high hydrostatic pressure processing juices. *Food Chem.* **2019**, *289*, 215–222. [CrossRef]
32. Goodner, K.L. Practical retention index models of OV-101, DB-1, DB-5, and DB-Wax for flavor and fragrance compounds. *LWT Food Sci. Technol.* **2008**, *41*, 951–958. [CrossRef]
33. Babushok, V.I.; Linstrom, P.J.; Zenkevich, I.G. Retention Indices for Frequently Reported Compounds of Plant Essential Oils. *J. Phys. Chem. Ref. Data* **2011**, *40*, 043101. [CrossRef]
34. An, K.; Liu, H.; Fu, M. Identification of the cooked off-flavor in heat-sterilized lychee (*Litchi chinensis* Sonn.) juice by means of molecular sensory science. *Food Chem.* **2019**, *301*, 125282. [CrossRef] [PubMed]
35. Pino, J.A.; Mesa, J. Contribution of volatile compounds to mango (*Mangifera indica* L.) aroma. *Flavor Fragr. J.* **2006**, *21*, 207–213. [CrossRef]
36. Van Gemert, L. *Compilations of Odour Threshold Values in Air, Water and Other Media*; BACIS: Zeist, The Netherlands, 2003.

6

Assessment of Antibiotic and Pesticides Residues in Breast Milk of Syrian Refugee Lactating Mothers

Nadia Smadi [1,†], Adla Jammoul [2,†] and Nada El Darra [1,*]

1 Faculty of Heath Sciences, Beirut Arab University, Tarik El Jedidah–Beirut, P.O.Box: 115020, Riad EL Solh, Beirut 1107 2809, Lebanon
2 Food Department, Lebanese Agricultural Research Institute, Fanar P.O. Box 2611, Beirut 1107 2809, Lebanon
* Correspondence: n.aldarra@bau.edu.lb
† These authors contributed equally.

Abstract: Occupational exposures and current diet are both sources of environmental contaminants that can be transferred in the mother's body. These chemicals can definitely penetrate to the developing foetus and the nursing infant from contaminated breast milk during the lactation period. Nowadays, one of the special interests is the exposure of new-borns to toxic chemicals such as pesticides and antibiotics reported in human milk due to their potential harms, especially developmental deficits in early childhood. The aim of our current study was to assess the occurrence of pesticide residues and antibiotic residues contamination in breast milk collected from Syrian refugee lactating mothers residing in North Lebanon Camps. A total of 120 breast milk samples (40 in triplicate) were collected from camps in Akkar, North Lebanon using an electrical pump. A survey was administrated to determine socio-demographic characteristics, dietary and smoking habits and medical history of participating lactating mothers. The milk samples were analysed for the presence of antibiotic residues and pesticide residues using liquid and gas chromatography tandem mass spectroscopy (LC-MS/MS) and Gas Chromatography-Tandem Mass Spectrometry (GC-MS/MS). This study reported the absence of antibiotic residues in 96.66% of our samples (n = 120) and the presence of pesticides residues in only 5% of our total breast milk sample. Our results considered the breast milk collected from Syrian refugee lactating mothers as safe from chemical contamination. It is worth conducting more studies on other Syrian refugee camps to test the effect of camp living conditions on breast milk safety.

Keywords: breast milk; antibiotic residues; pesticide residues; LC-MS/MS; GC-MS/MS

---

1. Introduction

Breastfeeding is admitted to present numerous beneficial health effects and it is vastly considered as the best suitable food for the baby [1]. The World Health Organization recommends to exclusively breastfeed for six months after birth [2]. However, the European recommendation varies among countries between four and six of exclusive breastfeeding [3].

In Lebanon and Syria, two Middle Eastern Arab countries sharing almost the same socio-demographic characteristics including religion and language, a percentage of 10% was shown for an exclusive breastfeeding rate at six months [4]. According to the United Nations Children Fund [5], the average of breastfed Syrian infants under six months of age was found to be 43%. The absence of support for breastfeeding by social workers and health care providers were considered the main reasons for low prevalence of exclusive breastfeeding. It could be also due to psychological reasons such as mother's emotional stress and the perceived breast-milk inadequacy [6].

Breast milk is an important source of energy and nutrients for children's health [7]. However, many studies have been conducted to assess the chemical contaminants in breast milk associated with its health effects on the infant and the mother. This contamination could occur due the exposure of the nursing mother on a daily basis to chemical pollution of the environment. These environmental chemicals are released from several basic activities and different sources such as water, air, food and manufactured products [8].

Few research studies have reported the presence of antibiotic and pesticide residues in breast milk. It is worth mentioning that studies screening antibiotic residues are rare on the international scale. The latest study conducted in Turkey [9], detected quinolone and beta-lactam residues in human milk samples, with an occurrence of 8/34 (23.5%) and 29/34 (85.2%), respectively. It is to be noted that the lactating mothers enrolled in this study do not have antibiotic history. Antibiotic residues in human milk were thought to derive from the mothers' food such as chicken, meat and dairy products [10].

Additionally, some studies have screened the presence of pesticide residues in breast milk especially organochlorine pesticides [11–15]. The most important study was carried out in China where a wide range of concentrations for 23 organochlorine pesticide (OCP) compounds were detected in 24 pooled samples of breast milk. The dichlorodiphenyltrichloroethane (DDT) contamination in breast milk was the highest with a mean value 582.8 ± 362.7 ng/g. The explanation for such high contamination was mainly from the mother's dietary intake since positive correlation was observed between concentration of DDT in human milk and consumption of animal origin food in the Chinese population [16].

However, to our knowledge no studies have been conducted on Syrian refugee lactating mothers residing in camps especially where the living and environmental conditions are very poor. Moreover, our study is the first conduced in Lebanon, screening antibiotics and pesticide residues in breast milk. Thus, the aim of this study was to assess the occurrence of antibiotic and pesticide residues in the breast milk of Syrian refugee lactating women residing in Lebanon camps. It investigated as well the socio-demographic and nutritional factors associated with milk contamination.

2. Materials and Methods

2.1. *Study Population and Area*

This cross-sectional study was conducted in Syrian refugee camps located in Lebanon, North Lebanon city—Akkar camps. The participants in this study were lactating Syrian refugee mothers residing in camps. In total, 40 women were randomly recruited and invited to participate in the study by providing mature breast milk samples.

2.2. *Ethical Approval*

The study protocol and ethics of this study were approved by the institutional review board (IRB) code: 2019H-0087-HS-M-0325 (17 January 2019), of Beirut Arab University prior to approaching the participants and a written, signed informed consent was obtained from all participants. All women who volunteered to participate in this study were informed about the purpose of the study and agreed to be part of it.

2.3. *Questionnaire*

A questionnaire was used to collect data about the socio-demographic characteristics of participating lactating mothers. The questionnaire was adopted from different studies [14,16–20]. It consisted of eight subsections collected from different sources, addressing general information about the lactating mothers and the new-borns, dietary habits of lactating mothers, smoking status, geographic characteristics, pesticides spraying, medical history and supplements intake as well as the drinking water intake.

The questionnaire was translated into Arabic to ensure accuracy of the translation, back translation from Arabic to English was conducted by another translator. Both English versions were the same. The Arabic version of the questionnaire was pilot-tested on a sample of 30 mothers. The questionnaire was administered twice within a period of two weeks between each sitting. To determine test–retest reliability, a paired *t*-test was performed to compare mean scores at T_1 (before) and T_2 (after two weeks). Pearson's correlation was calculated between T_1 and T_2, with a consistency interclass correlation (ICC) [21]. The values of ICC were considered excellent if >0.75 [22].

2.4. Sample Collection

During December 2018 and February 2019, 50 nursing mothers were contacted but only 45 agreed to answer the survey and participated in our study. The 45 nursing mothers that were visited for the survey were contacted again and only 40 of them agreed to donate breast milk by triplicate over three weeks.

Breast milk collection was done in the morning an hour after the previous breastfeeding. Samples (25–40 mL) were collected using a sterile, single-use electrical pump (Chico, Mod.06836, China). A polypropylene pump container was sterilized with 70% alcohol solution to prevent cross contamination between samples. Each sample was placed into a separate 100 mL dark glass (to preserve the samples from light), numbered then transported on ice (at 4 °C) to the laboratory, where they were transferred into urine cups, numbered and stored at −20 °C until analysis.

2.5. Chemicals and Reagents

The reagents used were of analytical grade. The antibiotic standards belonged to three different families: sulphonamides (sulfamethazine), tetracyclines (tetracycline, oxytetracycline) and beta-lactam antibiotics (ampicillin); in addition, 161 kinds of pesticide were purchased from Sigma-Aldrich (St. Louis, MO, USA). The standards used present high purity grades (>99%). Individual stock solutions were prepared at 1000 µg/L in acetonitrile and stored at −20 °C.

The working standard solutions of a concentration of 10 µg/L each were prepared as dilution of the stock solution in water/methanol (50:50, *v*/*v*). The working standard solutions were stored at −20 °C. HPLC-grade water, HPLC-grade acetonitrile, and magnesium sulphate ($MgSO_4$) were also supplied by Sigma-Aldrich.

2.6. Sample Extraction for Antibiotics Residues Analysis

Using liquid chromatography–mass spectrometry, a multi-class method was developed for identifying and quantifying three antibiotics belonging to two different chemical classes (tetracyclines and beta-lactams). Afterwards, this method was optimized for the detection of antibiotics in breast milk.

100 µL from 1 % formic acid was added to a 2 mL of breast milk then vortexed for 60 s and kept in the dark for 10 min. Then 500 µL of EDTA (0.5%) was added and shaken vigorously for 1 min. Next, 8 mL TCA (5%) was added then shaken for 10 min before subjecting the tubes to centrifugation at 5000 rpm at 4 °C for 10 min. The supernatant obtained was then subjected to filtration through a 0.45 µm polyvinylidene fluoride (PVDF) filter for further liquid chromatography tandem mass spectroscopy (LC-MS/MS) analysis.

2.7. Sample Extraction for Pesticide Residues Analysis: QuEChERS Extraction

To ensure a quick and easy sample treatment, a modified QuEChERS (quick, easy, cheap, effective, rugged and safe) approach, originally developed as a powerful sample preparation, was used to analyse hundreds of pesticides in different kinds of food.

Acetonitrile (10 mL) was added to 10 mL of breast milk and shaken vigorously before adding magnesium sulphate $MgSO_4$ (4 g), sodium chloride NaCl (1 g), sodium dibasic citrate (1 g) and sodium tribasic citrate (0.5 g) and then shaken for 3 min. The tubes were then subjected to centrifugation at 5000 rpm at 4 °C for 5 min.

The aqueous supernatant (5 mL) was mixed with dispersive Enhanced Matrix Removal (EMR) dissolved in 5 mL water then shaken vigorously for 5 min. Another centrifugation was conducted at 5000 rpm at 4 °C for 5 min.

The aqueous supernatant (5 mL) obtained was transferred in a new Polish EMR Lipid (MgSO4 and NaCl tube, then shaken vigorously for 3 min before being centrifuged at 5000 rpm at 4 °C for 5 min. Afterwards, a filtration was made through a 0.22 μm polyvinylidene fluoride (PVDF) filter for further LC-MS/MS and GC/MSMS analysis.

2.8. LC-MS/MS Equipment

LC-MS/MS analyses was conducted using an Agilent 6430 LC/MS (Agilent technologies, Santa Clara, USA) for antibiotic and pesticide residues detection, and LC-MS/MS analyses were performed on LC-NexeraX2 Shimadzu 8045 LC/MS (Kyoto, Japan). The mass spectrometer was operated with an electron spray ionisation (ESI) in multiple reaction monitoring (MRM) mode, ionspray voltage 4 kV, nitrogen for desolvation and dried gas 11 L/min.

The quantification of the three antibiotics families and 161 kinds of pesticide in 120 breast milk samples was done through measuring peak areas in the multiple reaction monitoring chromatogram, and comparing them with the relevant matrix-matched calibration curves. A calibration curve ranging between 1 μg/L and 500 μg/L was carried out to verify linearity.

The performance of the analytical method was evaluated by checking the identification criteria for the presence of two transitions at the same retention time, the signal to noise ratio ≥ 10, the relative retention time of the analyte within a tolerance of 2.5% and the relative ion intensities ratio within a tolerance defined by the EU commission decision 2002/657/EC. The MRM parameters are shown in Table 1.

Table 1. Multiple reactions monitoring (MRM) acquisition condition for each antibiotic used.

Antibiotic Family	Antibiotic	Precursor Ion (*m/z*)	Collision Energy for Precursor Ion (eV)	Product Ion (*m/z*)	Collision Energy for Production Ion (eV)	Cone Voltage (V)	Retention Time (min)
Sulphonamides	Sulfamethazine	279.0 > 186.1	17	279.0 > 184.0	23	Default	5.012
Tetracyclines	Tetracycline	445.1 > 410.25	20	445.1 > 427.15	12		4.886
	Oxytetracycline	461.1 > 443.3	14	461.1 > 426.15	18		4.934
Beta-Lactam	Ampicillin	350.3 > 106.05	20	350.3 > 159.9	13		4.944

2.9. LC/MS/MS Parameters

The separation of the pesticide residues was performed using a C18 analytical column (zorbax 2.1 mm inner diameter × 150 mm length, 3.5 μm particle size; Agilent technologies, Santa Clara, CA, USA) and a C18 analytical column (Shim-pack GIST 2.1 mm inner diameter × 100 mm length, 3 μm particle size; Japan) for antibiotic residues separation. The separation of pesticides was accomplished at 40 °C. The flow rate and injection volume were 0.4 mL/min and 10 μL, respectively. The mobile phases used were (A) 5 mM ammonium acetate and 0.1% formic acid in water, and (B) 5 mM ammonium acetate in methanol. The gradient elution program started with 100% A for 5 min, then decreased to 50% for 13 min, then to 0% for 2 min and returned back to the initial conditions for 5 min. The final run time of the method was 29 min.

The separation of sulfamides, tetracyclines and beta-lactam was accomplished at 40 °C. The flow rate and injection volume were 0.3 mL/min and 10 μL, respectively. The mobile phases used were (A) 5 mM ammonium acetate and 0.1% formic acid in water, and (B) 5 mM ammonium acetate in methanol. The gradient elution program was as follow: B (40%) (2 min), B (90%) (9 min), B (10%) (9 min and 1 s), the final run time of the method was 13 min.

2.10. GC/MSMS Parameters

GC-MS/MS analyses were carried out with an Agilent 7890A GC equipped with 7693 Agilent auto-sampler and 7000C triple quadrupole GC/MS system. The separation was performed on a HP-5MS Agilent technologies (Santa Clara, CA, USA), 0.25 mm × 30 m, 0.25 μm, and helium (purity 99.996%)

was used as a carrier gas at a constant pressure of 11 psi. The inlet temperature was set at 250 °C; the mode of inlet was splitless; the injection volume was 1 μL. The column temperature program was as follows: the initial temperature was maintained at 70 °C for 1 min, increased to 160 °C at a rate of 50 °C/min, raised to 200 °C at 2 °C/min, and then at 16 °C/min up to 280 °C, and held there for 7.2 min. The total run time was 35 min. The mass spectrometer was operated with an electron impact (EI) source in multiple reaction monitoring (MRM) mode. The electron energy was 70 eV, and the source, transfer line and quadrupole temperatures were set at 280 °C and 150 °C, respectively. In order to prevent instrument damage, the solvent delay was set at 4.5 min.

2.11. Recovery Test

In-house validation was performed by fortifying the blank matrix at a level of 50 μg/L. The extraction was performed by the methods described in the Section 2.6 and 2.7. The spiked and blank samples were then analysed by LCMS/MS. Recovery was calculated by comparing the analysed concentrations with spiked concentrations.

3. Results and Discussion

3.1. Validation of the Arabic Version of the Questionnaire

The results of test–retest reliability are presented in Table 2. The paired t-test showed that the mean scores of the Arabic questionnaire did not vary significantly between T_1 and T_2. A statistically significant inter-item correlation, $p < 0.05$, was noted between T_1 and T_2. An excellent consistency was also noted, with an ICC varying between 0.998 and 1.00 [22].

Table 2. Mean and standard deviations, Pearson's correlation coefficients and intraclass correlation for the scores of the Arabic and Arabic versions (n = 30).

Scale	Mean Scores		Paired t-Test	Correlation between Scores at T1 and T2	Intraclass Correlation (ICC)	
	Arabic T_1	Arabic T_2				
	Mean ± SD		p-Value	Correlation Coefficient	ICC	95% CI
Dietary Habits of Lactating Mothers	71.2 ± 10.21	71.43 ± 10.23	0.032	0.998	0.998	0.996–0.999
Smoking Status	5.16 ± 0.94	5.14 ± 0.92	0.000	0.999	0.999	0.998–1.000
Demographic Characteristics	3.33 ± 0.711	3.32 ± 0.743	0.001	1.000	1.000	1.000
Pesticides	2.866 ± 0.860	2.867 ± 0.862	0.000	1.000	1.000	1.000
Medical History and Supplements Intake	6.566 ± 1.381	6.565 ± 1.380	0.000	0.999	0.999	0.998–1.000

3.2. Survey Results

The results of the questionnaire filled by 40 Syrian refugee nursing mothers is presented in Table 3. All the participants had an age below 30 years. The number of years of residency in North Lebanon camps for all lactating mothers was between one year and five years. All the mothers (100%) had an educational degree below secondary and they were not employed.

While assessing the smoking habits, it was shown that 100% of nursing mothers had never smoked before or after pregnancy since Syrian culture inhibits women from smoking, but they were randomly exposed to second hand smoking from their husbands. Lactating mothers were also asked about their supplement intake, 100% indicated that they were taking vitamins and iron supplements during their pregnancy period, however they stopped consuming these supplements after their child's birth in order to save money.

Table 3. Survey results of participating mothers.

Characteristics	Frequency	Percent
Residence years in Tripoli Lebanon camps (1–5 years)	40	100
Age (<30 years)	40	100
Gestational Age (= 9 months)	40	100
Lactation time (>= 120 days)	40	100
Level of education (Below secondary)	40	100
Occupation Not employed (Housekeeper)	40	100
Number of newborn (=1 child)	40	100
Infant Gender		
Male	16	40
Female	24	60
Age of newborn		
=< 5months	11	27.5
>5 months	29	72.5
Birth weight (g)		
=< 3.5 kg	39	97.5
> 3.5kg	1	2.5
Irregular newborn Sleep pattern	40	100
Colic Crying of newborn	40	100
Fish intake		
Never	40	100
Sea food intake		
Never	40	100
Cereals intake		
Twice a week	39	97.5
Daily	1	2.5
Potatoes intake		
Once a week	7	17.5
Twice a week	2	5.0
More than twice a week	18	45
Daily	13	32.5
Fresh Vegetables intake		
Once a week	26	65
Twice a week	14	35
Milk intake		
Never	39	97.5
Daily	1	2.5
Dairy Product intake		
Never	11	27.5
Once a week	19	47.5
> once a week	4	10
Twice a week	6	15
Pasta intake		
Once a week	15	37.5
More than Once a week	5.0	5.0
Twice a week	37.5	37.5
More than Twice a week	20.0	20.0
Rice intake		
More than Twice a week	10	25.0
Daily	30	75.0

Table 3. *Cont.*

Characteristics	Frequency	Percent
Grains		
Once a week	4	10.0
Twice a week	13	32.5
More than Twice a week	23	57.5
Soft Drink intake		
Never	12	30.0
Once a week	20	50.0
More than Once a week	2	5.0
Twice a week	6	15.0
Jam intake		
Never	27	67.5
Once a week	13	32.5
Coffee intake		
Never	23	57.5
Once a week	9	22.5
Twice a week	8	20.0
Tea intake		
Once a week	1	2.5
More than Once a week	1	2.5
Twice a week	11	27.5
More than Twice a week	12	30.0
Daily	15	37.5
Fruits intake		
Once a week	22	55.0
Twice a week	18	45.0
Salty Snack intake		
Never	19	47.5
Once a week	21	52.5
Chocolate intake		
Never	21	52.5
Once a week	17	42.5
Twice a week	2	5.0
Meat and Poultry intake		
Never	40	100
Eggs intake		
Never	16	40.0
Once a week	13	32.5
More than Once a week	6	15.0
Twice a week	5	12.5
Beverages		
Never	11	27.5
Once a week	17	42.5
More than Once a week	3	7.5
Twice a week	9	22.5
Smoking status		
Before pregnancy	40	100
No	40	100
During pregnancy	40	100
No		
Random smoke exposure		
Yes		

Table 3. *Cont.*

Characteristics	Frequency	Percent
Nearby Waste Disposal		
Yes	40	100
Nearby Cultivation Activity		
Yes	40	100
Spray indoor to prevent mosquitoes		
Yes	40	100
Spray outdoor with pesticides		
Yes	40	100
Vitamin supplement in pregnancy		
Yes	40	100
Iron supplement in pregnancy		
Yes	40	100
Vitamin_supplement_in_postpartum_2_months		
No	40	100
Iron_supplement_in_postpartum_2_months		
No	40	100
Antibiotic intake in pregnancy		
No	40	100
Antibiotic intake after pregnancy		
No	40	100
Anemia at any time		
Yes	40	100
Drinking water		
Well artesian water	40	100
Water bottles intake		
1–2 bottle	40	100

When questioned on their dietary habits, it was clearly noticed that 100% of the participants never consumed fish and sea food due to poverty. The consumption of meat and poultry was also negligible for the same reason. With respect to their location of residence, 100% were living near a waste disposal site and cultivation activity since the camp was located in an agricultural area.

3.3. Assessment of Antibiotic Residues in Breast Milk

3.3.1. LC-MS/MS Method Performance for Antibiotic Residues

An in-house method was verified as per the criteria specified in EU commission decision 2002/675/EC to quantify antibiotics residues. The validation parameters were fixed by spiking blank powdered milk at a level of 50 ng/g. The measured parameters were specificity, linear range, repeatability, reproducibility, accuracy and limit of quantification (LOQ).

The antibiotic residue chromatograms of the reference standards used as well as the calibration curves are shown in Figure 1. The calibration curves were created from six matrix-calibration standards which were injected in each batch in the range of 1 to 50 ng/g. The calibration curves showed good linearity, characterized by a high correlation coefficient ($R^2 > 0.99$).

The precision of the method was determined using the spiked standards at the 50 µg/L level. The results for repeatability ranged from 5 and 18.8% (Table 4). The limit of quantification is considered as the lowest quantified level with $S/N \geq 10$ in presence of the two transitions at the same retention time. The LOQ was calculated to be 2 µg/kg for all tested antibiotics. The mean recoveries of the residues for the spiked samples ranged between 97% and 108% (Table 4). These values presented high mean recoveries within the acceptable range (80%–120%) except for ampicillin, which presented a low mean recovery, although it was still higher than 50%. These values were within the acceptable ranges (50%–120%) recommended by AOAC, 2002 [23].

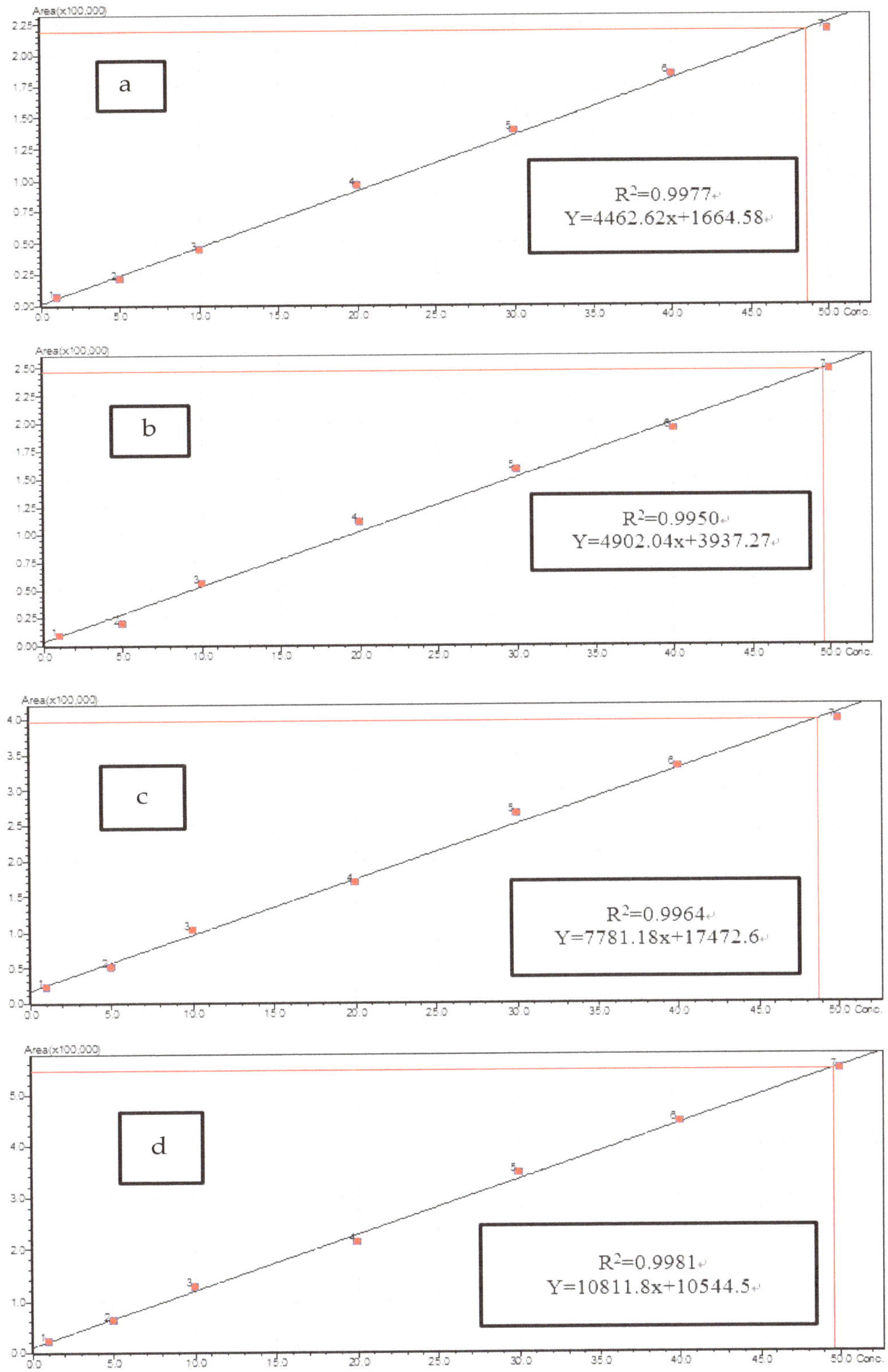

Figure 1. Calibration curves for (**a**) oxytetracycline, (**b**) tetracycline, (**c**) ampicillin and (**d**) sulfamethazine 1, 5, 10, 20, 30, 40 and 50 μg/L.

Table 4. Results of in-house verification of the LC-MS/MS method for the antibiotics considered in this study, with the standard deviation (STDEV), relative standard deviation of repeatability (RSD) and limits of quantification (LOQ).

Antibiotic Family	Antibiotic	Spiking Level (50 μg/L)	Mean Recoveries (%)	STDEV	RSD (%)	LOQ (μg/L)
Sulphonamides	Sulfamethazine	49.075	98.15	5.72	5.82	2 μg/L
Tetracyclines	Tetracycline	53.85	107.71	6.96	6.46	
	Oxytetracycline	48.664	97.328	18.22	18.72	
Beta-lactam	Ampicillin	26.475	52.95	0.627	1.18	

3.3.2. Occurrence of Antibiotic Residues in Breast Milk Samples

The method developed was applied to the determination of four antibiotics from three different chemical classes (sulphonamides, tetracyclines, and beta-lactams) in 120 breast milk samples collected from 40 lactating mothers residing in Lebanon's camp. In order to validate the results, an internal quality control was carried out on every batch of samples. Moreover, the retention time, quantification and confirmation transitions and relative ion intensities of the detected ion of the human milk samples were compared to those of the corresponding calibration standards in the same batch to identify the detected analytes.

While assessing the antibiotic residues in the breast milk samples (Table 5), the results showed that only 3.33% of the samples were contaminated with antibiotic residues. It is to be noted that none of the samples contained more than one antibiotic residues. The antibiotic residue contaminating the four samples was oxytetracycline. Of the samples, 96.67% presented a total absence of antibiotic residues. This result is well correlated with the survey, that shows that all the mothers did not receive antibiotics during pregnancy nor lactation period. Additionally, the survey results showed that 100% of the nursing mothers do not consume meat and chicken, that are known to be sources of antibiotic residues, due to the misuse of antibiotics in farms [10,24].

Table 5. Occurrence of sulphonamides, tetracyclines, and β-lactams in the 120 breast milk samples.

Breast Milk Samples (n = 120)	Sulphonamides	Tetracyclines		β-Lactams
	Sulfamethazine	Tetracycline	Oxytetracycline	Ampicillin
Mean * (μg/L)	0	0	5.04	0
min (μg/L)	0	0	0	0
max (μg/L)	0	0	6	0
n positive	0	0	3	0
% positive	0	0	2.5	0

* mean value of four contaminated samples.

3.3.3. Mean Concentrations of Antibiotic Residues for the Different Families in Breast Milk Samples

Table 5 represents the summary of multi-antibiotic residues occurrence in 120 human milk samples. The assessment of the sulphonamides, beta-lactam families and tetracyclines showed that three antibiotic residues were not detected in all milk samples, i.e., tetracyclines, sulfamethazine and ampicillin. However, as mentioned before, oxytetracycline was detected in four samples (3.3%). These four samples belong to three mothers; one of them reported one antibiotic residue concentration between the LOQ (= 2 μg/L) and LOD (= LOQ/3 = 2/3 = 0.66μg/L) followed by a second concentration higher than the LOQ in the week after. The other two samples exceeded the LOQ.

Hence, the percentage of positive samples (>LOQ = 2 μg/L) is 2.5%, since the fourth sample presented a mean of 0.64 μg/L which is lower than LOQ and LOD.

The mother that reported the highest value of oxytetracycline in her breast milk (6 μg/L) was the only mother that consumed milk on a daily basis. The mean value of oxytetracycline (5.04 μg/L) is

2.5 times higher than the LOQ (2 μg/L). The low oxytetracycline concentration in only three breast milk samples is derived from the nursing mother's food since they did not use any kind of antibiotics after pregnancy. However antibiotic residues can be found in dairy products since antibiotics are usually used in farms of dairy cattle for the treatment of diseases such as mastitis [25]. Probably the source of oxytetracycline is from nursing mothers' dairy product intake because our survey findings indicate the contaminated breast milk samples belong to 15% of mothers consuming dairy products twice per week. The presence of oxytetracycline in dairy products, could be due to the misuse of this antibiotic in farms and lack of abidance to the recommended withdrawal times [26].

3.4. Assessment of Pesticide Residues in Breast Milk

3.4.1. LC-MS/MS Method Performance for Pesticide Residuess

The pesticide residue chromatograms of the reference standards used as well as the calibration curves are shown in Figure 2. The calibration curves were created from nine matrix-calibration standards which were injected in each batch in the range from 1 to 500 ng/mL. The calibration curves showed good linearity, characterized by a high correlation coefficient (R^2 > 0.99).

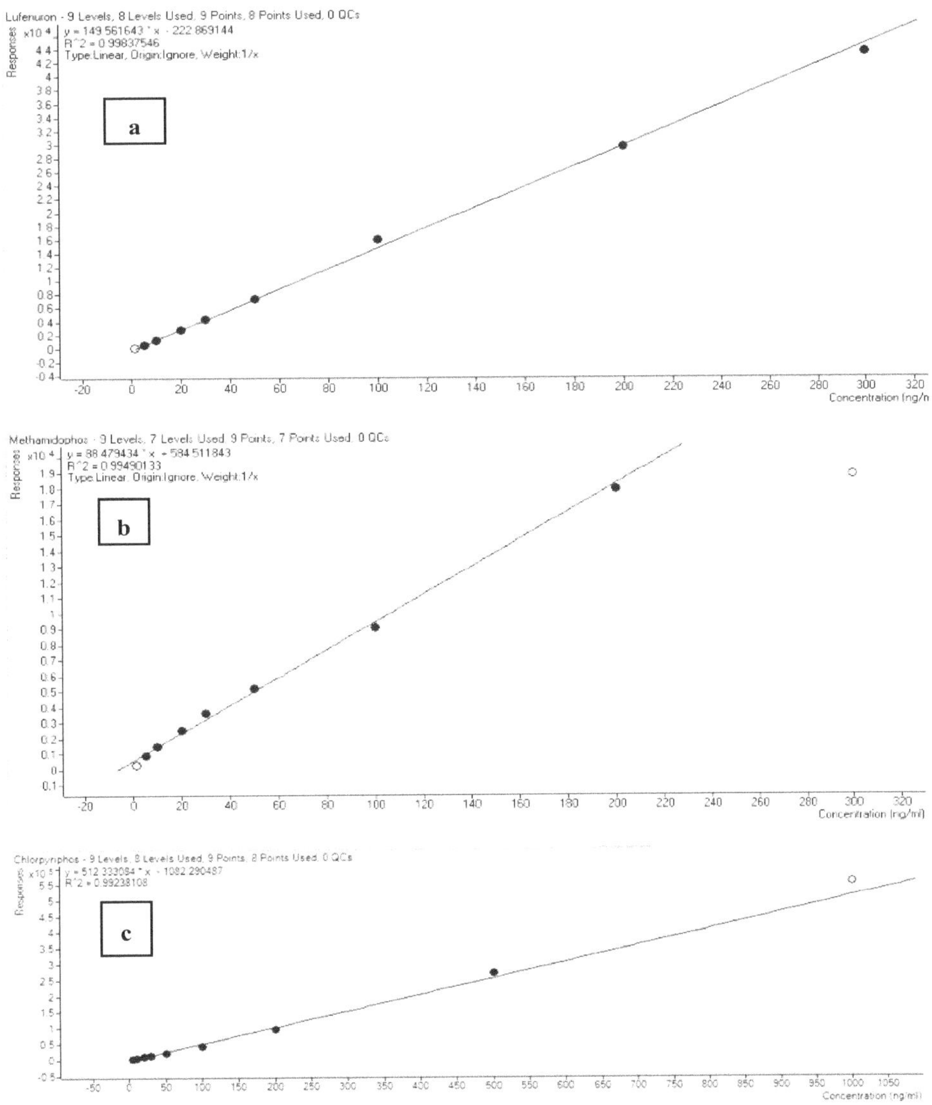

Figure 2. LC standard calibration curves for (**a**) lufeneron, (**b**) methamidophos and (**c**) chlorpyriphos at 5, 10, 20, 30, 50, 100, 200, 500 μg/L.

The precision of the method was determined using two spiked levels of 0.05 and 0.1 mg/kg. The limit of quantification was considered as the lowest quantified level with S/N ≥ 3 in presence of the two transitions at the same retention time. The LOQ was calculated to be 5 μg/kg for all tested pesticides. The mean recoveries of the residues for the spiked samples was between the acceptable range (80%–120%).

3.4.2. Occurrence of Pesticide Residues in Breast Milk Samples

While assessing the pesticide residues in the breast milk samples, the results showed that only 5% of the samples were contaminated with pesticide residues. The samples 95% were non-contaminated. It is to be noted that none of the samples contained more than one pesticide residue.

3.4.3. Mean Concentrations of Pesticide Residues in Breast Milk Samples

Table 6 represents the summary of multi-pesticide residues occurrence in 120 human milk samples. The assessment of the 161 pesticides showed that only three pesticide residues were detected in six samples, lufeneron, methamidophos and chlorpyriphos. Hence, the percentage of positive samples (>LOQ = 5 μg/L) was 4.16%. Only four samples were contaminated with lufeneron, the fifth sample was contaminated with methamidophos and the final sample was contaminated with chlorpyriphos with a concentration of 12.32 μg/L. The recovery tests results were at one level (10 μg/L) for methamidophos (86.5%), chlorpyriphos (103.6%) and lufeneron (105.2%).

Table 6. Occurrence in the 120 breast milk samples of lufeneron, methamidophos and chlorpyriphos.

Breast Milk Samples (*n* = 120)	**Lufeneron**	**Methamidophos**	**Chlorpyriphos**
Mean * (μg/L)	5.8754	2.198	2.05
min (μg/L)	5.1208	0	0
max (μg/L)	12.0447	13.1927	12.32
n positive	4	1	1
% positive	3.33	0.83	0.83

* mean values of six contaminated samples.

While assessing the pesticide residues in breast milk, the percentage of positive samples was only 5% in which methamidophos presented the highest concentration (13.1927 μg/L) in one sample only. The only mean value that exceeded the LOQ (5 μg/L) was reported for lufeneron pesticide (5.874 μg/L). The present results can be explained by nursing mother exposure to pesticides from fruit and vegetable intake [27], since the percentage of mothers consuming fruit and vegetables twice per week were 55% and 35% respectively, to the contrary of other studies conducted in several countries where the percentage of pesticide contamination in breast milk is high [14].

4. Conclusions

The contamination of breast milk is a critical problem since it affects the health of both the mother and her infant. In the last few decades, numerous studies reported chemical contamination in breast milk, especially pesticide residues, but antibiotic residue studies on breast milk remain rare. These contaminations are usually associated with socio-demographic status and dietary habits of nursing mothers. This study conducted for the first time, in Syrian refugee camps, presented an absence of antibiotic residues in the majority of the samples and the presence of pesticide residues in only 5% of our total breast milk sample. These findings consider the breast milk collected from Syrian refugee lactating mothers safe from the chemical contaminants screened. This could be due to the poor living and environmental conditions in the camp. It is worth conducting more studies on other Syrian refugee camps to test the effect of the camp living conditions on breast milk safety.

Author Contributions: N.S. declares that she is a Msc student. She contributed in the sampling, running the experiments and the data acquisition and data analysis. A.J. contributed in the design and the conception of the work, interpretation and the analysis of the data as the manuscript writing and submission. N.E.D. declares that she is the supervisor of the Msc student. She contributed in the design and the conception of the work, interpretation and the analysis of the data as well as the manuscript writing and submission. She is the corresponding author.

Acknowledgments: The authors thankfully acknowledge Michel Afram, the director of the Lebanese Agricultural Research Institute for his continuous support.

References

1. Hörnell, A.; Lagström, H.; Lande, B.; Thorsdottir, I. Breastfeeding, introduction of other foods and effects on health: A systematic literature review for the 5th Nordic Nutrition Recommendations. *Food Nutr. Res.* **2013**, *57*, 20823. [CrossRef] [PubMed]
2. WHO. 2008. Available online: https://www.who.int/maternal_child_adolescent/documents/9789241596664/en/ (accessed on 26 April 2019).
3. Agostoni, C.; Bergman, R.; Bresson, J.-L.; Michaelsen, K.F.; Przyrembel, H.; Sanz, Y.; Tomé, D. Scientific opinion on the appropriate age for introduction of complementary feeding of infants: EFSA panel on dietetic products, nutrition and allergies (NDA). *EFSA J.* **2009**, *7*, 1423.
4. Hamade, H.; Naja, F.; Keyrouz, S.; Hwalla, N.; Karam, J.; Al-Rustom, L.; Nasreddine, L. Breastfeeding knowledge, attitude, perceived behavior, and intention among female undergraduate university students in the Middle East: The case of Lebanon and Syria. *Food Nutr. Bull.* **2014**, *35*, 179–190. [CrossRef] [PubMed]
5. UNICEF. UNICEF. At a Glance: Syrian Arab Republic. 2013. Available online: http://www.unicef.org/infobycountry/syria_statistics.htmL (accessed on 26 April 2019).
6. Forste, R.; Weiss, J.; Lippincott, E. The decision to breastfeed in the United States: Does race matter? *Pediatrics* **2001**, *108*, 291–296. [CrossRef] [PubMed]
7. Grote, V.; Verduci, E.; Scaglioni, S.; Vecchi, F.; Contarini, G.; Giovannini, M.; Koletzko, B.; Agostoni, C. Breast milk composition and infant nutrient intakes during the first 12 months of life. *Eur. J. Clin. Nutr.* **2016**, *70*, 250. [CrossRef]
8. WHO. *Indicators for Assessing Infant and Young Child Feeding Practices*; WHO: Geneva, Switzerland, 2007.
9. Dinleyici, M.; Yildirim, G.K.; Aydemir, O.; Kaya, T.B.; Bildirici, Y.; Carman, K.B. Human milk antibiotic residue levels and their relationship with delivery mode, maternal antibiotic use and maternal dietary habits. *Eur. Rev. Med. Pharmacol. Sci.* **2018**, *22*, 6560–6566.
10. Jammoul, A.; El Darra, N. Evaluation of Antibiotics Residues in Chicken Meat Samples in Lebanon. *Antibiotics* **2019**, *8*, 69. [CrossRef]
11. Anderson, H.A.; Wolff, M.S. Environmental contaminants in human milk. *J. Expo. Sci. Environ. Epidemiol.* **2000**, *10*, 755. [CrossRef]
12. Szyrwińska, K.; Lulek, J. Exposure to specific polychlorinated biphenyls and some chlorinated pesticides via breast milk in Poland. *Chemosphere* **2007**, *66*, 1895–1903. [CrossRef]
13. Zhou, J.; Zeng, X.; Zheng, K.; Zhu, X.; Ma, L.; Xu, Q.; Zhang, X.; Yu, Y.; Sheng, G.; Fu, J. Musks and organochlorine pesticides in breast milk from Shanghai, China: Levels, temporal trends and exposure assessment. *Ecotoxicol. Environ. Saf.* **2012**, *84*, 325–333. [CrossRef]
14. Rojas-Squella, X.; Santos, L.; Baumann, W.; Landaeta, D.; Jaimes, A.; Correa, J.C.; Sarmiento, O.L.; Ramos-Bonilla, J.P. Presence of organochlorine pesticides in breast milk samples from Colombian women. *Chemosphere* **2013**, *91*, 733–739. [CrossRef]
15. Müller, M.H.B.; Polder, A.; Brynildsrud, O.B.; Karimi, M.; Lie, E.; Manyilizu, W.B. Organochlorine pesticides (OCPs) and polychlorinated biphenyls (PCBs) in human breast milk and associated health risks to nursing infants in Northern Tanzania. *Environ. Res.* **2017**, *154*, 425–434. [CrossRef]
16. Zhou, P.; Wu, Y.; Yin, S.; Li, J.; Zhao, Y.; Zhang, L.; Chen, H.; Liu, Y.; Yang, X.; Li, X. National survey of the levels of persistent organochlorine pesticides in the breast milk of mothers in China. *Environ. Pollut.* **2011**, *159*, 524–531. [CrossRef]

17. Örün, E.; Yalçın, S.S.; Aykut, O.; Orhan, G.; Morgil, G.K.; Yurdakök, K.; Uzun, R. Breast milk lead and cadmium levels from suburban areas of Ankara. *Sci. Total Environ.* **2011**, *409*, 2467–2472. [CrossRef]
18. Cantú-Cornelio, F.; Aguilar-Toalá, J.E.; de León-Rodríguez, C.I.; Esparza-Romero, J.; Vallejo-Cordoba, B.; González-Córdova, A.F.; García, H.S.; Hernández-Mendoza, A. Occurrence and factors associated with the presence of aflatoxin M1 in breast milk samples of nursing mothers in central Mexico. *Food Control* **2016**, *62*, 16–22. [CrossRef]
19. Radonić, J.R.; Kocić Tanackov, S.D.; Mihajlović, I.J.; Grujić, Z.S.; Vojinović Miloradov, M.B.; Škrinjar, M.M.; Turk Sekulić, M.M. Occurrence of aflatoxin M1 in human milk samples in Vojvodina, Serbia: Estimation of average daily intake by babies. *J. Environ. Sci. Health Part B* **2017**, *52*, 59–63. [CrossRef]
20. Bassil, M.; Daou, F.; Hassan, H.; Yamani, O.; Abi, J.; Attieh, Z.; Elaridi, J. Chemosphere Lead, cadmium and arsenic in human milk and their socio-demographic and lifestyle determinants in Lebanon. *Chemosphere* **2018**, *191*, 911–921. [CrossRef]
21. Weir, J.P. Quantifying test-retest reliability using the intraclass correlation coefficient and the SEM. *J. Strength Cond. Res.* **2005**, *19*, 231–240.
22. Fleiss, J.L.; Wallenstein, S.; Chillton, N.W.; Goodson, J.M. A re-examination of within-mouth correlations of attachment level and of change in attachment level. *J. Clin. Periodontol.* **1988**, *15*, 411–414. [CrossRef]
23. Horwitz, W. *AOAC Guidelines for Single Laboratory Validation of Chemical Methods for Dietary Supplements and Botanicals*; AOAC International: Gaithersburg, MD, USA, 2002; pp. 12–19.
24. Ramatla, T.; Ngoma, L.; Adetunji, M.; Mwanza, M. Evaluation of Antibiotic Residues in Raw Meat Using Different Analytical Methods. *Antibiotics* **2017**, *6*, 34. [CrossRef]
25. Olatoye, I.O.; Daniel, O.F.; Ishola, S.A. Screening of antibiotics and chemical analysis of penicillin residue in fresh milk and traditional dairy products in Oyo state, Nigeria. *Vet. World* **2016**, *9*, 948–954. [CrossRef]
26. Obaidat, M.M.; Salman, A.E.B.; Davis, M.A.; Roess, A.A. Major diseases, extensive misuse, and high antimicrobial resistance of Escherichia coli in large-and small-scale dairy cattle farms in Jordan. *J. Dairy Sci.* **2018**, *101*, 2324–2334. [CrossRef]
27. Elgueta, S.; Moyano, S.; Sepúlveda, P.; Quiroz, C.; Correa, A. Pesticide residues in leafy vegetables and human health risk assessment in North Central agricultural areas of Chile. *Food Addit. Contam. Part B* **2017**, *10*, 105–112. [CrossRef]

7

Aflatoxin Contamination of Milk Marketed in Pakistan

Agha Waqar Yunus [1,*], Nida Imtiaz [1], Haider Khan [1], Mohammed Nawaz Mohammed Ibrahim [2] and Yusuf Zafar [3]

1 Animal Sciences Institute, PARC National Agricultural Research Center, Park Road, Islamabad 45500, Pakistan; nidaimtiaz234@yahoo.com (N.I.); haiderkhan_1@hotmail.com (H.K.)
2 International Livestock Research Institute, ILRI-Pakistan, Park Road, Islamabad 45500, Pakistan; m.ibrahim@cgiar.org
3 Pakistan Agricultural Research Council, G-5/1, Islamabad 44000, Pakistan; y_zafar@yahoo.com
* Correspondence: aghawaqaryunus@yahoo.com

Abstract: A longitudinal one-year study was conducted to determine aflatoxin M_1 levels in different types of milk marketed in Pakistan. Processed and raw liquid milk from 21 sources, two milk powder and six tea whitener brands were sampled on monthly basis from Islamabad. The aflatoxin M_1 levels in liquid milk were lower ($p < 0.05$) in summer (April to July) compared with the levels in winter (January, November and December). The mean aflatoxin M_1 levels were 254.9, 939.5, and 1535.0 ng/L in UHT, pasteurized, and raw milk, respectively (differing at $p < 0.001$). The mean toxin level in powdered milk after reconstitution was 522.1 ng/L. Overall, 12.9, 41.0, 91.9 and 50.0% of the UHT, pasteurized, raw and powdered milk samples, respectively, exceeded the Codex maximum tolerable limit of 500 ng of aflatoxin M_1/L. It was estimated that consumers of raw and processed milk were exposed to 11.9 and 4.5 ng aflatoxin M_1, respectively, per kg of body weight daily. The study indicates potential aflatoxin M_1 exposure risks for the consumers of raw milk in the country. The levels of the toxin though comparatively lower in milk powder, requires attention as this type of milk is consumed by infants.

Keywords: aflatoxin M_1; milk; pasteurized; Pakistan; UHT

Key Contribution: The mean AFM_1 levels in raw milk were found to be three times higher than the Codex maximum tolerable limit of 500 ng of AFM_1/L. The overall mean aflatoxin M_1 level in processed milk (UHT, pasteurized and powdered milk) was 63% lower than the raw milk, with only the UHT milk qualifying the Codex limit. Compared with the local milk powder brands, UHT milk could be used to reduce the daily aflatoxin M_1 exposure in infants.

---

1. Introduction

Aflatoxins are the toxic secondary metabolites of various *Aspergillus* spp. that commonly contaminate food and feed ingredients. The aflatoxins encountered in agricultural commodities include aflatoxin B_1, B_2, G_1 and G_2. In contaminated foodstuffs, the percentage of aflatoxin B_1 (AFB_1) in total aflatoxins is over 90%. Once ingested by animals, AFB_1 is also carried to milk in the form of the toxic metabolite aflatoxin M_1 (AFM_1) [1]. All of these toxins are known to exert potent hepatotoxic, immunotoxic and carcinogenic effects in animals and humans consuming the contaminated food. Due to high carcinogenicity, aflatoxins are the only group of mycotoxins for which legislation and control protocols are in place, even in many developing countries [2]. The toxicity of aflatoxins is known to be higher in younger age groups (infants, children and young animals). Monitoring the levels of AFM_1 in milk and baby foods is therefore more critical. Consequently, the levels of the mycotoxin allowed in milk are lower than the levels allowed in other foodstuffs. The EU further restricts the levels allowed in infant milk formula to half of the levels allowed in milk.

The maximum tolerable limit of AFM_1 in liquid milk is 500 ng/L in the USA and in the Codex standards, while only 50 ng/L in the EU. In Pakistan, the maximum tolerable limit of AFM_1 is 10 μg/kg in milk powder while no particular legislation has been made for liquid milk. This is despite the fact that specific monsoon conditions in the country favor mycotoxin development in food and feedstuffs, pushing Pakistan into a high risk area. The studies conducted in Pakistan also show that 25 to 90% of milk samples [3–6] could be contaminated with AFM_1.

There have been notable differences in the AFM_1 levels in milk reported by different authors from Pakistan. In this regard, Muhammad et al. [4] reported 17,380 ng/L as the mean AFM_1 level in milk sampled from Lahore in the year 2007, with 81% samples exceeding the 500 ng AFM_1/L limit. Contrary to this, Iqbal et al. [6] reported 64 ng/L mean AFM_1 level in milk sampled in the year 2011 in the urban areas of Punjab province, with 15% samples exceeding the 500 ng AFM_1/L limit. These differences in the AFM_1 contamination level reported by various authors could be due to different seasons, different feeds used by farmers in different areas, and different methods of AFM_1 quantification. Overall, such differences make it impractical to infer risk of exposure for the consumers of milk in other cities. The present study was therefore conducted as a longitudinal one year study to determine the AFM_1 levels in various types of milk, primarily processed, available in Islamabad the capital city of Pakistan. To the best of our knowledge, there is no previous longitudinal study on AFM_1 contamination in processed milk in Pakistan. Also, AFM_1 contamination of milk has not been previously investigated in Islamabad city. Data on processed milk from one city are however applicable to milk consumers in other cities because processing companies collect milk from farmers located in different areas and distribute it to consumers in all cities of Pakistan.

2. Results and Discussion

2.1. Aflatoxin Contamination of Liquid Milk

The seasonal trend regarding aflatoxin contamination of liquid milk (average of UHT, pasteurized, and raw milk) is presented in Figure 1. The levels of AFM_1 in liquid milk were lower ($p \leq 0.021$) in the summer months of April to June compared with the levels in the winter months of January, November and December. This seasonal variation in AFM_1 levels is in accordance with previous findings in which comparatively lower levels of AFM_1 were found in raw milk sampled during summer months in the country [7,8].

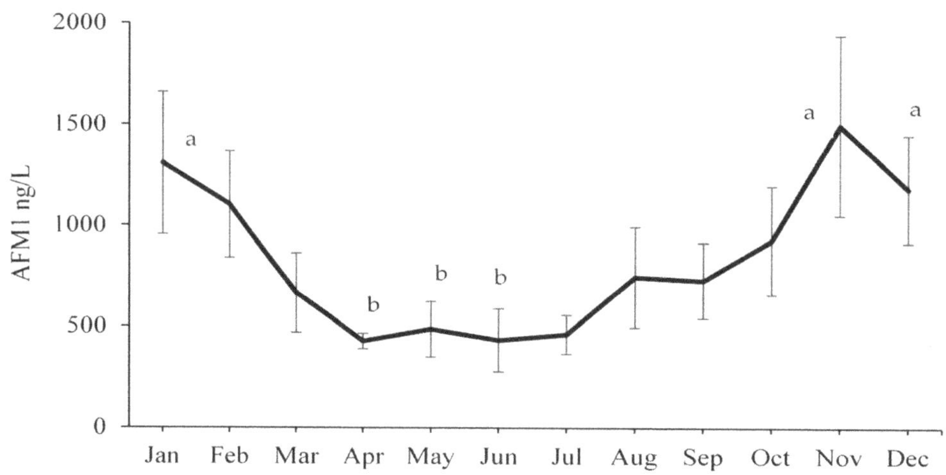

Figure 1. Average aflatoxin M_1 levels ± SEM, in liquid milk during different months (means bearing different alphabets differ significantly at $p < 0.05$).

The data regarding AFM_1 levels in different types of liquid milk are presented in Table 1. In case of UHT milk, the mean AFM_1 level was 254.9 ng/L, with 12.9% samples exceeding 500 ng/L, the maximum allowed limit in the USA. The maximum AFM_1 level noted was 1536 ng/L. Only five out of 85 UHT milk samples (5.8%) had AFM_1 levels less than 50 ng of AFM_1/L, the maximum allowed limit in the EU. Overall, the UHT milk samples had lower ($p < 0.001$) AFM_1 levels compared with the pasteurized and raw milk samples.

Table 1. Aflatoxin M_1 levels (ng/L) [1] in liquid milk samples during the year 2016.

Month	UHT Milk	Pasteurized Milk	Raw Milk	*p* Value	*F* Value
Jan	434.0 ± 517.5 [b]	1861.1 ± 1855.9 [ab]	2422.3 ± 198.6 [a]	0.043	4.12
Feb	135.5 ± 34.4 [b]	1613.9 ± 905.0 [a]	2155.7 ± 603.9 [a]	0.000	18.25
Mar	48.3 ± 37.8 [c]	556.0 ± 476.5 [b]	1908.7 ± 367.3 [a]	0.000	40.32
Apr	413.2 ± 201.2	334.0 ± 151.6	549.2 ± 38.3	0.055	3.49
May	109.5 ± 48.2 [b]	553.8 ± 770.5 [ab]	994.6 ± 539.7 [a]	0.029	4.41
Jun	61.4 ± 26.1 [b]	355.2 ± 351.1 [b]	1123.3 ± 1015.2 [a]	0.022	5.07
Jul	213.5 ± 150.4	565.3 ± 666.6	628.3 ± 134.4	0.169	1.98
Aug	218.9 ± 82.1	1148.3 ± 1820.0	886.2 ± 289.0	0.289	1.34
Sep	278.2 ± 124.1	813.8 ± 1073.7	1365.9 ± 459.2	0.078	3.04
Oct	279.5 ± 99.5 [b]	913.3 ± 1238.3 [b]	2231.7 ± 847.1 [a]	0.008	6.89
Nov	441.1 ± 184.8 [b]	1385.6 ± 1469.5 [b]	3796.3 ± 2858.1 [a]	0.009	6.40
Dec	395.1 ± 155.5 [b]	1570.9 ± 1457.5 [a]	1894.4 ± 1184.4 [a]	0.040	3.92
n	85	78	55		
Mean	254.9 ± 223.9 [c]	939.5 ± 1164.6 [b]	1535.0 ± 1234.5 [a]	0.000	31.74
Median	198.4	368.5	1037.6	-	-
Max	1536.0	4808.0	7460.7	-	-
Min	n.d.	32.8	1912	-	-

[1] Arithmetic means ± standard deviation. [abc] Means bearing different superscripts differ significantly within a row ($p < 0.05$). n.d. = less than detection limit.

In case of pasteurized milk, the mean AFM_1 level was 939.5 ng/L and only 41% samples were below the limit of 500 ng AFM_1/L. Only four pasteurized milk samples (5.1%) were below the limit of 50 ng AFM_1/L. Compared with the UHT milk, the pasteurized milk had higher ($p < 0.001$) AFM_1 levels. Raw milk was found to have higher ($p < 0.001$) AFM_1 levels than the UHT and pasteurized milk. None of the 34 raw milk samples qualified the EU limit of 50 ng AFM_1/mL milk. Only five raw milk samples (9.1%) had AFM_1 levels lower than the USA standard of 500 ng/L of milk, indicating potential health risks for the consumers of raw milk. Mean AFM_1 level in raw milk in the country has been reported to be in the range of 46 to 340 ng/L by various authors, which is lower than the mean level being reported here [5,8,9]. However, the mean AFM_1 level in raw milk collected from peri-urban farms in Lahore was found in one report to be 17,380 ng/L which is very high compared to the present findings [4]. The differences in various reports indicate that the AFM_1 contamination may vary in different seasons and areas.

In general, the milk sampled from rural areas of Pakistan, where more green fodder is available, has been found to have lower AFM_1 levels compared to the milk collected from urban areas [6]. This could also explain the lower levels found in UHT milk in the present study which is in general collected from rural areas followed by testing for AFM_1 levels before packaging. The farmers are usually paid according to the quality of the milk. The UHT milk companies also have the option to either not buy or discard the milk having higher AFM_1 contamination. Compared to the UHT milk, the companies marketing pasteurized milk have their own dairy farms and thus do not have the option to discard milk. It is therefore important to educate these companies to control aflatoxin in the dairy feeds, coupled with legislation, for a market advantage.

2.2. *Aflatoxin Contamination of Tea Whiteners and Milk Powder*

The data regarding AFM_1 levels in tea whiteners are presented in Table 2. The mean AFM_1 level in tea whiteners was only 98.9 ng/L, with 2.1% samples (1 out of 48) exceeding the 500 ng AFM_1/L limit. The mean AFM_1 levels in different months did not differ statistically (p = 0.081). Tea whiteners are usually made using both the milk and non-milk constituents, hence it is logical to expect a lower AFM_1 contamination in this commodity. It should however be noted that tea whiteners are not a substitute of milk, and the present results do not imply use of such products as a replacement of milk.

Table 2. Aflatoxin M_1 levels (ng/L) in tea whiteners.

Month	Tea Whiteners
Jan	138.8 ± 195.7
Feb	49.4 ± 83.3
Mar	0.6 ± 0.9
Apr	178.2 ± 173.5
May	16.4 ± 6.2
Jun	32.1 ± 24.4
Jul	69.4 ± 89.3
Aug	120.2 ± 55.4
Sep	60.9 ± 40.7
Oct	38.5 ± 30.5
Nov	120.8 ± 80.8
Dec	339.4 ± 357.3
p value	0.081
F value	1.849
n	48
Overall Mean	98.9 ± 161.8
Median	40.6
Max	932.6
Min	n.d.

Data presented as arithmetic means ± standard deviation; n.d. = less than detection limit.

The mean AFM_1 level in powdered milk after reconstitution was 522.1 ng/L i.e., close to the maximum tolerable limit of 500 ng AFM_1/L. In total, 50.0% of the powdered milk samples had AFM_1 levels lower than the 500 ng/L limit. Only two powder milk samples (10.0%) passed the EU standard of 50 ng AFM_1/L. It appears from these results that the powdered milk was a safer option for consumers compared to the pasteurized and raw milk. However, these results need to be examined in consideration of the fact that the studied milk powder brands are used as infant milk formula. As infants are more sensitive to the hazards of AFM_1, effective measures should be in place to ensure that all powdered milk brands marketed in Pakistan meet at least the 500 ng AFM_1/L limit.

2.3. *Miscellaneous Quality Variables of Milk Samples*

Miscellaneous quality variables were studied to identify any case of adulteration which would result in dilution of the natural AFM_1 content of milk and making comparison biased. Data regarding fat, total solids, and SNF content of milk samples are presented in Table 3. Fat percentage was lower ($p \leq 0.029$) in the UHT milk compared with the pasteurized milk samples in the month of January. The UHT milk samples were also found to have lower ($p \leq 0.037$) total solids compared with both the pasteurized and raw milk samples in January. These data are indicative of fat removal by milk processors to make milk uniform regarding fat content. The SNF contents of different types of milk did not differ during the study duration. Furthermore, no addition of starch, cane sugar, urea, detergent and hydrogen peroxide was found in liquid milk samples. These data indicate that no particular adulteration occurred in the liquid milk samples included in the study.

Table 3. Quality variables of liquid milk included in the study.

Mon	Fat (%)				Total Solids (%)				SNF (%)			
	UHT	Pasteur	Raw	p	UHT	Pasteur	Raw	p	UHT	Pasteur	Raw	p
Jan	3.21 ± 0.35^{b}	4.12 ± 0.13^{a}	3.84 ± 0.69^{a}	0.014	11.00 ± 1.22^{b}	12.56 ± 0.18^{a}	12.22 ± 1.02^{a}	0.027	7.78 ± 0.96	8.44 ± 0.23	8.38 ± 1.46	0.498
Feb	3.49 ± 0.29	4.12 ± 0.13	3.86 ± 0.64	0.077	11.62 ± 0.93	12.50 ± 0.28	11.79 ± 1.39	0.361	8.13 ± 0.69	8.37 ± 0.31	7.93 ± 1.48	0.758
Mar	3.66 ± 0.38	4.58 ± 0.13	3.90 ± 0.90	0.061	11.73 ± 0.87	12.31 ± 0.34	12.11 ± 0.62	0.331	8.07 ± 0.78	7.72 ± 0.37	8.22 ± 1.03	0.591
Apr	3.58 ± 0.23	3.72 ± 0.38	3.59 ± 0.69	0.868	11.66 ± 0.63	12.00 ± 0.45	11.83 ± 1.18	0.808	8.08 ± 0.57	8.27 ± 0.26	8.25 ± 1.24	0.911
May	3.36 ± 0.46	3.90 ± 0.28	4.09 ± 0.85	0.103	11.49 ± 0.98	12.00 ± 0.72	12.53 ± 0.89	0.106	8.12 ± 0.61	8.08 ± 0.51	8.44 ± 1.45	0.778
Jun	3.46 ± 0.27	3.76 ± 0.50	4.01 ± 0.94	0.302	11.73 ± 0.68	12.21 ± 0.62	12.26 ± 0.87	0.364	8.24 ± 0.44	8.45 ± 0.28	8.25 ± 1.29	0.914
Jul	3.63 ± 0.63	3.96 ± 0.44	4.14 ± 1.06	0.484	11.78 ± 1.16	12.37 ± 0.35	12.21 ± 1.04	0.547	8.15 ± 0.71	8.40 ± 0.33	8.07 ± 1.37	0.838
Aug	3.36 ± 0.18	3.10 ± 0.97	3.88 ± 0.61	0.100	11.58 ± 0.85	11.25 ± 1.33	11.97 ± 1.15	0.513	8.21 ± 0.67	8.15 ± 0.53	8.10 ± 1.41	0.978
Sep	3.56 ± 0.32	4.00 ± 0.45	4.01 ± 0.91	0.336	11.84 ± 0.93	12.53 ± 0.55	12.19 ± 1.34	0.548	8.38 ± 0.65	8.53 ± 0.23	8.15 ± 1.68	0.848
Oct	3.52 ± 0.22	3.96 ± 0.53	3.97 ± 1.14	0.494	12.59 ± 0.70	13.15 ± 0.78	12.56 ± 1.12	0.484	9.07 ± 0.60	9.19 ± 0.25	8.58 ± 1.28	0.438
Nov	3.54 ± 0.29	4.00 ± 0.71	4.16 ± 0.80	0.193	11.18 ± 1.55^{b}	12.98 ± 0.85^{a}	12.10 ± 0.72^{ab}	0.040	7.64 ± 1.41	9.00 ± 0.24	7.93 ± 0.87	0.090
Dec	3.41 ± 0.34	3.76 ± 0.32	4.04 ± 0.63	0.069	11.46 ± 1.18	11.81 ± 0.56	12.04 ± 0.79	0.474	8.04 ± 0.99	8.05 ± 0.49	8.01 ± 1.19	0.996

Data presented as arithmetic means ± standard deviation; UHT = milk with ultraheat treatment, Pasteur. = pasteurized milk. ^{abc} Means bearing different superscripts differ significantly within a row under each quality variable ($p < 0.05$).

In the case of tea whiteners, one brand was found to be positive for neutralizer/soda in April and then from June through December. Another brand of tea whitener was found to be positive for detergent only in January.

2.4. Exposure to Aflatoxin through Milk

The users of raw and processed milk (UHT and pasteurized) were estimated to be exposed to 11.9 and 4.5 ng AFM_1 per kg body weight daily, respectively (Table 4). The overall daily AFM_1 intake was found to be 11.6 ng per kg body weight, which is very high compared with 3.1 ng/kg mean estimated previously for consumers in Karachi [10]. The higher exposure estimated in the present study could be due to the higher contamination levels noted in Islamabad, and the use of weighted AFM_1 mean in our study. The mean AFM_1 in milk if not corrected for relative consumption would have resulted in an estimate of 6.4 ng/kg day.

Table 4. Estimated daily AFM_1 intake in users of different types of milk.

Type of Milk	AFM_1 (ng/L)	Daily Milk Consumption [1] (mL)	Average Body Weight [2] (kg)	Daily Intake (ng/kg bw)
Children:				
Milk powder for 1 year old	353.1	500	9.3	19.1
Milk powder for 5 year old	728.7	500	18.5	19.7
Adults:				
UHT	254.9	466	60.0	2.0
Pasteurized	939.5	466	60.0	7.3
Raw	1535.0	466	60.0	11.9
Overall liquid	1487.4	466	60.0	11.6

[1] Milk consumption for children as recommended by the milk powder manufacturer. Milk consumption in case of adult is the per head milk available in Pakistan [11]. [2] Average body weight of children taken from Royal College of Pediatrics and Child Health (UK-WHO) charts. Average weight of Pakistani children of one year and five years of age are 5.6 and 15.3, respectively [12]. bw = body weight.

The presently estimated AFM_1 intake for children using milk powder is alarmingly high despite the acceptable levels of AFM_1 in milk powder for one-year-old children. Compared to adults, the consumption of milk per kg body weight is higher in case of children which results in higher contribution of aflatoxin from milk. It may be noted that UHT milk was found to have the lowest AFM_1 levels in the present study. If UHT milk was used for calculation of daily exposure of children, then the figures for one- and five-year-old children had been 13.8 and 6.9 ng per kg body weight. This shows the need to reduce the levels of AFM_1 in milk powder produced in the country, especially in the milk for the five-year age group. In a separate study (unpublished data), we found that the AFM_1 levels in imported milk powders was less than 60 ng/L and would result in an exposure of only 3.2 ng/kg body weight daily in one-year-old children. This implies that during legislation, the levels of AFM_1 allowed in milk for one-year-old and younger children should be kept lower than 500 ng/L.

3. Conclusions

This study, though conducted on a limited scale, identifies raw milk as less suitable for human consumption in the capital city Islamabad due to high AFM_1 levels. The processed liquid milk was found to be a safer option for consumers. In all types of milk, the levels of AFM_1 were lower during summer months. Reducing the levels of AFM_1 in milk for infants, while strictly following the 500 ng AFM_1/L limit in milk for older children are suggested as ways to reduce the exposure of children to AFM_1.

4. Materials and Methods

4.1. Sampling

All the regularly available locally produced processed milk brands were collected once a month from supermarkets in the twin cities of Islamabad-Rawalpindi in the year 2016. Each time, 21 different liquid milk samples were collected including eight UHT milk brands, seven pasteurized milk brands, and raw milk from six different suppliers. Additionally, two brands of milk powder—each for a different age group of children—and six brands of tea whiteners were sampled on monthly basis.

As UHT and powdered milk have long expiry time, it was ensured that a different and fresh batch was collected on each sampling. The collected liquid milk samples were transported in chilled form to the laboratory where these were subsampled and stored at −20 °C until AFM_1 analysis. Identity of the collected samples was masked from analysts to make the study blind.

4.2. Aflatoxin M_1 Analysis

The milk samples were analyzed for AFM_1 contamination using ELISA kits (AM1-E01, Immunolab GmbH, Kassel, Germany) following protocols specified by the manufacturer. The kit had a quantification range of 10 to 1000 ng of AFM_1/L. Samples were analyzed several times in different dilutions until the AFM_1 levels in the diluted samples fell within the quantification range of the ELISA kit. Limit of detection (LOD) was found to be 4.4 ng AFM_1/L, while recovery of AFM_1 was 86.9% at 500 ng of AFM_1/L. Results were not corrected for recovery.

The kit was validated before start of this study [13]. In addition, external standards were run with each microtiter plate to ensure reliability of the AFM_1 quantification. Two types of reference materials were used to make external standards. The first was a 44 ng of AFM_1/L reference skim milk powder (RMBD-248, EU Joint Res Center, IRMM, Geel, Belgium) which was used to make 4.4 and 44 ng of AFM_1/L standards. The second reference material was a purified 9.786 μg of AFM_1/mL solution (46319U, Supelco, Bellefonte, PA, USA) which was diluted with ddH_2O to make 5, 50, 500 ng of AFM_1/L secondary standards for use with each microtiter plate.

The ELISA kits were read on an ELISA reader (BDSL Immunoskan MS 355, Labsystems, Vantaa, Finland). For quantification of AFM_1, a software based on four parametric curve estimation provided by the kit manufacturer was used.

4.3. Miscellaneous Milk Quality Variables

Raw milk in the country is often adulterated with water, urea, starch, and neutralizers, etc [14]. Comparison of adulterated milk with pure milk is not justified because addition of water and non-milk constituents result in milk dilution and lowering of the AFM_1 contents. Therefore, all the liquid milk samples were analyzed for milk fat, total solids, and solids-not-fat (SNF) to identify any notable case of adulteration. Milk fat was determined following the Gerber method and using a Gerber Centrifuge (Funke Dr. N. Gerber, Berlin, Germany) [15]. Total solids were determined using Quevenne lactometer and following the principles of lactodensitometry as detailed by Ling [16]. Solids-not-fat were calculated as difference between total solids and milk fat.

Qualitative tests were performed to detect adulterants and stabilizers including starch, cane-sugar, urea, detergent, and hydrogen peroxide following methods described by Sharma et al. [17].

4.4. Calculations and Statistics

The exposure to AFM_1 through a specific milk type was estimated using a formula:

$$\begin{aligned}&\text{Average daily AFM}_1\text{ intake (ng/kg body weight) through a specific type of milk}\\&= [(\text{mean AFM}_1\text{ ng/L in the specific type of milk})\\&\times(\text{milk availability per head per day in Pakistan})]/(\text{average body weight})\end{aligned} \quad (1)$$

where, milk availability was taken as 170.2 L/head/year [11], and average body weight was taken as 60 kg. Daily consumption of milk powder in case of children was taken as recommended by the milk powder manufacturer. To calculate the exposure level for one and five year old children, average body weight was taken from Royal College of Pediatrics and Child Health (UK-WHO) charts.

In case of overall AFM_1 intake through liquid milk, the mean AFM_1 levels were normalized using a formula, keeping in view only 5% share of processed milk in the overall milk market:

$$\text{Average daily AFM}_1\text{ intake through liquid milk (ng/kg body weight)} \\ = [(\text{mean AFM}_1\text{ in raw milk} \times 95) + (\text{mean AFM}_1\text{ in processed milk} \times 5)]/100 \quad (2)$$

Means were compared using ANOVA and LSD in IBM SPSS Statistics 20 (IBM Corp., Armonk, New York, NY, USA, 2011). Values of AFM_1 less than the LOD were considered as zero for calculating monthly means.

Author Contributions: Conceptualization, A.W.Y., and M.N.M.I.; Methodology, A.W.Y. and H.K.; Aflatoxin M_1 Analysis, N.I. and A.W.Y.; Milk Quality Analyses, H.K.; Data Curation, A.W.Y.; Writing-Original Draft Preparation, A.W.Y.; Writing-Review & Editing, A.W.Y., M.N.M.I. and Y.Z.; Visualization, A.W.Y., M.N.M.I. and Y.Z.; Supervision, A.W.Y. and M.N.M.I.; Project Administration and Funding Acquisition, A.W.Y. and M.N.M.I.

Acknowledgments: The authors highly acknowledge the support given by Aman Ullah, Umer Farooq, Eitazaz Ehsan, Zahid and other staff members of various laboratories of Animal Sciences Institute, PARC National Agricultural Research Center, Islamabad during analyses of milk samples. Also, the contribution of Johanna Lindhal (ILRI scientist) for her editorial comments is appreciated.

References

1. Yunus, A.W.; Razzazi-Fazeli, E.; Böhm, J. Aflatoxin B1 in affecting broiler's performance, immunity, and gastrointestinal tract: A review of history and contemporary issues. *Toxins* **2011**, *3*, 566–590. [CrossRef] [PubMed]
2. Food and Agriculture Organization (FAO). *Worldwide Regulations for Mycotoxins in Food and Feed in 2003*; FAO Food and Nutrition Paper 81; Food and Agriculture Organization of the United Nations: Rome, Italy, 2004.
3. Raza, R. Occurrence of aflatoxin M_1 in the milk marketed in the city of Karachi, Pakistan. *J. Chemical Soc. Pakistan* **2006**, *28*, 155–157.
4. Muhammad, K.; Tipu, M.Y.; Abbad, M.; Khan, A.M.; Anjum, A.A. Monitoring of aflatoxin M_1 in market raw milk in Lahore City, Pakistan. *Pakistan J. Zool.* **2010**, *42*, 697–700.
5. Iqbal, S.Z.; Asi, M.R.; Ariño, A. Aflatoxin M_1 contamination in cow and buffalo milk samples from the North West Frontier Province (NWFP) and Punjab provinces of Pakistan. *Food Add. Contam. B* **2011**, *4*, 282–288. [CrossRef] [PubMed]
6. Iqbal, S.Z.; Asi, M.R.; Selamat, J. Aflatoxin M1 in milk from urban and rural farmhouses of Punjab, Pakistan. *Food Add. Contam. B* **2014**, *7*, 17–20. [CrossRef] [PubMed]
7. Hussain, I. Micro-Analysis of Aflatoxin M_1 in Dairy Products at Trace Levels and Its Elimination. Ph.D. Thesis, Institute of Chemistry, University of the Punjab, Punjab, Pakistan, 2009.
8. Younus, M.; Abbas, T.; Rafique, K.; Sajid, M.; Aslam, M.; Zafar, M. *Analyses of Selected Heavy Metals and Aflatoxin M1 in Milk for Human Consumption in Jhang City, Pakistan*; Working Paper 12; International Food Policy Research Institute (IFPRI): Washington DC, USA, 2013.
9. Sadia, A.; Jabbar, M.A.; Deng, Y.; Hussain, E.A.; Riffat, S.; Naveed, S.; Arif, M. A survey of aflatoxin M_1 in milk and sweets of Punjab, Pakistan. *Food Control.* **2012**, *26*, 235–240. [CrossRef]
10. Jawaid, S.; Talpur, F.T.; Afridi, H.I. Contamination profile of aflatoxin M_1 residues in milk supply chain of Sindh Pakistan. *Toxicol. Rep.* **2015**, *2*, 1418–1422. [CrossRef] [PubMed]

11. Government of Pakistan. *Economic Survey 2016–2017*; Ministry of Finance, Government of Pakistan: Islamabad, Pakistan, 2017; p. 194.
12. Aziz, S.; Noor-ul-Ain, W.; Majeed, R.; Khan, M.A.; Qayum, I.; Ahmed, I.; Hosain, K. Growth centile charts (anthropometric measurement) of Pakistani pediatric population. *J. Pakistan Med. Assoc.* **2012**, *62*, 367–377.
13. Imtiaz, N.; Yunus, A.W. Comparison of some ELISA kits for aflatoxin M_1 quantification. *J. AOAC Int.* **2019**, *101*. [CrossRef]
14. Barham, G.S.; Khaskheli, M.; Soomro, A.H.; Nizamani, A.A. Screening of adulterants and extent of added water I market milk at central zone of Sindh, Pakistan. *IOSR J. Environ. Sci. Toxicol. Food Technol.* **2014**, *8*, 77–83. [CrossRef]
15. James, C.S. Determination of the fat content of dairy products by the Gerber method. In *Analytical Chemistry of Foods*; Blackie Academic and Professionals: Glasgow, UK, 1995; pp. 93–95.
16. Ling, T.R. *Textbook of Dairy Chemistry*, 3rd ed.; Chapman and Hall: London, UK, 1963; Volume 2.
17. Sharma, R.; Rajput, Y.S. *Detection of Adulterants in Milk—A Laboratory Manual*; Karnal National Dairy Research Institute: Karnal, India, 2012.

8

Quantitative Analysis of Spectinomycin and Lincomycin in Poultry Eggs by Accelerated Solvent Extraction Coupled with Gas Chromatography Tandem Mass Spectrometry

Bo Wang [1,2], Yajuan Wang [2,3], Xing Xie [4], Zhixiang Diao [2,3], Kaizhou Xie [2,3,*], Genxi Zhang [2,3], Tao Zhang [2,3] and Guojun Dai [2,3]

1 College of Veterinary Medicine, Yangzhou University, Yangzhou 225009, China; dz120180009@yzu.edu.cn
2 Joint International Research Laboratory of Agriculture & Agri-Product Safety, Yangzhou University, Yangzhou 225009, China; yzwyj163@gmail.com (Y.W.); yzdzx163@gmail.com (Z.D.); gxzhang@yzu.edu.cn (G.Z.); zhangt@yzu.edu.cn (T.Z.); daigj@yzu.edu.cn (G.D.)
3 College of Animal Science and Technology, Yangzhou University, Yangzhou 225009, China
4 Institute of Veterinary Medicine, Jiangsu Academy of Agricultural Sciences, Key Laboratory of Veterinary Biological Engineering and Technology, Ministry of Agriculture, Nanjing 210014, China; xiexing@jaas.ac.cn
* Correspondence: kzxie@yzu.edu.cn

Abstract: A method based on accelerated solvent extraction (ASE) coupled with gas chromatography tandem mass spectrometry (GC-MS/MS) was developed for the quantitative analysis of spectinomycin and lincomycin in poultry egg (whole egg, albumen and yolk) samples. In this work, the samples were extracted and purified using an ASE350 instrument and solid-phase extraction (SPE) cartridges, and the parameters of the ASE method were experimentally optimized. The appropriate SPE cartridges were selected, and the conditions for the derivatization reaction were optimized. After derivatization, the poultry egg (whole egg, albumen and yolk) samples were analyzed by GC-MS/MS. This study used blank poultry egg (whole egg, albumen and yolk) samples to evaluate the specificity, sensitivity, linearity, recovery and precision of the method. The linearity (5.6–2000 μg/kg for spectinomycin and 5.9–200 μg/kg for lincomycin), correlation coefficient (≥0.9991), recovery (80.0%–95.7%), precision (relative standard deviations, 1.0%–3.4%), limit of detection (2.3–4.3 μg/kg) and limit of quantification (5.6–9.5 μg/kg) of the method met the requirements for EU parameter verification. Compared with traditional liquid–liquid extraction methods, the proposed method is fast and consumes less reagents, and 24 samples can be processed at a time. Finally, the feasibility of the method was evaluated by testing real samples, and spectinomycin and lincomycin residues in poultry eggs were successfully detected.

Keywords: poultry eggs; spectinomycin; lincomycin; ASE; GC-EI/MS/MS

1. Introduction

Spectinomycin and lincomycin are aminoglycoside and lincosamide antibiotics, respectively, and they have synergistic and complementary effects on each other's antibacterial spectra and antibacterial mechanisms. Spectinomycin is an inhibitor of bacterial protein synthesis and acts on the 30S subunit of ribosomes, and its antibacterial mechanism mainly involves preventing the binding of messenger ribonucleic acid and ribosomes, thereby hindering the synthesis of proteins and resulting in bactericidal effects [1]. The antibacterial mechanism of lincomycin mainly consists of binding to the bacterial ribosomal 50S subunit, which inhibits peptide acyltransferase, hinders the synthesis of bacterial proteins and results in bactericidal effects [2]. Spectinomycin has strong antibacterial activity against Gram-negative bacteria and weak activity against Gram-positive bacteria, whereas lincomycin has no effect on Gram-negative bacteria but a strong antibacterial effect on Gram-positive bacteria. Therefore, spectinomycin and lincomycin are usually used in combination to treat infections with Gram-positive bacteria and Gram-negative bacteria and are widely used to treat piglet diarrhea and infection by *Mycoplasma hyopneumoniae* and *Mycoplasma pneumoniae*, which cause chronic respiratory diseases in chickens [3–6]. However, spectinomycin can damage the eighth cranial nerve, exert kidney toxicity and block neuromuscular transmission; lincomycin has strong side effects, damages the gastrointestinal tract and liver, and even causes anaphylactic shock and death. Thus, China and the European Union (EU) have listed spectinomycin as a banned drug for poultry eggs and set maximum residue limits (MRLs) of 300–5000 μg/kg for spectinomycin and 50–1500 μg/kg for lincomycin in animal-derived foods [7,8]. The MRLs in the United States are 100–4000 μg/kg for spectinomycin in chicken and cattle muscle and liver and 100–600 μg/kg for lincomycin in pig muscle and liver. In addition, the presence of the latter two drugs in other animal-derived foods has been banned [9]. Japan has set MRLs for spectinomycin of 500–5000 μg/kg and for lincomycin of 200–1500 μg/kg in animal-derived foods [10]. Thus, it is important to develop fast and efficient analytical methods to detect spectinomycin and lincomycin in poultry eggs.

To date, many methods have been used to measure spectinomycin and lincomycin in animal-derived foods and animal feedstuffs, including fluorescent latex immunoassay (FLI) [11], micellar electrokinetic capillary chromatography combined with ultraviolet detection (MEKC-UVD) [12], enzyme-linked immunosorbent assay (ELISA) [13], high-performance liquid chromatography with electrochemical detection (HPLC-ECD) [14,15], HPLC with fluorescence detection (FLD) [16], HPLC-UVD [17,18], HPLC with evaporative light-scattering detection (ELSD) [19,20], hydrophilic interaction chromatography with mass spectrometry (HILIC-MS) [21], HILIC tandem MS (MS/MS) [22], HPLC-MS [23], HPLC-MS/MS [24–28], gas chromatography–nitrogen phosphorus detection (GC-NPD) [29,30] and GC-MS [30]. The FLI, ELISA, ECD, FLD, UVD and ELSD methods have low sensitivity, specificity, recovery and precision and have many limitations. Molognoni et al. [22] developed a liquid–liquid extraction (LLE) method combined with HILIC-MS/MS for the determination of spectinomycin, halquinol, zilpaterol and melamine residues in animal feedstuffs with good recovery and precision. Juan et al. [27] reported an accelerated solvent extraction (ASE) method for the trace analysis of macrolide and lincosamide antibiotics in meat and milk using HPLC-MS/MS, and the method was fast, sensitive and automatic, making it suitable for the determination of macrolide and lincosamide residues in meat and milk. Tao et al. [30] established an ASE approach for extracting lincomycin and spectinomycin residues from swine and bovine tissues using GC-NPD and GC-MS. ASE is an automated extraction technology that is widely used for veterinary drug residue detection in animal food because of its advantages, such as rapid analysis, low organic solvent use and batch sample processing. Compared to liquid chromatography, gas chromatography has been reported less frequently for the detection of spectinomycin and lincomycin in animal foods. Moreover, the use of single-stage GC-MS has several difficult limitations, such as the inability to effectively exclude sample matrix-derived interferences, to confirm false positives and quasi-deterministic parameters and to quantify target compounds. However, a gas chromatography–tandem mass spectrometry (GC-MS/MS) method can effectively address these issues and accurately quantify target compounds. Thus, an ASE-GC-MS/MS method was developed to determine spectinomycin and lincomycin residues in poultry eggs. The method parameters were validated according to the EU [31] and the Food and Drug Administration (FDA) [32] validation requirements.

2. Materials and Methods

2.1. Chemicals and Reagents

Spectinomycin (97.9% standard) and lincomycin (98.9% standard) were purchased from the Food and Drug Control Agency (Beijing, China). *N*,O-bis(Trimethylsilyl)trifluoroacetamide (BSTFA, >99.0%

standard) was obtained from Sigma-Aldrich (St. Louis, MO, USA). Sodium dodecyl sulfonate (SDS, ≥99.0% standard) was purchased from Sangon Biotech (Shanghai, China). Acetonitrile and methanol (HPLC grade) were acquired from Merck (Fairfield, OH, USA). Analytical-grade phosphoric acid (H_3PO_4), sodium hydroxide, acetic acid, n-hexane, potassium dihydrogen phosphate (KH_2PO_4) and trichloroacetic acid (TCA) were obtained from Sinopharm Chemical Reagent Co. (Shanghai, China). Ultrapure water was obtained from a PURELAB Option-Q synthesis system (ELGA Lab Waters, High Wycombe, Bucks, UK).

Standard stock solutions of spectinomycin and lincomycin at 1 mg/mL were prepared in pure methanol. The standard working solutions were obtained by diluting the standard stock solutions with pure methanol according to the test needs.

2.2. *GC-MS/MS Analysis*

The GC-MS/MS system consisted of a Trace 1300 gas chromatograph, a TSQ 8000 triple quadrupole tandem mass spectrometer and a Triplus RSH automatic sample injector, and the TraceFinder 3.0 software was used for the analysis (Thermo Fisher Corp., Waltham, MA, USA). GC separation was performed using the following temperature program: 160 °C for 1 min; a ramp at 25 °C/min to 250 °C, followed by a 1 min hold; and a ramp at 15 °C/min to 300 °C, followed by a 5 min hold. A Thermo Fisher TG-5MS amine column (30 m × 0.25 mm; inside diameter (i.d.), 0.25 μm) was used. The GC was operated in splitless mode with a carrier gas (helium, 99.999% standard, 60 psi) flow rate of 1.0 mL/min. The injector temperature was held at 280 °C, and the injection volume was 1.0 μL.

The MS/MS system was equipped with an electron impact (EI) source and used in full scan mode and selected reaction monitoring (SRM) mode. The typical MS parameters were as follows: ionization voltage, 70 eV; ion source temperature, 280 °C; and transfer line temperature, 280 °C. The retention times and relevant MS parameters are presented in Table 1.

Table 1. Retention times and relevant mass spectrometry (MS) parameters for the analytes.

Analyte	Retention Time (min)	Molecular Weight (*m*/*z*)	Mass Transitions (*m*/*z*)	Collision Energy (eV)
Spectinomycin	6.93	332.15	201.1 > 75.0 *	16
			201.1 > 185.1	8
Lincomycin	10.53	406. 21	126.1 > 42.0 *	22
			126.1 > 82.0	22

Note: * Ion pair used for quantification.

2.3. *Preparation of the Samples*

Considering that consumers have separate uses for whole eggs, albumens and yolks in hen, duck and goose eggs, we studied the elimination of spectinomycin and lincomycin residues in whole eggs, albumens and yolks. Because pigeon and quail eggs are relatively small, consumers generally use these as whole eggs. Thus, blank hen, duck and goose eggs were collected as whole eggs, albumen and yolk samples, and blank pigeon and quail eggs were collected as whole egg samples. Blank hen, duck and goose eggs (whole eggs, albumens and yolks) as well as pigeon and quail eggs (whole eggs) were separately homogenized, divided and frozen. In this work, LLE and ASE were used to extract the poultry egg samples, which were then cleaned up by SPE and finally derivatized.

2.3.1. Liquid–Liquid Extraction

Homogenized poultry eggs (2.0 ± 0.02 g) were precisely weighed and then added to 10 mL of 0.01 M KH_2PO_4 solution (pH 4.0). The sample was vortexed for 5 min at 2000× *g*, homogenized ultrasonically for 10 min and then centrifuged for 10 min at 8000× *g*. The extraction solution was collected, and the sample was extracted again. The two extracts were combined, and 5 mL of n-hexane was added. The mixture was vortexed for 5 min at 2000× *g* and then centrifuged for 10 min at 8000× *g*.

After degreasing twice with n-hexane, the extract was added to 5 mL of 3% TCA solution, vortexed for 5 min at 2000× *g* and then centrifuged for 10 min at 8000× *g*. The liquid–liquid extraction procedure was performed according to the National Food Safety Standard (GB 29685-2013) [33].

2.3.2. Accelerated Solvent Extraction

Homogenized poultry eggs (2.0 ± 0.02 g) and 4.0 g of diatomaceous earth were fully ground, and then sample preparation was performed. The fat-removal and extraction parameters for the ASE350 instrument (Thermo Fisher Scientific Co. Ltd., Waltham, MA, USA) were as follows: 60 °C, 1500 psi, a static extraction time of 5 min and a nitrogen purge time of 60 s. One extraction was performed with a total solvent rinse of 40% and n-hexane to remove the fat, and two extractions were then performed with a total solvent rinse of 50% and 0.01 M KH_2PO_4 solution (pH 4.0) to extract the analytes, after which the sample extract was collected.

2.3.3. Solid-Phase Extraction

After the sample was processed by LLE or ASE, 10% NaOH solution was added to the extract to adjust the pH to 5.8 ± 0.2, 2 mL of 0.2 M SDS solution was added, and the sample was then vortexed for 1 min. After standing for 15 min, the extract was cleaned up by SPE with an Oasis PRiME HLB cartridge (3 mL/60 mg, Waters Corp., Milford, MA, USA) that had been activated and equilibrated by the addition of 3 mL of methanol, 3 mL of ultra-pure water and 3 mL of 0.02 M SDS solution. After 20 mL of the extracts was added to the Oasis PRiME HLB cartridge at a constant rate (2.0 mL/min) and allowed to completely pass through the cartridge, 9 mL of ultrapure water was added in three portions for rinsing. Finally, 6 mL of methanol was used to elute the two target compounds.

2.4. Derivatization Reaction

After the extract was dried under a stream of nitrogen at 40 °C, 200 µL of BSTFA and 100 µL of acetonitrile were sequentially added to the sample, which was then vortexed for 1 min. Then, the mixture was placed in a 75 °C oven for 60 min. After the derivatization reaction was complete, the mixture was cooled to room temperature and dried under a stream of nitrogen at 40 °C. Finally, 2 mL of n-hexane was added to the sample to dissolve the residue, and the resulting solution was vortexed for 1 min and passed through a 0.22 µm organic phase needle filter into the GC-MS/MS system.

2.5. Quality Parameters

Seven spiked concentration levels for the two analytes were used to establish the linear regression equations: the limit of quantification (LOQ) and 50, 100, 500, 1000, 1500 and 2000 µg/kg for spectinomycin and the LOQ and 10, 20, 50, 100, 150 and 200 µg/kg for lincomycin. The peak areas as a function of the analyte concentration were used to establish standard working curves. The correlation coefficients (R^2 values) were determined and should all have been ≥0.9991. The other parameters were tested according to the EU [31] and the FDA requirements [32], and the TraceFinder 3.0 software (Thermo Fisher Corp., Waltham, MA, USA) was used for the analysis.

3. Results and Discussion

3.1. Optimization of the ASE Conditions

Due to the complexity of the matrices of animal-derived foods, the detection of veterinary drug residues in such foods usually requires sample pretreatment involving extraction and clean-up to avoid clogging the chromatography column and contaminating the instrument. Several methods, such as LLE [22,24,26], solid-phase extraction (SPE) [16,21], core-shell molecularly imprinted solid-phase extraction (CSMISPE) [18] and ASE [27,30], have been developed for the extraction of spectinomycin and lincomycin from animal tissues, meat, milk and animal feedstuffs as well as from swine, calf and chicken plasma. Compared with the LLE, SPE and CSMISPE methods, the ASE method has the advantages of a short extraction time, lower consumption of organic reagents and batch sample processing. Therefore, in this study, the ASE method was used to extract spectinomycin and lincomycin from poultry eggs, and the analyte recoveries were compared with those for the LLE method.

Tao et al. [30] used a 0.01 M KH_2PO_4 solution as an extractant to successfully extract spectinomycin and lincomycin from animal tissues. Based on the chemical properties of spectinomycin and lincomycin, a 0.01 M KH_2PO_4 solution was also selected as the extractant in the present study. In this experiment, the pH of the 0.01 M KH_2PO_4 solution was adjusted with H_3PO_4, and the effects of different pH values (3.0–5.5) on the response values of the two compounds were compared. When the 0.01 M KH_2PO_4 solution (pH 4.0) was used as the extractant, the response values of spectinomycin and lincomycin were the highest (Figure 1a). Thus, the 0.01 M KH_2PO_4 solution (pH 4.0) was finally selected as the extractant in this study. At 1500 psi, the effects of the temperature (40 °C, 60 °C, 80 °C, 100 °C and 120 °C), the amount of extractant (40%, 50%, 60%, 70% and 80% of the extraction cell volume), and the number of extractions (1 and 2 static cycles) on the recovery of spectinomycin and lincomycin from poultry eggs were compared. Firstly, using the 0.01 M KH_2PO_4 solution (pH 4.0) as the extractant and the optimal conditions for ASE extraction temperature were tested under 1500 psi, and the optimal extraction temperature was determined to be 60 °C (Figure 1b). Secondly, under the conditions of 1500 psi, 60 °C and using the 0.01 M KH_2PO_4 solution (pH 4.0) as the extractant, we optimized the amount of extractant and the number of extractions, and a 50% extraction cell volume and two static cycles obtained the best response value (Figure 1c). Thus, the optimal extraction conditions for the ASE method were as follows (Figure 1): 60 °C, 1500 psi, a 0.01 M KH_2PO_4 solution (pH 4.0) as the extractant, 50% extraction cell volume, static extraction for 5 min, one degreasing cycle and two static cycles.

3.2. *Optimization of the SPE Conditions*

An ion-pair reagent can be combined with the analyte to form an ion-pair and become neutral so that the analyte molecules are retained on the chromatographic column. A test revealed that the ion-pair reagent was susceptible to pH-induced changes: slight changes in pH affected the ion-pair reagent and, consequently, the recoveries of the target compounds. To solve this problem, after LLE or ASE, 10% NaOH was added to the sample extract to adjust the pH (to 5.4, 5.6, 5.8, 6.0 and 6.2), and then 2 mL of 0.2 M SDS solution was added to change the polarity. Adjusting the pH of the extract to 5.8 ± 0.2, the response value of the target was slightly improved. SPE cartridges were used to isolate spectinomycin and lincomycin from poultry eggs. The effects of different ion-pair reagents (sodium hexane sulfonate, sodium heptane sulfonate, sodium octane sulfonate and sodium dodecyl sulfonate) on the recoveries of the target compounds were compared. Sodium dodecyl sulfonate yielded the highest responses for the quantitative ion pairs (spectinomycin: m/z 201.1 > 75.0, lincomycin: m/z 126.1 > 42.0), which resulted in higher analyte recovery (Figure 2). Therefore, a 0.02 M sodium dodecyl sulfonate solution was used to equilibrate the SPE cartridge. This study compared C_{18} cartridges (6 mL/500 mg, Agela Technologies, Tianjin, China), PCX cartridges (6 mL/500 mg, Agela Technologies), and Oasis PRiME HLB cartridges (3 mL/60 mg, Waters Corp) in terms of the target compound recoveries. The C_{18} cartridge (6 mL/500 mg) produced interferences and did not effectively clean up the samples. The PCX cartridge (6 mL/500 mg) resulted in poor peak shapes and recoveries of less than 70%. The Oasis PRiME HLB cartridge (3 mL/60 mg) effectively cleaned up the samples and yielded recoveries above 80%. The Oasis PRiME HLB cartridge is a new type of solid-phase extraction cartridge that can remove 99% of the phospholipid matrix interferences in the sample, which minimizes the matrix effect of mass spectrometry, resulting in more stable data, a longer column life cycle, less instrument maintenance and less downtime. Therefore, the Oasis PRiME HLB cartridge (3 mL/60 mg) was used for sample clean-up.

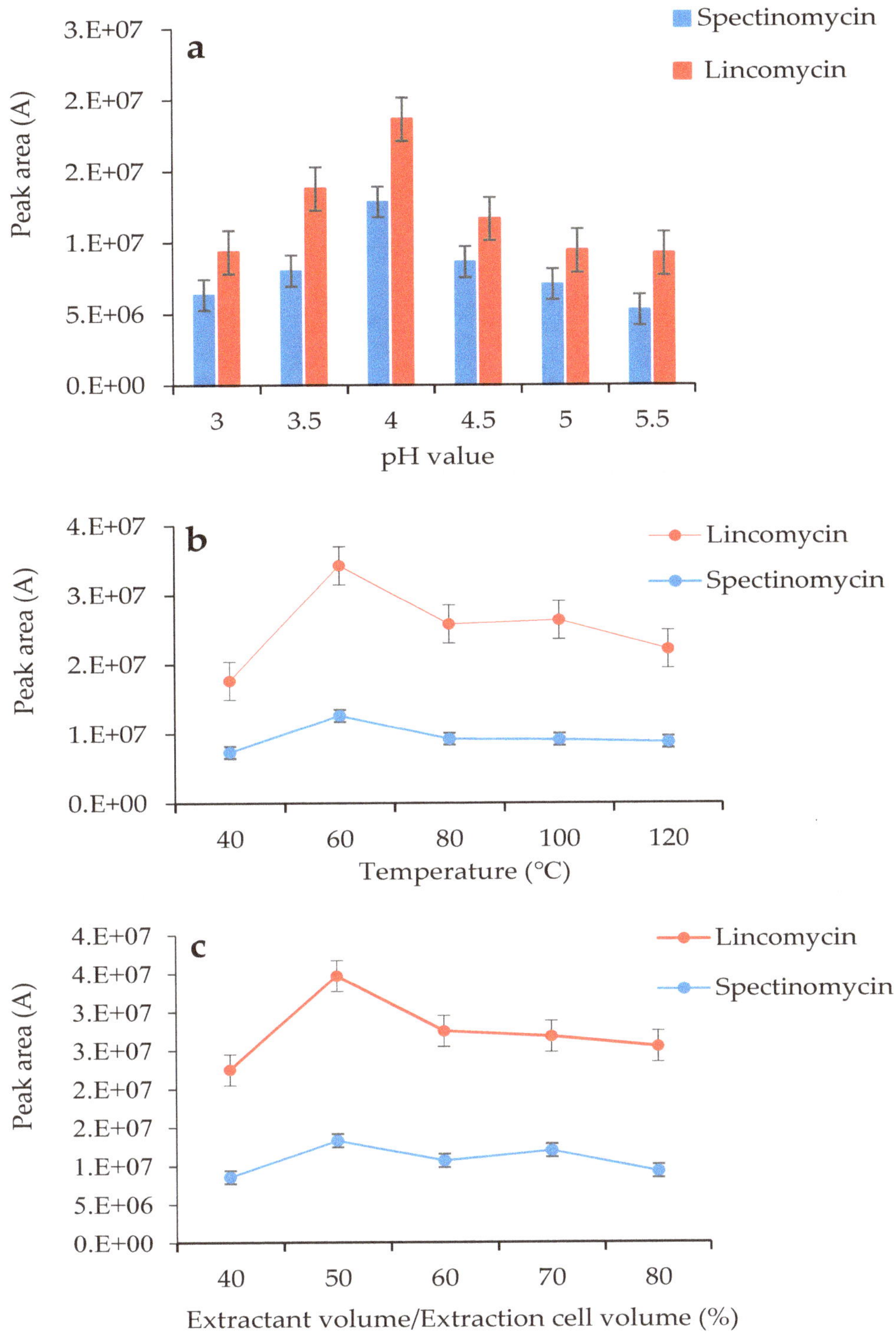

Figure 1. Effects of pH (**a**), temperature (**b**) and extractant volume (**c**) on the extraction efficiency of accelerated solvent extraction (ASE).

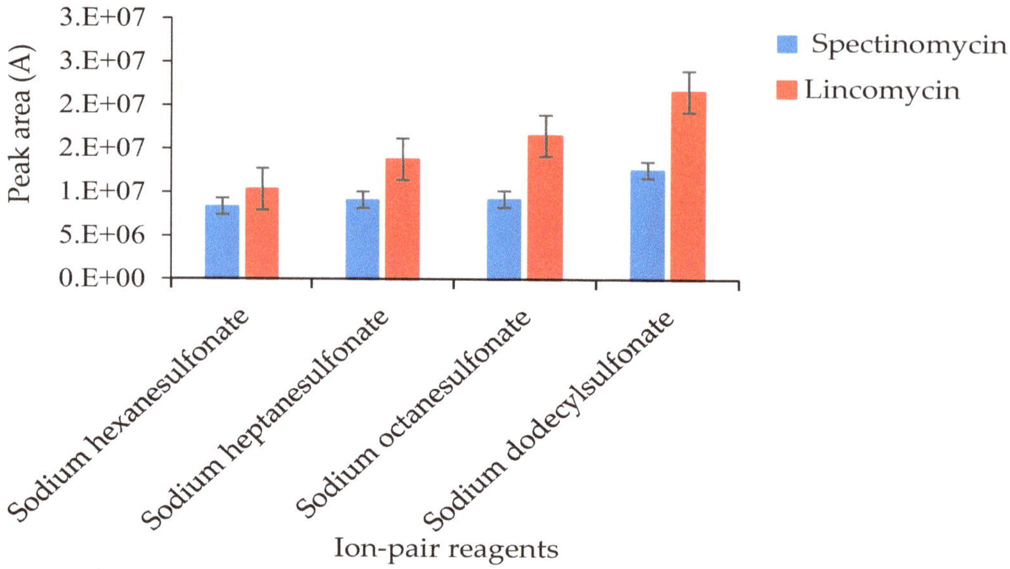

Figure 2. Effects of different ion-pair reagents on the recovery of spectinomycin and lincomycin.

After the optimization of the extraction and clean-up conditions, the effects of the LLE-SPE and ASE-SPE methods on the recoveries of spectinomycin and lincomycin from poultry eggs were compared. The results (Table 2) show that the recoveries for the ASE-SPE method were higher than those for the LLE-SPE method. Therefore, the ASE-SPE method was used to extract and clean up spectinomycin and lincomycin residues in poultry eggs.

Table 2. Comparison of the effects of the extraction method on the recoveries of 50 μg/kg spectinomycin and lincomycin in poultry eggs (%) (n = 6). Liquid–liquid extraction, LLE; solid-phase extraction, SPE.

Analyte	Matrix	Sample Preparation Method	
		LLE-SPE	ASE-SPE
Spectinomycin	Hen whole egg	80.5 ± 1.6	86.0 ± 1.5
	Albumen	81.4 ± 1.4	86.6 ± 2.0
	Yolk	81.2 ± 1.7	86.6 ± 2.1
	Duck whole egg	80.9 ± 3.4	86.1 ± 2.7
	Albumen	74.4 ± 1.4	85.7 ± 2.2
	Yolk	82.1 ± 3.4	87.1 ± 1.3
	Goose whole egg	79.2 ± 2.2	87.7 ± 1.7
	Albumen	81.1 ± 1.6	83.4 ± 1.4
	Yolk	80.9 ± 3.7	85.9 ± 1.2
	Pigeon whole egg	80.5 ± 2.7	87.7 ± 2.2
	Quail whole egg	79.0 ± 3.2	87.2 ± 1.9
Lincomycin	Hen whole egg	79.4 ± 1.6	84.8 ± 1.6
	Albumen	78.7 ± 2.7	85.8 ± 2.2
	Yolk	81.3 ± 3.3	84.8 ± 2.0
	Duck whole egg	75.4 ± 1.7	85.4 ± 2.6
	Albumen	76.0 ± 2.4	93.4 ± 2.3
	Yolk	83.5 ± 1.6	87.9 ± 2.1
	Goose whole egg	82.3 ± 3.7	87.9 ± 1.1
	Albumen	75.7 ± 2.0	89.1 ± 1.1
	Yolk	76.6 ± 2.2	84.6 ± 2.5
	Pigeon whole egg	83.2 ± 3.6	86.2 ± 2.2
	Quail whole egg	79.3 ± 1.2	86.4 ± 1.5

3.3. Optimization of the GC-MS/MS Analysis

Spectinomycin and lincomycin are highly polar compounds and cannot be detected directly by GC techniques. Usually, derivatization is required to reduce the polarity and boiling point of these compounds before GC detection. Tao et al. [30] reported the successful detection of spectinomycin and lincomycin in animal tissues by a GC method after derivatization by BSTFA. Thus, BSTFA was used as the derivatization reagent in the present work, and the above method of optimizing the ASE parameters was used to optimize the following derivatization conditions: the amount of BSTFA (100–700 μL), amount of acetonitrile (50–300 μL), temperature (35–95 °C) and time (30–90 min). The optimal derivatization conditions (Figure 3) were 75 °C, 60 min, 200 μL of BSTFA and 100 μL of acetonitrile, under which spectinomycin and lincomycin were derivatized to spectinomycin- trimethylsilyl (TMS) and lincomycin-TMS (Figures 4 and 5). After derivatization, BSTFA was removed by drying the sample under a stream of nitrogen. Excess BSTFA crystallizes easily and will plug and damage the column. TMS-derivatized products are easily hydrolyzed and stable for 24 h. Therefore, TMS-derivatized products should be analyzed by GC-MS/MS within 24 h.

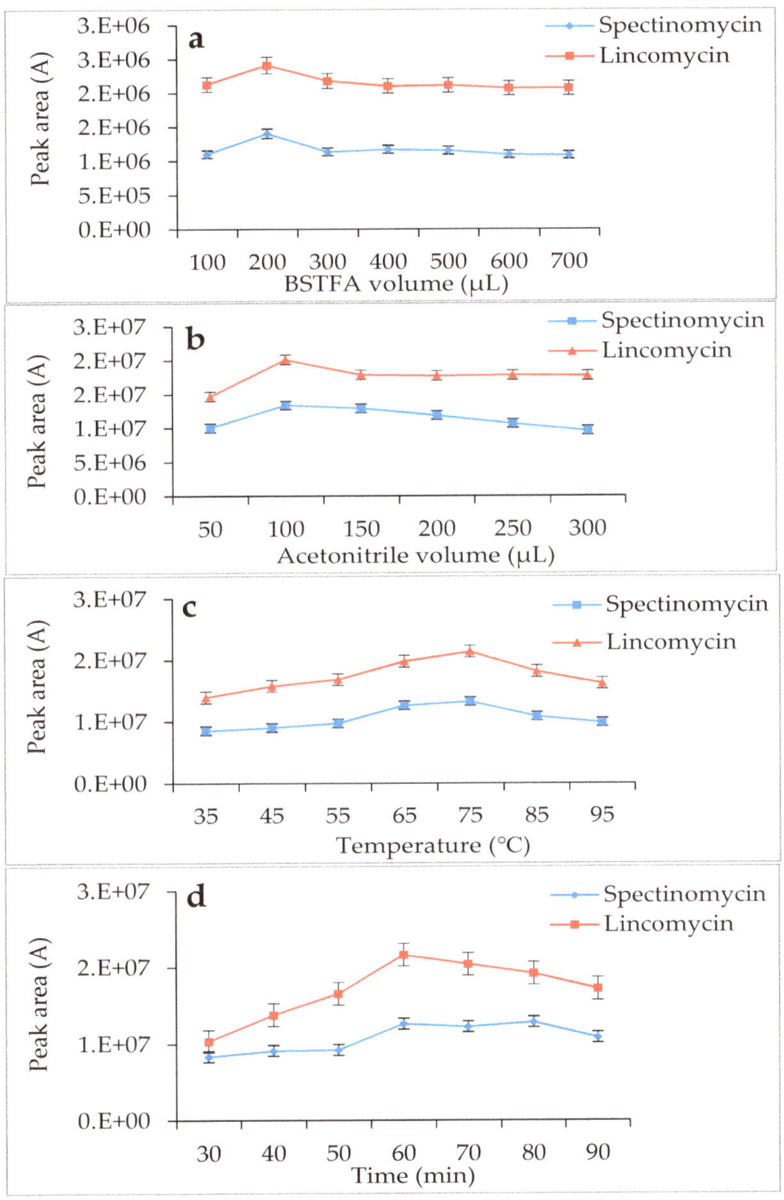

Figure 3. Effects of *N*,O-bis(Trimethylsilyl)trifluoroacetamide (BSTFA) volume (**a**), acetonitrile volume (**b**), temperature (**c**) and time (**d**) on the derivatization reaction.

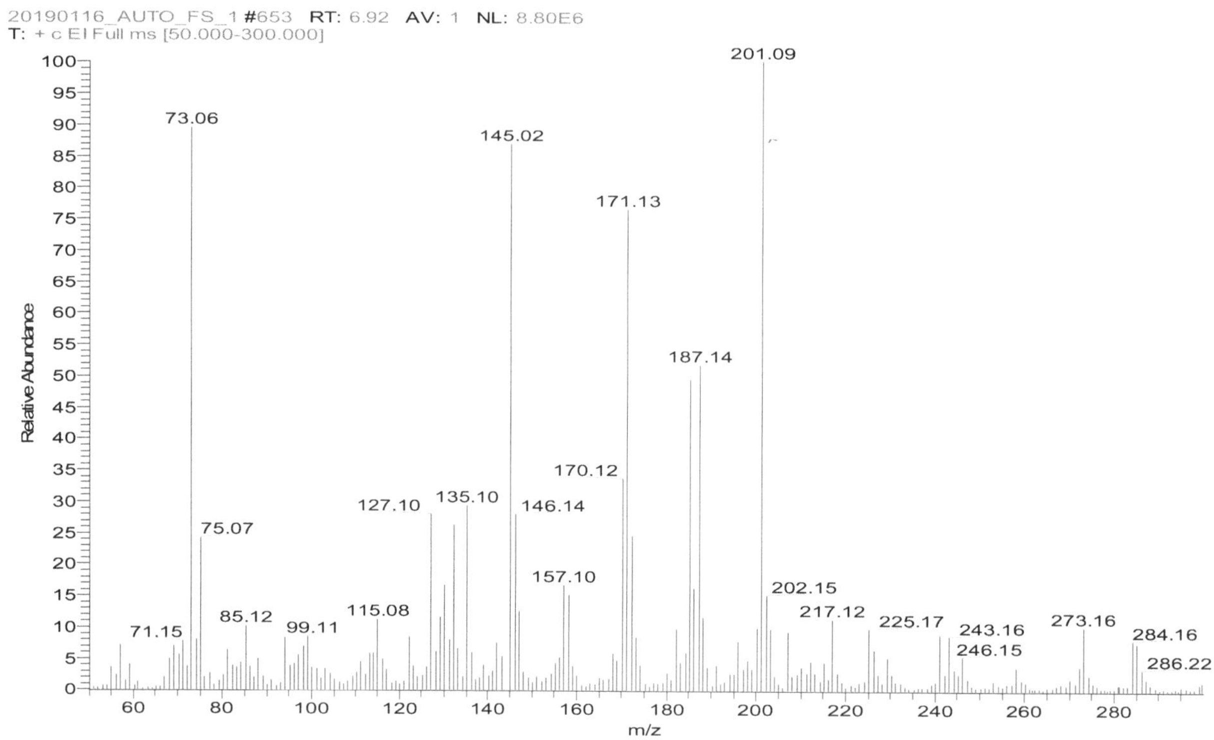

Figure 4. MS spectrum of spectinomycin-TMS.

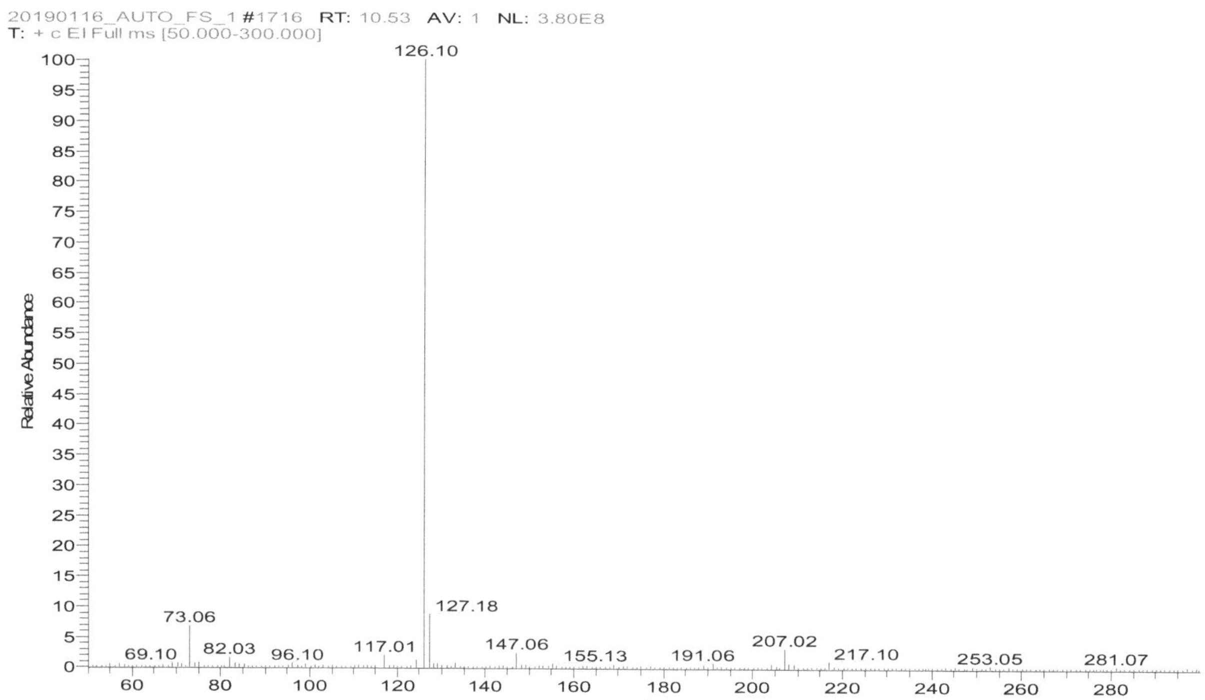

Figure 5. MS spectrum of lincomycin-TMS.

Several capillary columns, including DB-1 (30 m × 0.25 mm i.d., 0.25 μm), HP-5 (30 m × 0.25 mm i.d., 0.25 μm) and Rtx-5 (30 m × 0.25 mm i.d., 0.25 μm), have been reported for the detection of spectinomycin and lincomycin in animal-derived foods and were tested herein. According to previous reports [29,30], lincomycin and spectinomycin derivatives have moderate polarities and low boiling points, so nonpolar and moderately polar capillary columns are usually used to detect these two compounds. The inner surface of the moderately polar TG-5MS (30 m × 0.25 mm i.d., 0.25 μm) capillary column has been chemically treated to reduce the tailing of active basic compounds and increase the detection of amines. Therefore, a TG-5MS (30 m × 0.25 mm i.d., 0.25 μm) capillary column was selected to analyze spectinomycin and lincomycin residues in poultry eggs. Next, the oven temperature program was optimized to decrease the retention time (RT) of the target compounds (spectinomycin and lincomycin, 6.93 and 10.53 min) and shorten the total run time. Analysis was performed in full scan mode and SRM mode to identify precursor and product ions. In this study, two monitored ion pairs were selected for the qualitative and quantitative analysis of the target compounds. The derivatized products were analyzed under the optimized GC-MS/MS conditions. The total ion chromatogram (TIC) and extracted ion chromatograms (XICs) of a blank hen whole egg sample are shown in Figure 6. The TIC and XICs of the quantitative ions from the blank hen whole egg spiked with 50.0 μg/kg spectinomycin and 50.0 μg/kg lincomycin (Figure 7) showed that spectinomycin and lincomycin in hen whole eggs could be effectively separated with sharp peaks and no tailing.

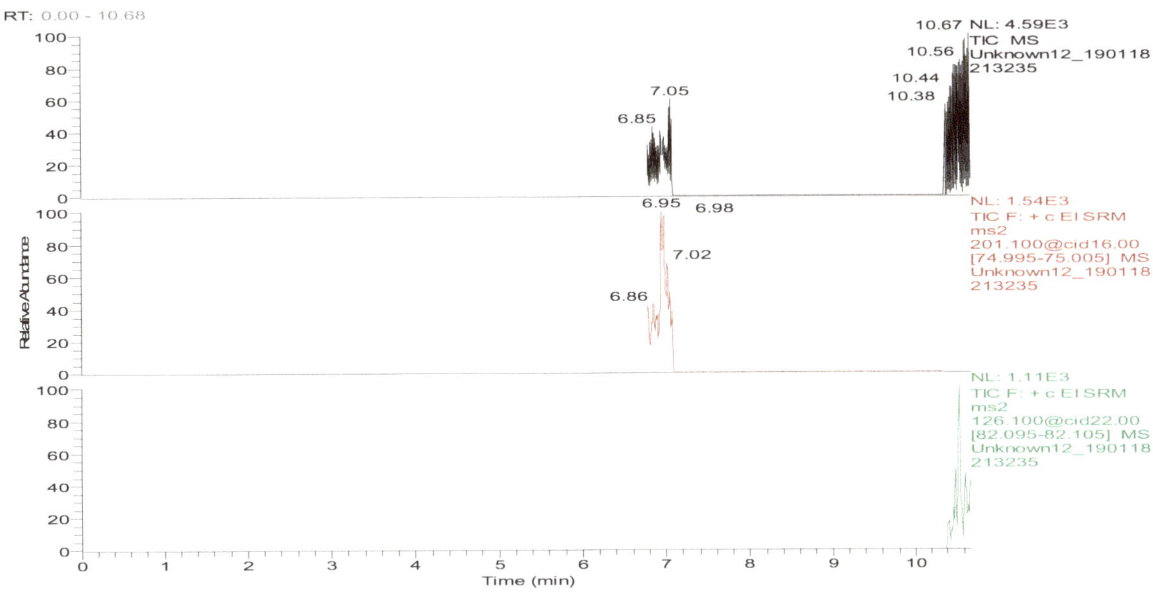

Figure 6. Total ion chromatogram and extracted ion chromatograms of a blank hen whole egg sample.

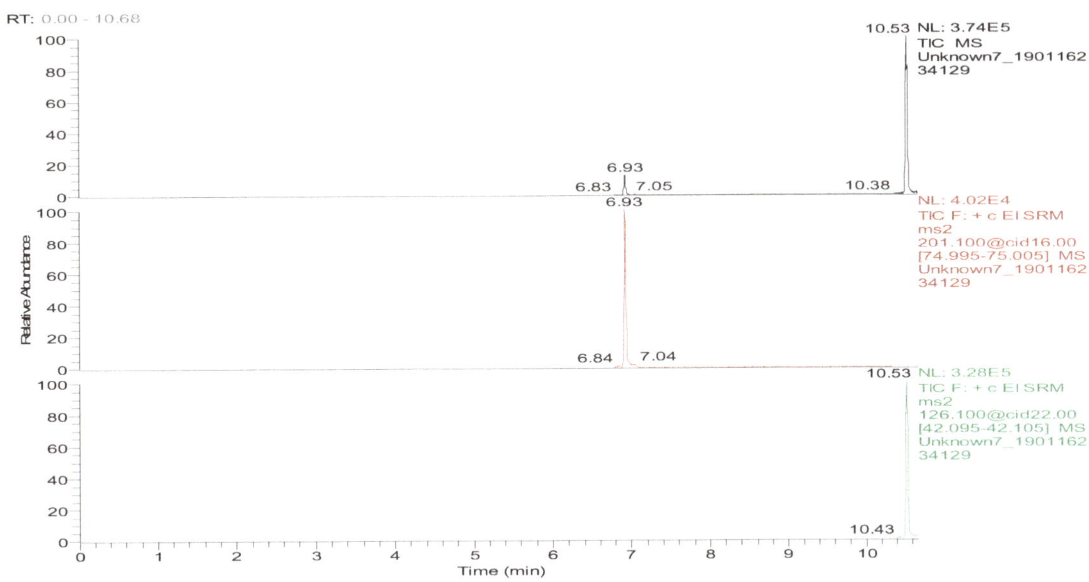

Figure 7. Total ion chromatogram and extracted ion chromatograms of a blank hen whole egg sample spiked with 50.0 μg/kg spectinomycin (retention time (RT), 6.93 min) and 50.0 μg/kg lincomycin (RT, 10.53 min).

3.4. Bioanalytical Method Validation

The specificity of the method for analyzing blank poultry eggs was determined by comparing Figures 6 and 7. Figure 6 shows that the blank hen whole egg sample did not contain spectinomycin and lincomycin. The blank poultry egg samples were extracted and cleaned up by the ASE-SPE method to obtain a blank matrix extract. The standard working solutions of spectinomycin and lincomycin and the reagents required for the abovementioned derivatization reaction were sequentially added to the blank matrix extract for derivatization. The standard curve was constructed from the GC-MS/MS analysis of the samples at the seven concentration levels. The linear ranges of spectinomycin and lincomycin were LOQ–2000 μg/kg and LOQ–200 μg/kg, respectively. The regression equation and determination coefficient data are listed in Table 3. According to the EU guidelines [31], the recovery and precision (intraday precision and interday precision) of the developed GC-MS/MS method were validated at the LOQ and at 0.5, 1.0 and 2 MRL (n = 6 at each level) for each drug in the poultry egg samples. In particular, 4000 μg/kg spectinomycin was added to the blank poultry egg sample; after extraction and purification by the ASE-SPE method, the sample was diluted with blank matrix extract 2-fold before the derivatization reaction was performed to ensure that the detected concentration of the sample was in the linear range. The measured concentration was multiplied by 2 to obtain the actual concentration of the original sample. By this method, the recovery and precision of measuring 4000 μg/kg spectinomycin in poultry eggs were evaluated. As shown in Tables 4 and 5, the recoveries of spectinomycin and lincomycin in the blank hen, duck and goose egg (whole egg, albumen and yolk) samples as well as in the pigeon and quail egg (whole egg) samples were 80.0%–95.7%, and the relative standard deviations (RSDs) were 1.0%–3.4%. In addition, the intraday RSDs were 1.9%–6.0%, and the interday RSDs were 2.2%–6.7%. These data indicate that the recovery and precision of the method meet the EU [31] and FDA [32] requirements for methodological parameters.

Table 3. Linearity, determination coefficient, limit of detection (LOD) and limit of quantification (LOQ) of spectinomycin and lincomycin in poultry eggs.

Analyte	Matrix	Regression Equation	Determination Coefficient (R^2)	Linear Range (μg/kg)	LOD (μg/kg)	LOQ (μg/kg)
Spectinomycin	Hen whole egg	$y = 1251x + 16346$	0.9993	6.0–2000	3.1	6.0
	Albumen	$y = 1298.4x + 34049$	0.9992	6.4–2000	3.0	6.4
	Yolk	$y = 1199.1x - 4540.7$	0.9997	5.6–2000	2.3	5.6
	Duck whole egg	$y = 1059.5x - 3709.3$	0.9999	6.3–2000	3.5	6.3
	Albumen	$y = 1113.1x - 4463.4$	0.9994	7.9–2000	3.8	7.9
	Yolk	$y = 1051.5x - 2573.8$	0.9993	8.0–2000	2.7	8.0
	Goose whole egg	$y = 1113.5x - 3245.8$	0.9996	7.1–2000	3.5	7.1
	Albumen	$y = 1179.6x - 3611.7$	0.9995	7.8–2000	3.2	7.8
	Yolk	$y = 1074.7x - 2777.5$	0.9996	6.7–2000	3.0	6.7
	Pigeon whole egg	$y = 1022.9x - 6310.6$	0.9991	8.0–2000	4.0	8.0
	Quail whole egg	$y = 1073.1x - 4316.6$	0.9998	7.6–2000	3.8	7.6
Lincomycin	Hen whole egg	$y = 10074x - 29019$	0.9994	8.4–200	3.1	8.4
	Albumen	$y = 7829.4x + 98669$	0.9992	6.7–200	2.6	6.7
	Yolk	$y = 13453x - 27487$	0.9994	5.9–200	2.5	5.9
	Duck whole egg	$y = 8127.3x - 16359$	0.9996	6.5–200	2.8	6.5
	Albumen	$y = 7772.3x - 312.45$	0.9992	7.3–200	3.0	7.3
	Yolk	$y = 7759.2x + 32016$	0.9992	8.0–200	4.3	8.0
	Goose whole egg	$y = 8825.4x + 2625.7$	0.9996	8.5–200	3.5	8.5
	Albumen	$y = 9240.1x + 6253.1$	0.9994	9.0–200	3.8	9.0
	Yolk	$y = 8381.4x + 27696$	0.9994	9.2–200	4.0	9.2
	Pigeon whole egg	$y = 8578.4x - 37597$	0.9993	9.5–200	3.9	9.5
	Quail whole egg	$y = 8736.2x - 61525$	0.9994	8.2–200	4.3	8.2

Table 4. Recovery and precision of spectinomycin and lincomycin spiked in blank poultry eggs (whole egg, $n = 6$).

Analyte	Matrix	Spike Level (μg/kg)	Recovery (%)	RSD (%)	Intraday RSD (%)	Interday RSD (%)
Spectinomycin	Hen whole egg	6.0	82.9 ± 1.5	1.8	3.0	4.2
		1000.0	84.5 ± 1.0	1.2	4.1	3.6
		2000.0 [a]	86.0 ± 1.5	1.7	4.0	4.1
		4000.0	95.7 ± 1.2	1.3	3.7	5.5
	Duck whole egg	6.3	80.9 ± 1.3	1.6	4.6	5.0
		1000.0	85.5 ± 1.5	1.8	3.6	4.5
		2000.0 [a]	86.1 ± 2.7	3.1	2.9	3.2
		4000.0	89.4 ± 1.8	2.0	2.9	3.4
	Goose whole egg	7.1	82.1 ± 1.7	2.1	3.2	5.0
		1000.0	86.8 ± 1.3	1.5	2.7	4.9
		2000.0 [a]	87.7 ± 1.7	1.9	2.5	3.0
		4000.0	90.6 ± 1.5	1.7	4.0	4.1
	Pigeon whole egg	8.0	82.3 ± 1.3	1.6	3.3	6.7
		1000.0	83.5 ± 1.1	1.3	3.1	4.6
		2000.0 [a]	87.7 ± 2.2	2.5	3.7	4.8
		4000.0	88.7 ± 1.3	1.5	2.5	3.7
	Quail whole egg	7.6	85.1 ± 1.4	1.6	3.1	6.0
		1000.0	85.3 ± 1.6	1.9	3.2	4.4
		2000.0 [a]	87.1 ± 1.9	2.2	3.0	3.9
		4000.0	94.8 ± 1.1	1.2	2.5	3.5
Lincomycin	Hen whole egg	8.4	82.7 ± 1.3	1.6	4.6	5.7
		25.0	85.4 ± 2.4	2.8	5.0	5.6
		50.0 [a]	84.8 ± 1.6	1.9	3.1	4.5
		100.0	87.1 ± 1.9	2.2	3.2	4.9
	Duck whole egg	6.5	81.8 ± 1.1	1.3	4.3	4.7
		25.0	86.7 ± 2.5	2.9	3.7	5.4
		50.0 [a]	85.4 ± 2.6	3.0	2.7	3.7
		100.0	86.1 ± 2.7	3.1	4.6	5.0
	Goose whole egg	8.5	80.1 ± 1.5	1.9	2.2	4.1
		25.0	84.0 ± 2.6	3.1	3.4	4.1
		50.0 [a]	87.9 ± 1.1	1.3	2.3	3.7
		100.0	93.5 ± 1.8	1.9	3.5	3.9
	Pigeon whole egg	9.5	81.4 ± 1.6	2.0	3.9	6.5
		25.0	82.2 ± 2.1	2.6	2.3	3.1
		50.0 [a]	86.2 ± 2.2	2.6	5.6	6.7
		100.0	91.3 ± 1.6	1.8	3.0	3.9
	Quail whole egg	8.2	80.0 ± 1.3	1.6	2.6	3.3
		25.0	86.7 ± 2.1	2.4	3.9	3.7
		50.0 [a]	86.4 ± 1.5	1.7	2.6	4.4
		100.0	87.3 ± 1.1	1.3	3.5	5.5

Note: [a] Maximum residue limit (MRL). RSD, relative standard deviation.

Table 5. Recovery and precision for spectinomycin and lincomycin spiked in blank poultry eggs (egg albumen and yolk, $n = 6$).

Analyte	Matrix	Spike Level (μg/kg)	Recovery (%)	RSD (%)	Intraday RSD (%)	Interday RSD (%)
Spectinomycin	Hen egg albumen	6.4	83.8 ± 2.4	2.9	3.0	4.6
		1000.0	84.2 ± 2.3	2.7	2.9	3.7
		2000.0 [a]	86.6 ± 2.0	2.3	2.8	3.6
		4000.0	93.0 ± 1.9	2.0	2.2	4.3
	Yolk	5.6	83.9 ± 2.5	3.0	3.5	3.7
		1000.0	84.1 ± 1.7	2.0	3.0	3.4
		2000.0 [a]	86.6 ± 2.1	2.4	3.7	5.5
		4000.0	94.5 ± 1.4	1.5	2.6	3.8
	Duck egg albumen	7.9	83.6 ± 1.3	1.6	2.7	4.7
		1000.0	84.7 ± 1.5	1.8	2.0	2.2
		2000.0 [a]	85.7 ± 2.2	2.6	3.0	3.9
		4000.0	87.3 ± 3.0	3.4	5.2	6.5
	Yolk	8.0	82.9 ± 1.9	2.3	2.8	3.5
		1000.0	85.3 ± 2.1	2.5	3.9	4.2
		2000.0 [a]	87.1 ± 1.3	1.5	2.8	3.3
		4000.0	90.3 ± 2.4	2.7	4.3	6.2
	Goose egg albumen	7.8	80.4 ± 1.5	1.9	2.0	3.7
		1000.0	84.0 ± 2.5	3.0	3.3	6.3
		2000.0 [a]	83.4 ± 1.4	1.7	2.5	3.1
		4000.0	88.1 ± 1.8	2.0	3.9	4.1
	Yolk	6.7	81.6 ± 2.4	2.9	3.1	3.4
		1000.0	82.4 ± 2.0	2.4	2.4	3.2
		2000.0 [a]	85.9 ± 1.2	1.4	2.0	3.5
		4000.0	89.4 ± 1.9	2.1	3.1	4.9
Lincomycin	Hen egg albumen	6.7	82.2 ± 1.2	1.5	2.8	3.4
		25.0	85.2 ± 1.6	1.9	3.2	4.1
		50.0 [a]	85.8 ± 2.2	2.6	3.4	4.5
		100.0	86.5 ± 2.5	2.9	3.5	5.7
	Yolk	5.9	82.7 ± 1.4	1.7	3.9	4.8
		25.0	85.4 ± 2.0	2.3	3.0	3.6
		50.0 [a]	84.8 ± 2.0	2.4	2.6	3.8
		100.0	91.2 ± 2.1	2.3	4.1	5.5
	Duck egg albumen	7.3	84.9 ± 1.5	1.8	1.9	3.5
		25.0	91.2 ± 2.4	2.6	6.0	5.5
		50.0 [a]	93.4 ± 2.3	2.5	3.7	4.3
		100.0	95.1 ± 2.2	2.3	3.2	4.9
	Yolk	8.0	85.2 ± 1.3	1.5	3.2	5.3
		25.0	86.5 ± 0.9	1.0	2.5	3.6
		50.0 [a]	87.9 ± 2.1	2.4	3.8	4.1
		100.0	89.1 ± 2.3	2.6	3.5	3.7
	Goose egg albumen	9.0	80.8 ± 1.6	2.0	2.8	5.3
		25.0	85.4 ± 1.2	1.4	2.6	4.2
		50.0 [a]	89.1 ± 1.1	1.2	2.7	3.5
		100.0	92.1 ± 1.2	1.3	3.9	3.6
	Yolk	9.2	80.2 ± 1.9	2.4	4.8	4.4
		25.0	85.6 ± 2.1	2.5	4.3	3.9
		50.0 [a]	84.6 ± 2.5	3.0	4.0	4.1
		100.0	87.9 ± 1.6	1.8	2.7	4.1

Note: [a] MRL.

Blank matrix extracts of the hen, duck and goose egg (whole egg, albumen and yolk) samples as well as of the pigeon and quail egg (whole egg) samples were prepared, and the spectinomycin and lincomycin standard working solutions were added to the blank matrix extract, derivatized and detected by GC-MS/MS. The concentrations corresponding to signal-to-noise (S/N) ratios of 3 and 10 for the target compounds were set as the limit of detection (LOD) and LOQ, respectively, of the target compounds in the hen, duck and goose egg (whole egg, albumen and yolk) samples as well as in the pigeon and quail egg (whole egg) samples. As shown in Table 3, the LODs of spectinomycin and lincomycin in the hen, duck, goose, pigeon and quail egg (whole egg) samples were 3.1, 3.5, 3.5, 4.0 and 3.8 μg/kg and 3.1, 2.8, 3.5, 3.9 and 4.3 μg/kg, respectively, and the LOQs of spectinomycin and lincomycin in the same poultry egg samples were 6.0, 6.3, 7.1, 8.0 and 7.6 μg/kg and 8.4, 6.5, 8.5, 9.5 and 8.2 μg/kg, respectively. The results for the LOD and LOQ of spectinomycin and lincomycin in the hen, duck and goose egg (albumen and yolk) samples are shown in Table 3. These LOQs and LOQs are relatively low, and the method is therefore highly sensitive and accurate.

3.5. Comparison of Different Detection Methods

Various analytical methods, including HPLC-FLD [16], HPLC-UVD [18], HILIC-MS/MS [22], HPLC-MS [23], HPLC-MS/MS [24,27], GC-NPD [30] and GC-MS [30], have been used to detect spectinomycin and lincomycin in meat, milk, feedstuffs, honey and animal tissues as well as in swine, calf and chicken plasma. Negarian et al. [18] established an HPLC-UVD method that showed better recovery (80.0%–89.0%) and precision (3.0%–3.9%) for the detection of lincomycin in milk and used CSMISPE to extract and clean up milk samples. Sin et al. [24] developed an LLE method to extract lincomycin from animal tissues and bovine milk. The average recoveries of lincomycin from animal tissues and bovine milk samples were 93.9%–107%, with a precision of 1.3%–7.8%. The LODs and LOQs of this method were 1.5–8.8 μg/kg and 25.0–50.0 μg/kg, respectively. Juan et al. [27] reported an ASE-HPLC-MS/MS method for the simultaneous determination of macrolide and lincosamide antibiotics in meat and milk. ASE is an automated technology that uses solvents at a relatively high pressure and a temperature below the critical points. Compared with LLE and SPE, ASE improves the work efficiency and reduces the amount of extractant required for analysis. Tao et al. [30] established GC-NPD and GC-MS methods for the determination of spectinomycin and lincomycin residues in animal tissues. Animal tissue samples were extracted by ASE, cleaned up with SPE cartridges and detected by GC-NPD and GC-MS. The average recoveries with the GC-NPD and GC-MS methods were 73.0%–97.0% and 70.0%–93.0%, and the RSDs were less than 17% and 21%, respectively. We compared the analysis time, sensitivity and recovery for spectinomycin and lincomycin analysis using different extraction and detection methods. As shown in Table 6, HPLC or GC with MS or MS/MS detection yielded higher sensitivity and precision than FLD, UVD and NPD.

ASE is an automated extraction technology that effectively improves the work efficiency, and 24 samples can be processed simultaneously in the same batch. In this study, LLE and ASE were used to effectively extract poultry egg samples. However, the LLE method is complicated and time- and reagent-consuming. After comparing sample pretreatment methods, we selected ASE for the extraction of spectinomycin and lincomycin residues from poultry eggs. Moreover, GC-MS/MS has higher sensitivity and precision than GC-MS. In this study, the parameters of ASE and GC-MS/MS were optimized to successfully detect spectinomycin and lincomycin in poultry eggs. The newly developed ASE-GC-MS/MS method provides new techniques and a scientific basis for the detection of spectinomycin and lincomycin residues in poultry eggs.

Table 6. Comparison of the present method with other methods.

Detection Method	Sample Preparation Method	Analyte	Animal-Derived Food	Analysis Time (min)	LOD (μg/kg)	LOQ (μg/kg)	Recovery (%)
HPLC-FLD [16]	SPE	Spectinomycin	Swine, calf and chicken plasma	12.0	-	-	91.0–104
HPLC-UVD [18]	CSMISPE	Lincomycin	Milk	9.0	20.0	80.0	80.0–89.0
HILIC-MS/MS [22]	LLE	Spectinomycin	Feedstuffs	10.0	-	-	80.0–92.0
HPLC-MS [23]	SPE	Lincomycin	Honey	10.0	7.0	10.0	102–105
HPLC-MS/MS [24]	LLE	Lincomycin	Animal tissues and milk	10.0	1.5–8.8	25.0–50.0	93.9–107
HPLC-MS/MS [27]	ASE	Lincomycin	Meat and milk	40.0	5.0–10.0	10.0–15.0	86.0–91.0
GC-NPD [30]	ASE	Spectinomycin Lincomycin	Animal tissues	13.0	8.1–9.4	16.4–21.4	73.0–97.0
GC-MS [30]	ASE	Spectinomycin Lincomycin	Animal tissues	15.0	1.9–3.1	4.7–5.7	70.0–93.0
GC-MS/MS	ASE	Spectinomycin Lincomycin	Poultry eggs	10.8	2.3–4.3	5.6–9.5	80.0–95.7

Note: "-" Not reported. FLD, fluorescence detection; UVD, ultraviolet detection; NPD, nitrogen phosphorus detection; CSMISPE, core-shell molecularly imprinted solid-phase extraction.

3.6. Real Sample Analysis

To evaluate the feasibility and accuracy of the newly developed method, we analyzed real samples using ASE-GC-MS/MS. One hundred and fifty commercial poultry eggs (30 hen eggs, 30 duck eggs, 30 goose eggs, 30 pigeon eggs and 30 quail eggs) were purchased from a local supermarket. Each poultry egg sample was processed in accordance with the sample pretreatment method described above and labeled, and each sample was detected and analyzed by the GC-MS/MS method. The target compounds were not detected in duck, goose, pigeon and quail eggs; only hen eggs were found to contain lincomycin residues (11.5 μg/kg less than the MRL). Therefore, the developed ASE-GC-MS/MS method can be applied to quantify spectinomycin and lincomycin in poultry egg samples.

4. Conclusions

In this study, we successfully developed a rapid, sensitive and specific ASE-GC-MS/MS method for the determination of spectinomycin and lincomycin residues in poultry egg samples. ASE is a promising technique for the preparation of animal-derived food samples. The developed method is accurate, has high recovery and precision, and fulfills the validation requirements of the Ministry of Agriculture of the People's Republic of China, the EU and the FDA. The analysis of real samples showed that this new method is feasible and can detect spectinomycin and lincomycin residues in poultry egg samples.

Author Contributions: Conceptualization, K.X.; Data Curation, B.W. and Y.W.; Formal Analysis, G.Z., T.Z. and G.D.; Funding Acquisition, B.W., X.X. and K.X.; Investigation, Y.W.; Methodology, B.W., Y.W., X.X. and K.X.; Resources, X.X. and Z.D.; Software, Z.D. and G.D.; Validation, Z.D., G.Z. and T.Z.; Writing of Original Draft, B.W. and Y.W.; and Writing of Review & Editing, B.W. All authors have read and agreed to the published version of the manuscript.

References

1. Butler, M.M.; Waidyarachchi, S.L.; Connolly, K.L.; Jerse, A.E.; Chai, W.; Lee, R.E.; Kohlhoff, S.A.; Shinabarger, D.L.; Bowlin, T.L. Aminomethyl spectinomycins as therapeutics for drug-resistant gonorrhea and chlamydia coinfections. *Antimicrob. Agents Chemother.* **2018**, *62*, e00325-18. [CrossRef] [PubMed]
2. Lin, A.H.; Murray, R.W.; Vidmar, T.J.; Marotti, K.R. The oxazolidinone eperezolid binds to the 50S ribosomal subunit and competes with binding of chloramphenicol and lincomycin. *Antimicrob. Agents Chemother.* **1997**, *41*, 2127–2131. [CrossRef] [PubMed]
3. Gao, P.; Hou, Q.; Kwok, L.-Y.; Huo, D.; Feng, S.; Zhang, H. Effect of feeding Lactobacillus plantarum P-8 on the faecal microbiota of broiler chickens exposed to lincomycin. *Sci. Bull.* **2017**, *62*, 105–113. [CrossRef]
4. Yu, Y.; Fang, J.-T.; Zheng, M.; Zhang, Q.; Walsh, T.R.; Liao, X.-P.; Sun, J.; Liu, Y.-H. Combination therapy strategies against multiple-resistant Streptococcus suis. *Front. Pharmacol.* **2018**, *9*, 489. [CrossRef]
5. Pokrant, E.; Maddaleno, A.; Lobos, R.; Trincado, L.; Lapierre, L.; San Martín, B.; Cornejo, J. Assessing the depletion of lincomycin in feathers from treated broiler chickens: A comparison with the concentration of its residues in edible tissues. *Food Addit. Contam.* **2019**, *36*, 1647–1653. [CrossRef]
6. Verrette, L.; Fairbrother, J.M.; Boulianne, M. Effect of cessation of ceftiofur and substitution with lincomycin-spectinomycin on extended-spectrum-β-lactamase/AmpC genes and multidrug resistance in escherichia coli from a canadian broiler production pyramid. *Appl. Environ. Microbiol.* **2019**, *85*, e00037-19. [CrossRef]
7. Ministry of Agriculture of the People's Republic of China. *Maxium Residue Level of Veterinary Drugs in Food of Animal Origin*; Notice no. 235 (Appendix 4); Ministry of Agriculture of the People's Republic of China: Beijing, China, 2002.

8. The European Medicines Agency. Pharmacologically Active Substances and their Classification Regarding Maximum Residue Limits in Foodstuffs of Animal Origin. In *Commission Regulation (EU)*; No. 37/2010 of 22 December 2009; The European Medicines Agency: Amsterdam, The Netherlands, 2010.
9. U.S. Food and Drug Administration. *CFR-Code of Federal Regulations Title 21 Part 556 Tolerances for Residue of New Animal Drugs in Food*; U.S. Food and Drug Administration: Rockville, MD, USA, 2014.
10. The Japan Food Chemical Research Foundation. *Maximum Residue Limits (MRLs) List of Agricultural Chemicals in Foods*; The Japan Food Chemical Research Foundation: Tokyo, Japan, 2015.
11. Medina, M.B. Development of a fluorescent latex immunoassay for detection of a spectinomycin antibiotic. *J. Agric. Food Chem.* **2004**, *52*, 3231–3236. [CrossRef]
12. Kowalski, P.; Konieczna, L.; Olędzka, I.; Plenis, A.; Bączek, T. Development and validation of electromigration technique for the determination of lincomycin and clindamycin residues in poultry tissues. *Food Anal. Methods* **2014**, *7*, 276–282. [CrossRef]
13. Cao, S.; Song, S.; Liu, L.; Kong, N.; Kuang, H.; Xu, C. Comparison of an enzyme-linked immunosorbent assay with an immunochromatographic assay for detection of lincomycin in milk and honey. *Immunol. Investig.* **2015**, *44*, 438–450. [CrossRef]
14. Schermerhorn, P.G.; Chu, P.-S.; Kijak, P.J. Determination of spectinomycin residues in bovine milk using liquid chromatography with electrochemical detection. *J. Agric. Food Chem.* **1995**, *43*, 2122–2125. [CrossRef]
15. Szúnyog, J.; Adams, E.; Liekens, K.; Roets, E.; Hoogmartens, J. Analysis of a formulation containing lincomycin and spectinomycin by liquid chromatography with pulsed electrochemical detection. *J. Pharm. Biomed. Anal.* **2002**, *29*, 213–220. [CrossRef]
16. Haagsna, N.; Scherpenisse, P.; Simmonds, R.J.; Wood, S.A.; Rees, S.A. High-performance liquid chromatographic determination of spectinomycin in swine, calf and chicken plasma using post-column derivatization. *J. Chromatogr. B Biomed. Sci. Appl.* **1995**, *672*, 165–171. [CrossRef]
17. Hamamoto, K.; Mizuno, Y.; Koike, R.; Yamaoka, R.; Takahashi, T.; Takahashi, Y. Residue analysis of spectinomycin in tissues of chicken and swine by HPLC. *J. Food Hyg. Soc. Jpn.* **2003**, *44*, 114–118. [CrossRef]
18. Negarian, M.; Mohammadinejad, A.; Mohajeri, S.A. Preparation, evaluation and application of core–shell molecularly imprinted particles as the sorbent in solid-phase extraction and analysis of lincomycin residue in pasteurized milk. *Food Chem.* **2019**, *288*, 29–38. [CrossRef]
19. Wang, M.J.; Hu, C.Q. Analysis of spectinomycin by HPLC with evaporative light-scattering detection. *Chromatographia* **2006**, *63*, 255–260. [CrossRef]
20. Zhou, J.; Zhang, L.; Wang, Y.; Yan, C. HPLC-ELSD analysis of spectinomycin dihydrochloride and its impurities. *J. Sep. Sci.* **2011**, *34*, 1811–1819. [CrossRef]
21. Peru, K.M.; Kuchta, S.L.; Headley, J.V.; Cessna, A.J. Development of a hydrophilic interaction chromatography–mass spectrometry assay for spectinomycin and lincomycin in liquid hog manure supernatant and run-off from cropland. *J. Chromatogr. A* **2006**, *1107*, 152–158. [CrossRef]
22. Molognoni, L.; de Souza, N.C.; de Sá Ploêncio, L.A.; Micke, G.A.; Daguer, H. Simultaneous analysis of spectinomycin, halquinol, zilpaterol, and melamine in feedingstuffs by ion-pair liquid chromatography–tandem mass spectrometry. *J. Chromatogr. A* **2018**, *1569*, 110–117. [CrossRef]
23. Thompson, T.S.; Noot, D.K.; Calvert, J.; Pernal, S.F. Determination of lincomycin and tylosin residues in honey using solid-phase extraction and liquid chromatography-atmospheric pressure chemical ionization mass spectrometry. *J. Chromatogr. A* **2003**, *1020*, 241–250. [CrossRef]
24. Sin, D.W.-M.; Wong, Y.-C.; Ip, A.C.-B. Quantitative analysis of lincomycin in animal tissues and bovine milk by liquid chromatography electrospray ionization tandem mass spectrometry. *J. Pharm. Biomed. Anal.* **2004**, *34*, 651–659. [CrossRef]
25. Sin, D.W.M.; Ho, C.; Wong, Y.C.; Ho, S.K.; Ip, A.C.B. Simultaneous determination of lincomycin and virginiamycin M1 in swine muscle, liver and kidney by liquid chromatography–electrospray ionization tandem mass spectrometry. *Anal. Chim. Acta* **2004**, *517*, 39–45. [CrossRef]
26. van Holthoon, F.L.; Essers, M.L.; Mulder, P.J.; Stead, S.L.; Caldow, M.; Ashwin, H.M.; Sharman, M. A generic method for the quantitative analysis of aminoglycosides (and spectinomycin) in animal tissue using methylated internal standards and liquid chromatography tandem mass spectrometry. *Anal. Chim. Acta* **2009**, *637*, 135–143. [CrossRef] [PubMed]

27. Juan, C.; Moltó, J.C.; Mañes, J.; Font, G. Determination of macrolide and lincosamide antibiotics by pressurised liquid extraction and liquid chromatography-tandem mass spectrometry in meat and milk. *Food Control* **2010**, *21*, 1703–1709. [CrossRef]
28. Maddaleno, A.; Pokrant, E.; Yanten, F.; San Martin, B.; Cornejo, J. Implementation and validation of an analytical method for lincomycin determination in feathers and edible tissues of broiler chickens by liquid chromatography tandem mass spectrometry. *J. Anal. Methods Chem.* **2019**, *2019*, 4569707. [CrossRef]
29. Luo, W.; Yin, B.; Ang, C.Y.W.; Rushing, L.; Thompson, H.C. Determination of lincomycin residues in salmon tissues by gas chromatography with nitrogen-phosphorus detection. *J. Chromatogr. B Biomed. Sci. Appl.* **1996**, *687*, 405–411. [CrossRef]
30. Tao, Y.; Chen, D.; Yu, G.; Yu, H.; Pan, Y.; Wang, Y.; Huang, L.; Yuan, Z. Simultaneous determination of lincomycin and spectinomycin residues in animal tissues by gas chromatography-nitrogen phosphorus detection and gas chromatography-mass spectrometry with accelerated solvent extraction. *Food Addit. Contam. A* **2011**, *28*, 145–154. [CrossRef]
31. The European Communities. Commission decision 2002/657/EC of 12 August 2002 implementing council directive 96/23/EC concerning the performance of analytical methods and the interpretation of results. *Off. J. Eur. Communities* **2002**, *221*, 8–36.
32. U.S. Department of Health and Human Services; Food and Drug Administration; Center for Drug Evaluation and Research; Center for Veterinary Medicine. *Guidance for Industry: Bioanalytical Method Validation*; US Department of Health and Human Services: Washington, DC, USA, 2001.
33. National Food Safety Standard. *GB 29685-2013: Determination of Lincomycin, Clindamycin and Spectinomycin in Animal Derived Food by Gas Chromatography-Mass Spectrometry Method*; Ministry of Agriculture of the People's Republic of China: Beijing, China, 2013.

9

Characterization of MS/MS Product Ions for the Differentiation of Structurally Isomeric Pesticides by High-Resolution Mass Spectrometry

Alberto Nuñez *, Yelena Sapozhnikova and Steven J. Lehotay

US Department of Agriculture, Agricultural Research Service, Eastern Regional Research Center, 600 East Mermaid Lane, Wyndmoor, PA 19038, USA; Yelena.Sapozhnikova@ars.usda.gov (Y.S.); steven.lehotay@ars.usda.gov (S.J.L.)

* Correspondence: alberto.nunez@ars.usda.gov

Abstract: Structural isomeric pesticides are used in agriculture and may be challenging to differentiate for accurate identification in pesticide monitoring programs. Due to structural similarity, isomeric pesticides are difficult to separate chromatographically, and thus, their accurate identification may rely solely on mass spectrometric analysis (MS). In this study, we challenged the ability of high-resolution quadrupole-orbitrap (Q-Orbitrap) mass spectrometry to produce and evaluate the tandem mass spectrometry (MS/MS) product ions for the selected five pairs of isomeric pesticides from different classes: Pebulate and vernolate, methiocarb and ethiofencarb, uniconazole and cyproconazole, sebuthylazine and terbuthylazine, and orbencarb and thiobencarb. The use of Q-Orbitrap instrument with a mass error <3 ppm allowed proposed elucidation of the product ion structures with consideration of the ion formulae, data interpretation, and literature searches. Product ions unique to pebulate, vernolate, methiocarb, ethiofencarb, and uniconazole were observed. Elucidation of the observed MS/MS product ion structures was conducted, and the fragmentation pathways were proposed. This information is valuable to increase selectivity in MS/MS analysis and differentiate isomeric pesticides, and thereby reduce the rates of false positives in pesticide monitoring programs.

Keywords: high-resolution mass spectrometry; Orbitrap; structure elucidation; pesticide residue analysis

1. Introduction

Numerous pesticides are used in agriculture worldwide to protect crops and increase harvest yields, with over 1.1 billion pounds used annually in the USA, and nearly six billion pounds (three million tons) usage worldwide [1]. Use of specific pesticides varies by different countries, and certain pesticides may be approved for applications in some countries but banned or no longer used in others. To protect consumer health, maximum residue limits (MRLs), or tolerances in the USA, are established and enforced by regulatory agencies around the world. MRLs depend on pesticide toxicity, physico-chemical properties, and application rates, which can range significantly for the same pesticide in different commodities [2]. Additionally, different countries may have different MRLs for the same pesticide in the same commodity. Hence, reliable identification and accurate measurement of pesticide residues in foods is important in global food trade because false positives and other forms of incorrect results can have detrimental economic and health consequences.

Common approaches for analysis of pesticide residues include liquid chromatography (LC) or gas chromatography (GC) with single or triple quadrupole for tandem mass spectrometry (MS) for separation and identification of targeted contaminants. In recent years, however, high-resolution accurate mass (HRAM) analyzers are becoming more common in routine analysis of pesticides and other contaminants [3,4]. During the MS/MS method development phase, precursor, and fragment/product ions are selected for analyte identification, and often, the most intense ions are selected. Many pesticides belonging to the same chemical classes (triazines, carbamates, *etc.*) produce the same product ions, and analyte/analyte interferences commonly complicate their accurate quantification. To achieve better selectivity and increase confidence in identification, specific fragments should be selected and studied when possible.

The use of HRAM analyzers in comparison with low resolution quadrupole analyzer instruments facilitates ion mass determination with error <5 ppm required for ion formula determination [5]. Combining this information with literature searches, databases, and data interpretation helps with the assignment of product ions structures. Currently, few published manuscripts on the analysis of contaminants report fragment/product ions with elucidation of the corresponding structures [6–10]. Yet, elucidation of ion structures is important to provide further support for elimination of false positive findings. This becomes even more important when dealing with identification of structurally isomeric pesticides, i.e., pesticides with the same molecular formula and weight, but different atom arrangement.

One example of isomeric pesticides are the carbamate herbicides, orbencarb and thiobencarb. Orbencarb (*S*-(2-chlorobenzyl) *N,N*-diethylthiocarbamate) was previously used on cereals, including barley, wheat, rye, maize, soybean, *etc.*, but not currently authorized for use in the European Union (EU) and not registered for use in the USA. Thiobencarb (*S*-(4-chlorobenzyl)-*N,N*-diethylthiocarbamate) differs from orbencarb only by Cl positioning on benzene ring (Table 1). It is used for weed control in paddy fields, mostly on rice, and according to Pesticide Action Network of North America, has an active status in the US products [11], but it is not approved for use in the EU. The MRL for thiobencarb is 10–200 ng/g depending on the country and commodity. As isomeric pesticides possess very similar properties, they are usually difficult or impossible to separate chromatographically. For example, both orbencarb and thiobencarb had the same retention time of 19.38 min and the same ions at *m/z* 258, 125, and 100 when analyzed by LC-MS/MS [12]; thus they were not differentiated by either LC separation or MS identification. MS/MS spectra of isomeric compounds are usually very similar and may have few distinct ions to tell them apart. In order to support these differentiations, it is important to show structures for those specific ions in support of the method of analysis. The use of HRAM is essential for the determination of ion formulas to be able to determine structures and fragmentation pathways [13,14].

In this study, we selected five pairs of isomeric pesticides (see Table 1): Pebulate and vernolate, methiocarb and ethiofencarb, uniconazole and cyproconazole, sebuthylazine and terbuthylazine, and orbencarb and thiobencarb, to test the hypothesis of applying HRAM for finding unique product ions to distinguish between the isomers. A Q-Orbitrap HRAM MS instrument with an error <3 ppm was utilized for identification of products ion formulae and the proposal of structures and pathways of fragmentation to differentiate isomeric pesticides. When necessary, the fragmentation patterns were based on product ions obtained from the MS^2 and their selected ions spectra (MS^3) obtained with a triple quadrupole instrument with a linear trap (Q-Trap) mass spectrometer.

Table 1. Isomeric structures of studied pesticides.

Name	CAS Number	Formula	Exact Mass	Structure
Vernolate	1929-77-7	$C_{10}H_{21}NOS$	203.1343	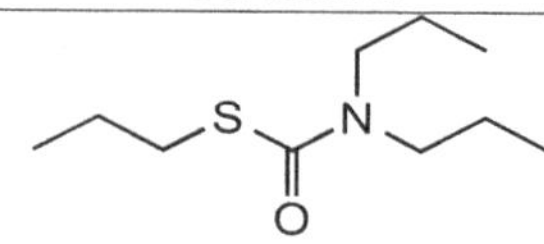
Pebulate	1114-71-2	$C_{10}H_{21}NOS$	203.1343	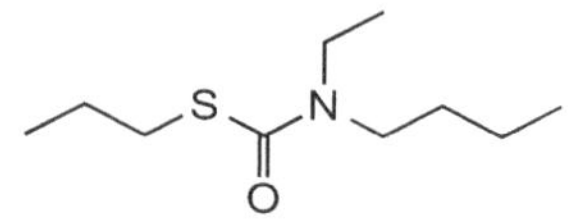
Cyproconazole	94361-06-5	$C_{15}H_{18}ClN_3O$	291.1138	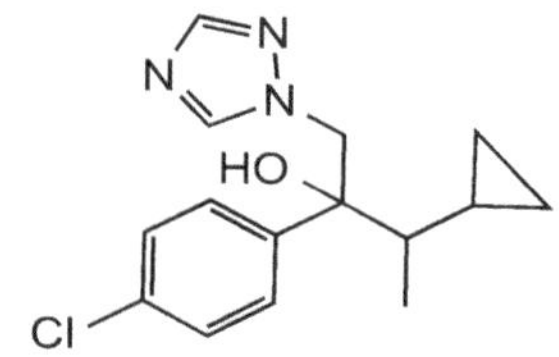
Uniconazole	83657-22-1	$C_{15}H_{18}ClN_3O$	291.1138	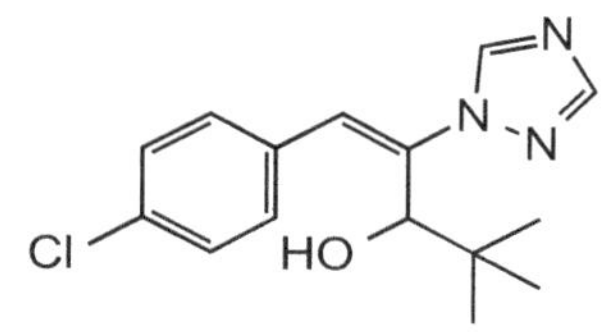
Methiocarb	2032-65-7	$C_{11}H_{15}NO_2S$	225.0823	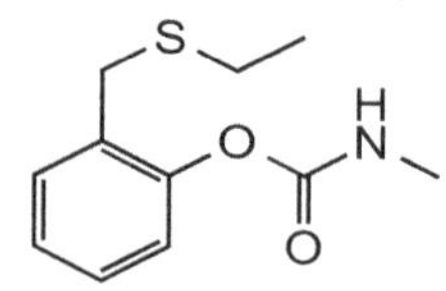
Ethiofencarb	29973-13-5	$C_{11}H_{15}NO_2S$	225.0823	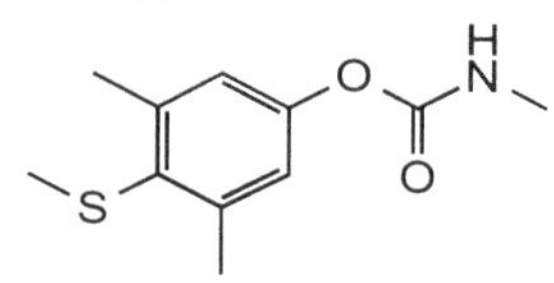
Thiobencarb	28249-77-6	$C_{12}H_{16}ClNOS$	257.0641	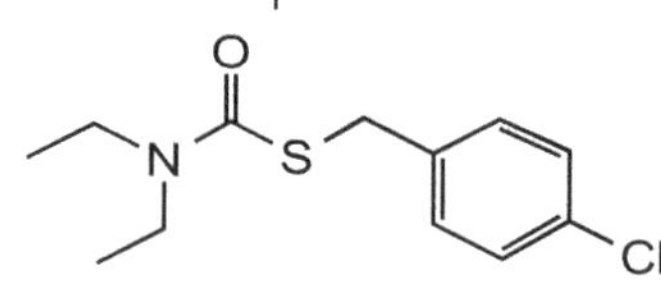
Orbencarb	34622-58-7	$C_{12}H_{16}ClNOS$	257.0641	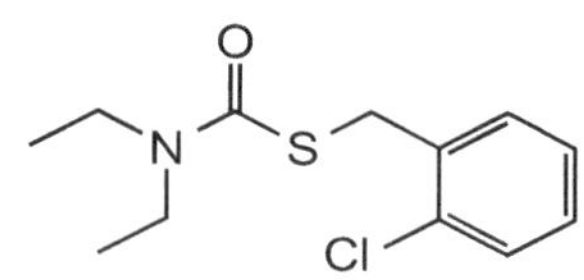
Sebuthylazine	7286-69-3	$C_9H_{16}ClN_5$	229.1094	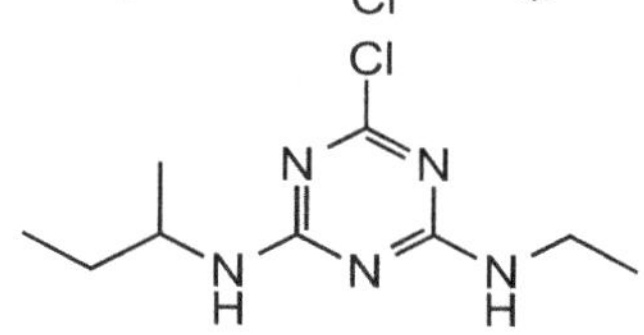
Terbuthylazine	5915-41-3	$C_9H_{16}ClN_5$	229.1094	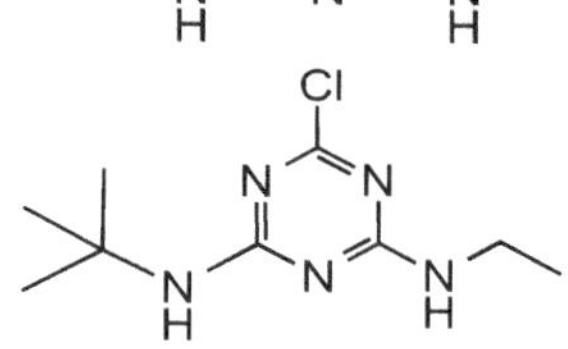

2. Materials and Methods

2.1. Reagents

Positive ion calibration solution for the Q-Orbitrap instrument was from Thermo Scientific (Rockford, IL, USA). Pesticide standards (purity > 95–99%) and formic acid (purity 98%) were obtained from Sigma Aldrich (St. Louis, MO, USA). Acetonitrile (MeCN), methanol (MeOH) and water were Optima LC-MS grade obtained from Fisher Scientific (Phillipsburg, NJ, USA), and deionized water was obtained from a Millipore (Bedford, MA, USA) Milli-Q system. Stock standard solutions were prepared at 2 mg/mL in MeOH, and working solutions at 0.5 µg/mL were prepared from stock solutions in 1:1 MeCN:H_2O (*v:v*) containing 0.1% formic acid.

2.2. Instrumentation and Conditions

Pesticide standard solutions (0.5 µg/mL) were infused at 10 µL/min using a Chemyx syringe pump model Fusion 101 into a Q-Exactive Plus (Thermo Scientific, Madison, WI, USA) mass spectrometer equipped with the electrospray ionization (ESI) probe (HESI-II) in positive mode to obtain MS/MS spectra. Collision energies were optimized and precursor ions were obtained with the mass error of <3 ppm. Capillary temperature was 300 °C with a spray voltage of 3.5 kV and nitrogen as sheath gas set to 60 arbitrary units of the HESI probe. Resolution was set to 140,000 (FWHM) for MS and MS/MS using normalized collision energy (NCE) adjusted to produce the ions of interest. The collision induced dissociation (CID) at the ion source was set at 5 V, except when a specific fragment ion was produced for subsequent MS/MS, in which case the CID was set to 30 V. The instrument was calibrated in the positive mode with a standard solution (Thermo Scientific) and the polysiloxanes *m/z* of 371.10124 and 445.12003 were used as lock masses in the MS mode.

Ion formation pathways were also investigated by infusion the selected pesticide standard solutions using a Sciex 6500 Q-Trap mass spectrometer (Sciex, Framingham, MA, USA) to obtain MS^2 and MS^3 spectrum of specific ions of interest. Ion source parameters were: Curtain gas at 10 psi, the ion spray potential was 5,000 V, the source temperature was 350 °C, entrance potential 10 V, collision gas (CAD) was set at medium, and ion source gases were 12 psi (GS1) and 10 psi (GS2).

3. Results and Discussion

Extensive compiled literature on mechanisms of fragmentation is available for electron ionization, also referred as odd electron ionization, but less is found regarding more generalized soft ionization, or even electron ionization, producing protonation, deprotonation, or alkyl adduct formation of the analytes. Recently, a very comprehensive compilation of drugs and pesticides MS/MS data interpretations has been published by Niessen and Correa [10], providing an excellent source for structural identification of MS/MS ions resulting from soft ionization mass spectrometry. In addition, availability of commercial software for spectra interpretation and public compound databases can help in elucidation of structures of product ions, but this approach should be considered with caution and some of these limitations have been described by Wright et al. [15]. Previously, we conducted structural characterization of product ions for 120 veterinary drugs with ESI quadrupole time-of-flight (Q-TOF) MS [7–9]. In the present study, a Q-Orbitrap instrument was used for characterization of product ions to differentiate isomeric pesticides. Occasionally, to better understand fragmentation pathway, product ions were selected for additional fragmentation. This was achieved by using a Q-Trap to obtain the MS^3 spectrum or, alternatively, by using adequate collision dissociation (CID) energy at the ion source of Q-Orbitrap to generate fragmentation for selection and subsequent MS/MS fragmentation. During the discussion, and with the purpose of simplification, the ion masses are used without considering the fractional masses that are reported in figures and schemes.

3.1. Pebulate and Vernolate

Vernolate (*S*-propyl-*N,N*-dipropylcarbamothioate) and pebulate (*S*-propyl-*N*-butyl-*N*-ethylcarbamothioate) are thiocarbamate herbicides. Both pesticides' applications ceased in the USA since 2001 and 2009, respectively [16], and in the EU in 2009 [17]. These pesticides are *S*-propyl thiocarbamates bearing a *N*-butyl-*N*-ethyl group (pebulate) and a *N,N*-dipropyl group (vernolate) substituent at the nitrogen. MS/MS spectra for these pesticides are presented in Figure 1, showing a very similar profile. The structure of some of these ions have been previously reported [18] and more recently described by Neissen and Castro [10]. The spectrum in Figure 1A for pebulate has two product ions at *m/z* 57 and 72 that are not observed for vernolate (Figure 1B). Additionally, vernolate has a significant ion at *m/z* 86 that is small for pebulate.

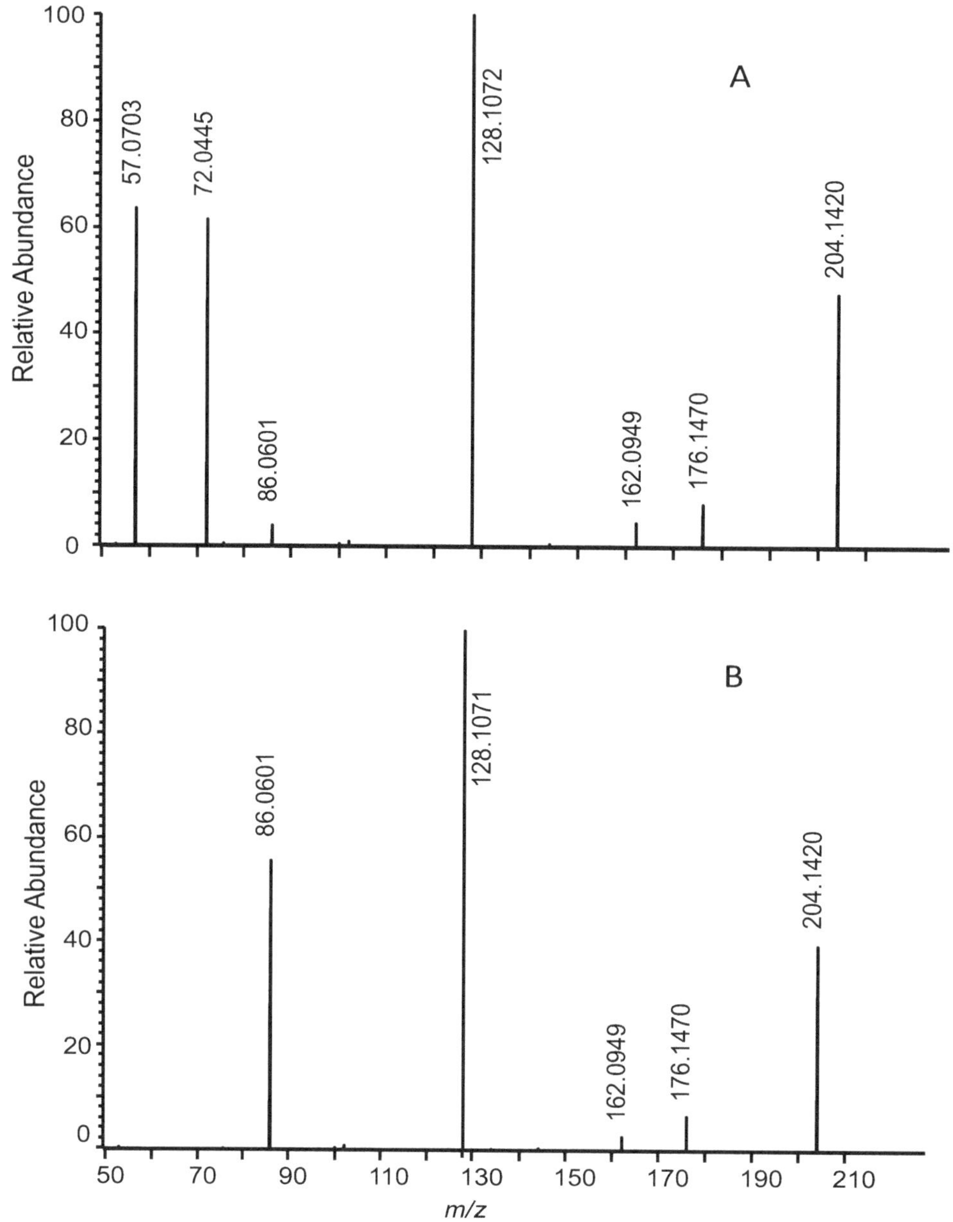

Figure 1. Positive ESI spectra obtained with an NCE 20 eV for: (**A**) pebulate and (**B**) vernolate.

Scheme 1 is the proposed fragmentation pathway for these pesticides indicating formation of two ions at *m/z* 128 and 162 that lead to product ions at *m/z* 57 and 72 for pebulate, and product ion at *m/z* 86 for vernolate. This ion resulted from the loss of butyl group but the equivalent ion resulting from the loss of ethyl group was not observed in the spectra. It is not clear if ion at *m/z* 162 can lead to those product ions, and is thus shown with a question mark in Scheme 1.

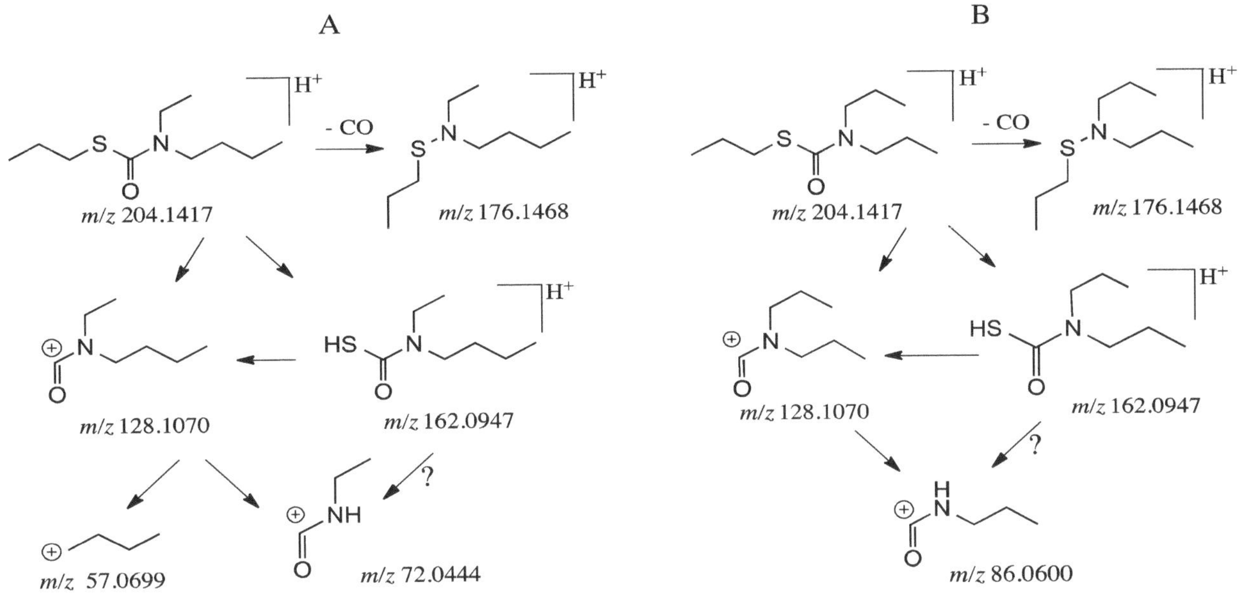

Scheme 1. Proposed fragmentation pathway for (**A**) pebulate and (**B**) vernolate.

The small ion at *m/z* 86 in the spectrum of pebulate is difficult to explain based on the fragmentation pattern observed. To further verify its origin, the *m/z* 162 ion was generated by adjusting CID energy at the ion source of Q-Orbitrap, and then selected for MS/MS. The resulting spectrum is shown in Figure 2. This approach generates a set of product ions that are presented in the pathway in Scheme 2. A careful inspection of the spectrum indicated the presence of a small ion at *m/z* 102 (Figure 2 insert). This ion is not expected for pebulate, nor is *m/z* 86 product ion, and consequently, this suggests that pebulate was contaminated with vernolate. This approach is not a true MS^3 because other isobaric ions could be present in the ion source, thus has to be used with caution, especially when low mass ions are selected. Additionally, the inclusion of solvent clusters in MS/MS high-energy collision cells can produce misleading product ions. However, exceptional mass accuracy and resolution of Q-Orbitrap are highly beneficial for assignments of ion formulae to better interpret fragments observed in this manner.

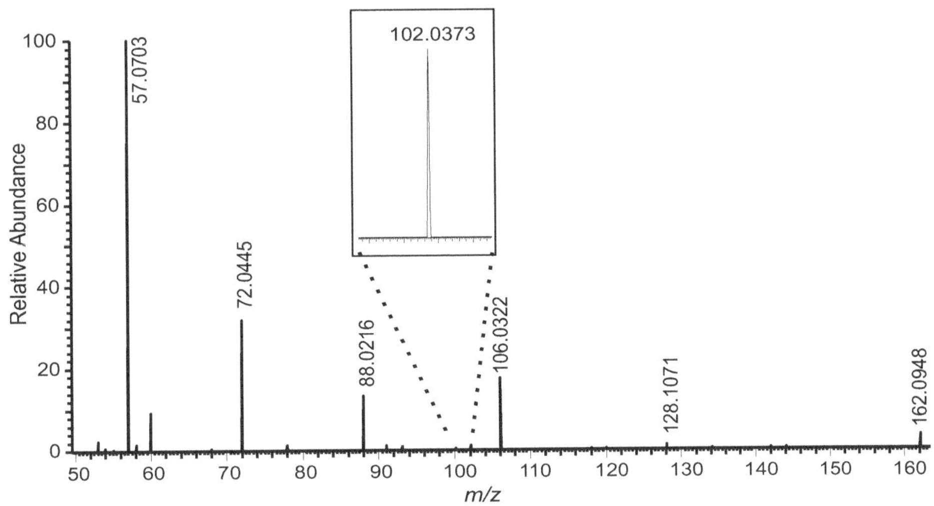

Figure 2. CID/HCD MS/MS spectrum of the selected ion at *m/z* 162 product ion of pebulate. The insert is the zoom region showing ion at *m/z* 102.

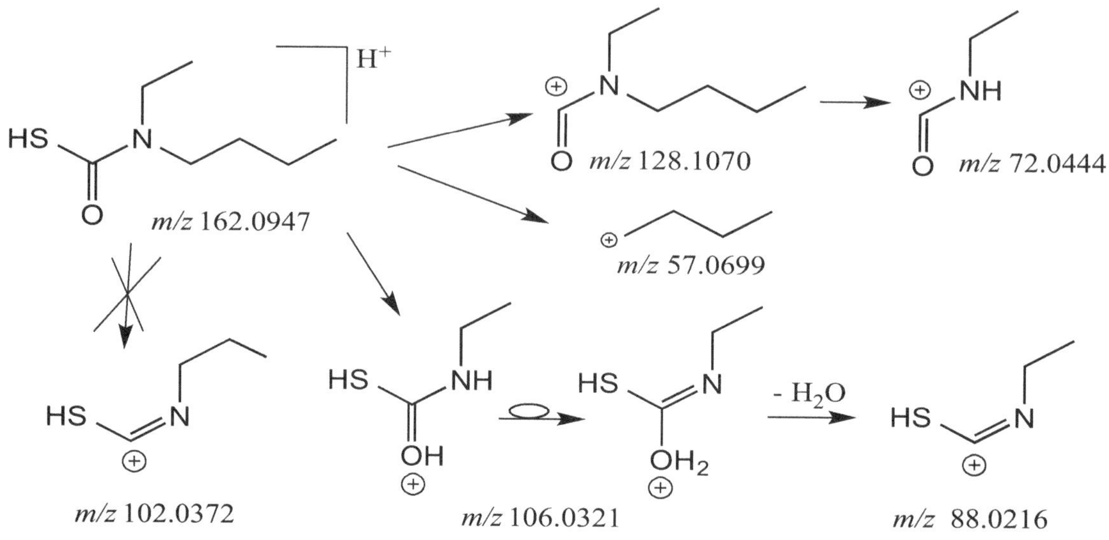

Scheme 2. Proposed fragmentation pathway for the CID generated product ion at *m/z* 162 of pebulate.

3.2. Methiocarb and Ethiofencarb

Methiocarb (3,5-dimethyl-4-methylthiophenyl-*N*-methylcarbamate) is a carbamate pesticide, currently approved for use in the EU. It is used for seed treatment, and applied to various agricultural crops, such as maize, lettuce, and fruits. It has an EU MRL of 50 ng/g in food of animal origin, 100 ng/g in grains, and 100–1000 ng/g in fruits and vegetables [17]. Methiocarb does not appear to be used in the USA, and no tolerance is listed in the USA MRL global database [19]. Ethiofencarb (α-ethylthio-*o*-tolyl methylcarbamate) is not approved for use in the EU, and no longer produced or used in the USA.

These two pesticides have MS/MS spectra producing a few fragments at 40–50 eV with product ions that can clearly differentiate between methiocarb and ethiofencarb as shown in Figure 3. However, the ions at *m/z* 107 and 169 are also observed at higher collision energies and only specific ions at *m/z* 121 for methiocarb and ion at *m/z* 164 can be used for differentiation. The pathway for the formation of these product ions is shown in Scheme 3, where ethiofencarb's loss of carbamate group leads to formation of ion at *m/z* 169 that subsequently leads to ion at *m/z* 121 after losing methanethiol [10]. Furthermore, the hydroxybenzyl cation at *m/z* 164 could rearrange to hydroxytropyllium ion [20].

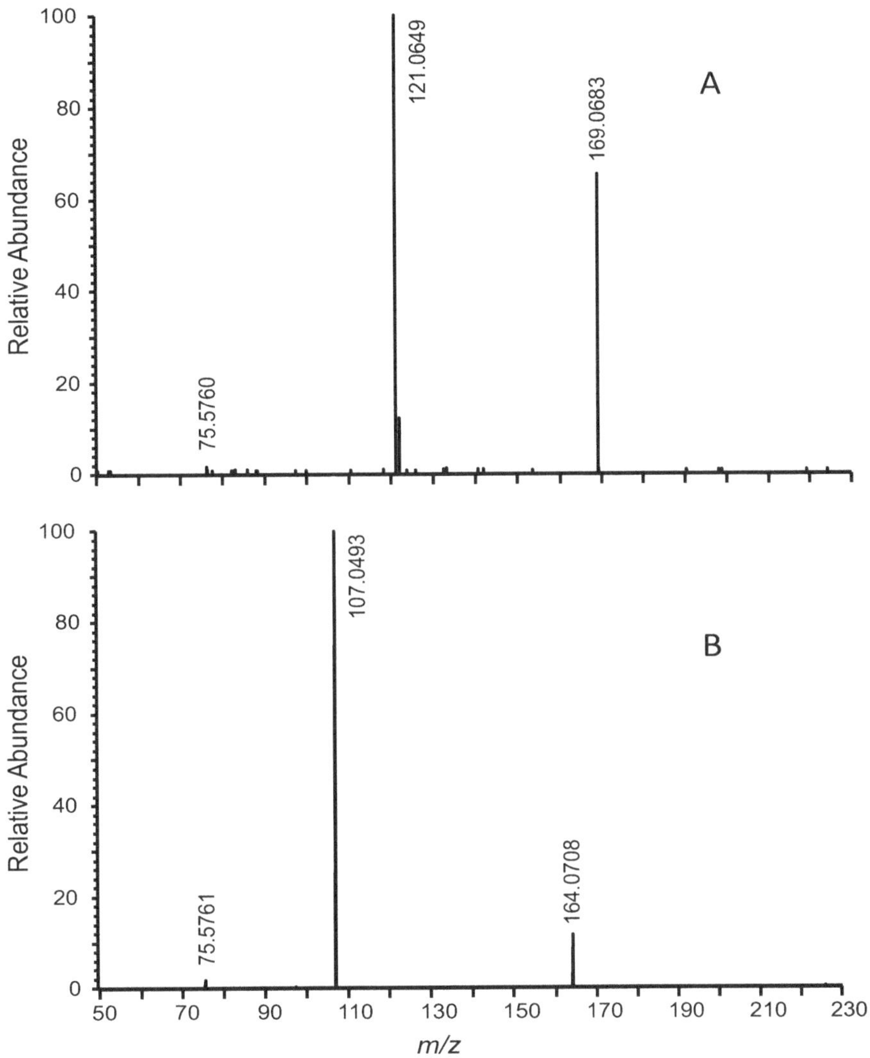

Figure 3. Positive ESI spectra obtained with an NCE 20 eV for (**A**) methiocarb and (**B**) ethiofencarb.

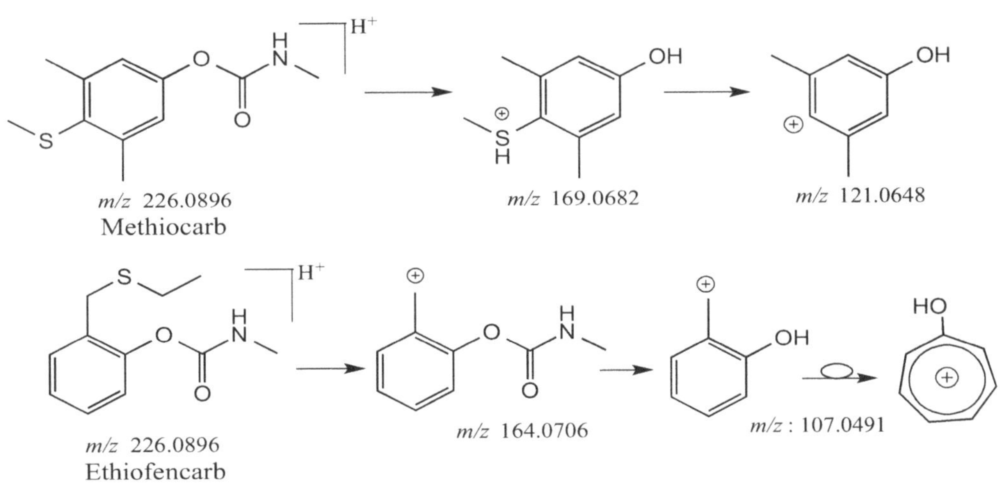

Scheme 3. Proposed pathway for the formation of the differentiation ions for methiocarb and ethiofencarb.

3.3. Uniconazole and Cyproconazole

Uniconazole ((*E*-3*S*)-1-(4-chlorophenyl)-4,4-dimethyl-2-(1,2,4-triazol-1-yl)- 1-penten-3-ol) and cyproconazole (2-(4-chlorophenyl)-3-cyclopropyl-1-(1H-1,2,4-triazol-1-yl)-2-butanol) are triazole fungicides. Uniconazole is not approved for use in the EU, but is used in the USA as plant growth regulator for flowering plants and vegetables (such as tomatoes, peppers, cucumber, and avocado) with the US tolerance of 10 ng/g. Cyproconazole, on the other hand, is approved for use in the USA and EU, and is widely used as a fungicide for cereals, vegetables and fruits, and nuts. Its MRL ranges from 10–1000 ng/g depending on the country and commodity.

In terms of chemical structure, cyproconazole and uniconazole are triazolyl derivatives with a monochlorobenzyl group as shown in Table 1. The most significant fragments for these compounds have been described by Niessen and Correa [10] and fragments include the loss of the triazolyl and the product ion corresponding to its protonated form ($[C_2H_4N_3]^+$). Another significant fragment is the chlorobenzyl ion. Ions like these have been reported as the tropylium cation, but rather high energy is needed to rearrange to this configuration and it formation is questioned [21,22]. Figure 4 shows the spectra for uniconazole and cyproconazole with product ions at *m/z* 70 for the protonated triazolyl cation and *m/z* 125 for the monochlorobenzyl cation. The spectra require 90 eV to produce ions that can be used to differentiate both monochlorobenzyl derivatives. Under this condition, loss of water is not observed, but this loss occurs at lower energy, most likely because the preferred site of protonation is at the triazolyl ring, leading to the elimination of this group (*m/z* 70).

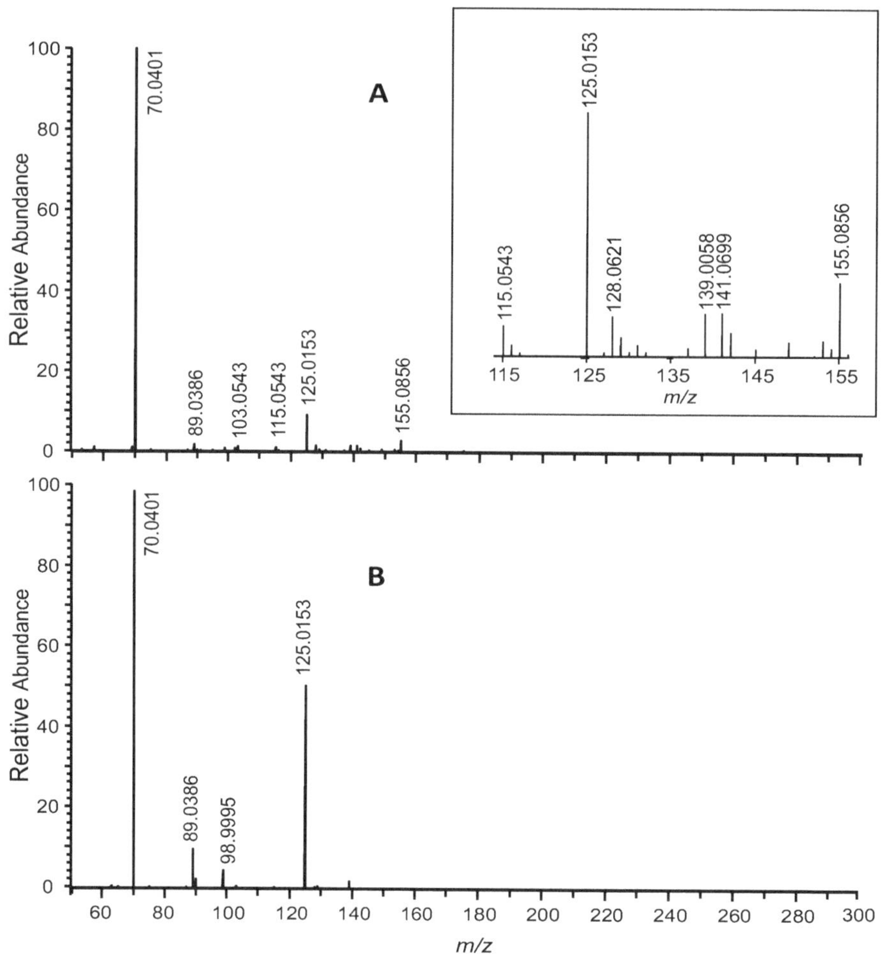

Figure 4. Positive ESI spectra obtained with an NCE 90 eV for: (**A**) uniconazole, with insert expanding the mass range between 115 and 155, and (**B**) cyproconazole.

Both spectra in Figure 4 have the same fragment peaks at different intensities, but uniconazole presented two distinctive product ions at *m/z* 155 and 115 at 70–90 eV that were not found for cyproconazole (Figure 4A insert). On the other hand, cyproconazole did not show any characteristic fragments. Q-Trap isolation and fragmentation of product ion at *m/z* 155 confirmed the subsequent ion at *m/z* 115. The fragmentation pathway required to explain these ions is complex and requires elimination of water, methyl, and triazole groups. The proximity of triazole to the methyl moiety in the terbuthyl group in a 3D model suggests that the methyl can migrate to the triazole and be eliminated as methyl triazole, leading to the ion at *m/z* 191, as shown in Scheme 4. The latter ion is very small and requires less energy to be observed, but seems to lead to the formation of product ion at *m/z* 155 after the loss of HCl. Positive charge in the structure will provide resonance stability to the ion. Loss of 1,2-propadiene from ion at *m/z* 155 produced a product ion at *m/z* 115. These two ions appear to be specific for identification of uniconazole, however, no specific ions were found for cyproconazole to differentiate it from uniconazole.

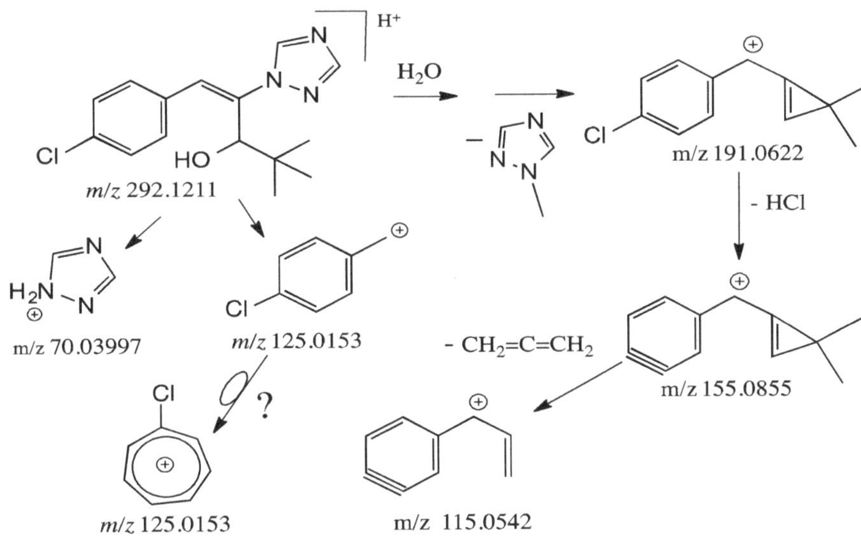

Scheme 4. Proposed pathway for the formation of the differentiation ions for uniconazole.

3.4. *Orbencarb and Thiobencarb*

These two pesticides differ in the position of chlorine group at the benzene ring (Table 1); consequently, no specific ions were observed for their differentiation. However, they form an intense product ion at *m/z* 125 (chlorobenzyl), which after loss of HCl produces *m/z* 89 (data not shown). Due to the ortho position of Cl in orbencarb and para position in thiobencarb, the product ion *m/z* 89 has different relative ion abundance ratios for these pesticides. Additionally, a product ion at *m/z* 100 corresponding to $[(CH_3\text{-}CH_2)_2N\text{-}C{=}O]^+$ is affected by the position of Cl, thus producing different ion intensities. Under controlled conditions, fragmentation of ions at *m/z* 125 and *m/z* 100 as precursors can provide additional evidence for distinction of these pesticides. The difference in intensities depends on collision energy, but when using the Q-Trap instrument, thiobencarb showed intensities of approximately 60% and 25% for ions at *m/z* 89 at *m/z* 100, respectively. The ion at *m/z* 89 is in agreement with a chlorobenzyl ion at *m/z* 125 that did not seem to rearrange into a chlorotropyllium ion (see Scheme 4) [21] because both pesticides would have the same possibility of forming this ion at *m/z* 89.

3.5. *Sebuthylazine and Terbuthylazine*

Sebuthylazine (2-*N*-butan-2-yl-6-chloro-4-*N*-ethyl-1,3,5-triazine-2,4-diamine) and terbuthylazine (2-*N*-tert-butyl-6-chloro-4-*N*-ethyl-1,3,5-triazine-2,4-diamine) are triazine group herbicides, very similar to the well-known herbicide, atrazine. No currently registered pesticide products containing sebuthylazine were found, and it is listed as obsolete in the EU. Terbuthylazine, on the other hand, is widely used in Europe and 45 other countries to control weeds in corn, potatoes, sorghum, pea, bean, grape, fruit, citrus, etc. with MRLs of 50–100 ng/g [15]. It has become a widely used triazine herbicide in Europe since atrazine was discontinued in 2004. Conversely, terbuthylazine is not registered in the USA [16].

Sebuthylazine and terbuthylazine are monochlorotriazines, and other well-known herbicides from this class include atrazine, propazine, and cyanazine. Fragmentation pathway of protonated monochlorotriazines has been previously extensively investigated and described [12,23]. MS/MS ions observed for both pesticides included *m/z* 174, 104, 132, and 96, similar to the ones previously reported, but no distinct product ions for differentiation of these isomers were observed.

4. Conclusions

Analysis of isomeric pesticides requires the identification of product ions that are unique to each isomeric form. The use of HRAM mass spectrometry provided the molecular ion formula that led to the proposal of pathways and structure for these distinctive product ions for pebulate-vernolate, uniconazole-cyproconazole, and methiocarb-ethiofencarb. Thiobencarb and orbencarb can only be differentiated according to the relative intensities of the ions at *m/z* 89 and 100, but for sebuthylazine and terbuthylazine no differentiation could be established. The importance of identifying specific ions and determining their structures provides further support for elimination of false positives, and consequently, avoiding wrongful prosecution by providing further evidence for the analytical results in legal proceedings.

Author Contributions: Conceptualization, A.N. and Y.S.; Methodology, A.N. and Y.S.; Software, A.N.; Validation, A.N., Y.S. and S.J.L.; Investigation, Y.S. and S.J.L.; Writing-Original Draft Preparation, A.N.; Writing-Review & Editing, A.N., Y.S., and S.J.L.

References

1. US EPA Pesticide Industry Sales and Usage, 2008–2012 Market Estimates. Available online: https://www.epa.gov/pesticides/pesticides-industry-sales-and-usage-2008-2012-market-estimates (accessed on 12 September 2018).
2. Winter, C.K. Pesticide tolerances and their relevance as safety standards. *Regul. Toxicol. Pharm.* **1992**, *15*, 137–150. [CrossRef]
3. Kaufmann, A.; Butcher, P.; Maden, K.; Walker, S.; Widmer, M. Comprehensive comparison of liquid chromatography selectivity as provided by two types of liquid chromatography detectors (high resolution mass spectrometry and tandem mass spectrometry): "Where is the crossover point?". *Anal. Chim. Acta* **2010**, *673*, 60–72. [CrossRef] [PubMed]
4. Wang, J.; Chow, W.; Chang, J.; Wong, J.W. Development and validation of a qualitative method for target screening of 448 pesticide residues in fruits and vegetables using UHPLC/ESI Q-Orbitrap based on data-independent acquisition and compound database. *J. Agric. Food Chem.* **2017**, *65*, 473–493. [CrossRef] [PubMed]
5. Volmer, D.A. Prerequisites for supplying complementary high-resolution mass spectrometry data in RCM publications. *Rapid Commun. Mass Spectrom.* **2010**, *24*, 3499–3500. [CrossRef]
6. Bethem, R.; Boison, J.; Gale, J.; Heller, D.; Lehotay, S.; Loo, J.; Musser, S.; Price, P.; Stein, S. Establishing the fitness for purpose of mass spectrometric methods. *J. Am. Soc. Mass Spectrom.* **2003**, *14*, 528–541. [CrossRef]
7. Geis-Asteggiante, L.; Nuñez, A.; Lehotay, S.J.; Lightfield, A.R. Structural characterization of product ions by electrospray ionization and quadrupole time-of-flight mass spectrometry to support regulatory analysis of veterinary drug residues in foods. *Rapid Commun. Mass Spectrom.* **2014**, *28*, 1061–1081. [CrossRef] [PubMed]

8. Nuñez, A.; Lehotay, S.J.; Geis-Asteggiante, L. Structural characterization of product ions by electrospray ionization and quadrupole time-of-flight mass spectrometry to support regulatory analysis of veterinary drug residues in foods. Part 2: Benzimidazoles, nitromidazoles, phenothiazines, and mectins. *Rapid Commun. Mass Spectrom,* **2015**, *29*, 719–729. [CrossRef] [PubMed]
9. Nuñez, A.; Lehotay, S.J.; Lightfield, A.R. Structural characterization of product ions by electrospray ionization and quadrupole time-of-flight mass spectrometry Part 3: Anthelmintics and thyreostats. *Rapid Commun. Mass Spectrom.* **2016**, *30*, 813–822. [CrossRef] [PubMed]
10. Niessen, W.M.A.; Correa, C.R.A. *Interpretation of MS-MS Mass Spectra of Drugs and Pesticides. Wiley Series on Mass Spectrometry*; Desiderio, D.M., Loo, J.A., Eds.; John Wiley and Sons, Inc.: Hoboken, NJ, USA, 2017; pp. 1–388.
11. Pesticide Action Network North Americ. Available online: http://www.pesticideinfo.org (accessed on 5 September 2018).
12. Wang, J.; Yang, C.; Beck, J.; Ghosh, D. *Streamlined Analysis of 400+ Pesticides in a Single Run Using the TSQ Quantum acCess MAX Mass Spectrometer and TraceFinder Software*; Thermo Scientific Application, Thermo Scientific: San Jose, CA, USA, 2016.
13. Marshall, A.G.; Hendrickson, C.L. High-Resolution Mass Spectrometers. *Annu. Rev. Anal. Chem.* **2008**, *1*, 579–599. [CrossRef] [PubMed]
14. Thurman, E.M.; Ferrer, I.; Zavitsanos, P.; Zweigenbaum, J.A. Analysis of isobaric pesticides in pepper with high-resolution liquid chromatography and mass spectrometry: Complementary or Redundant? *J. Agric. Food Chem.* **2013**, *61*, 2340–2347. [CrossRef] [PubMed]
15. Wright, P.; Alex, A.; Pullen, F. Prediction collision-induced dissociation spectra: Semi-empirical calculation as rapid and effective tool in software-aid mass spectral interpretation. *Rapid Commun. Mass Spectrom.* **2014**, *28*, 1127–1143. [CrossRef] [PubMed]
16. National Water Qaulity Assessment Project, Pesticide National Synthesis Project, Estimated Annual Agricutural Pesticide Use. Available online: https://water.usgs.gov/nawqa/pnsp/usage/maps/ (accessed on 5 September 2018).
17. EU Pesticide Database. Available online: http://ec.europa.eu/food/plant/pesticides/eu-pesticides-database/public/ (accessed on 10 September 2018).
18. Staub, R.E.; Spark, S.E.; Quistad, G.B.; Casida, J.E. S-methylation as a bioactivation mechanism for mono- and dithiocarbamate pesticides as aldehyde dehydrogenases inhibitors. *Chem. Res. Toxicol.* **1995**, *8*, 1063–1069. [CrossRef] [PubMed]
19. United States Department of Agriculture, Agricultural Marketing Service, Pesticide Database Program. Available online: https://apps.ams.usda.gov/pdp (accessed on 5 September 2018).
20. Cassady, C.J.; Freiser, B.S.; Russell, D.H. Structural determination of $[C_7H_7O]^+$ ions in the gas pahse by ion cyclotron resonance spectrometry. *Org. Mass Spectrom.* **1983**, *18*, 378–387. [CrossRef]
21. Pico, Y.; la Farre, M.; Soler, C.; Barcelo, D. Identification of unknown pesticides in fruit using ultra-performance liquid chromatography-quadrupole time-of-flight mass spectrometry Imazil as a case study of quantification. *J. Chromatogr. A* **2007**, *1176*, 123–134. [CrossRef] [PubMed]
22. Chai, Y.; Wang, L.; Sun, H.; Guo, C.; Pan, Y. Gas-pase chemistry of benzyl cations in dissociation of *N*-benzyliminium ios studied by mass spectrometry. *J. Am. Soc. Mass Spectrom.* **2012**, *23*, 823–833. [CrossRef] [PubMed]
23. Nelieu, S.; Stobiecki, M.; Kerhoas, L.; Einhorn, J. Screening and characterization of atrazine metabolites by liquid chromatography/tandem mass spectrometry. *Rapid Commun. Mass Spectrom.* **1994**, *8*, 945–952. [CrossRef]

10

Tissue Distribution and Elimination of Ciguatoxins in *Tridacna maxima* (*Tridacnidae*, Bivalvia) Fed *Gambierdiscus polynesiensis*

Mélanie Roué [1,*,†], Hélène Taiana Darius [2,†], André Ung [2], Jérôme Viallon [2], Manoella Sibat [3], Philipp Hess [3], Zouher Amzil [3] and Mireille Chinain [2]

1 Institut de Recherche pour le Développement (IRD)—UMR 241 EIO, PO box 53267, 98716 Pirae, Tahiti, French Polynesia
2 Institut Louis Malardé (ILM), Laboratory of Toxic Microalgae—UMR 241-EIO, PO Box 30, 98713 Papeete, Tahiti, French Polynesia; tdarius@ilm.pf (H.T.D.); aung@ilm.pf (A.U.); jviallon@ilm.pf (J.V.); mchinain@ilm.pf (M.C.)
3 IFREMER, Phycotoxins Laboratory, F-44311 Nantes CEDEX, France; manoella.sibat@ifremer.fr (M.S.); philipp.hess@ifremer.fr (P.H.); zouher.amzil@ifremer.fr (Z.A.)
* Correspondence: melanie.roue@ird.fr
† These authors contributed equally to this work.

Abstract: Ciguatera is a foodborne disease caused by the consumption of seafood contaminated with ciguatoxins (CTXs). Ciguatera-like poisoning events involving giant clams (*Tridacna maxima*) are reported occasionally from Pacific islands communities. The present study aimed at providing insights into CTXs tissue distribution and detoxification rate in giant clams exposed to toxic cells of *Gambierdiscus polynesiensis*, in the framework of seafood safety assessment. In a first experiment, three groups of tissue (viscera, flesh and mantle) were dissected from exposed individuals, and analyzed for their toxicity using the neuroblastoma cell-based assay (CBA-N2a) and liquid chromatography-tandem mass spectrometry (LC-MS/MS) analyses. The viscera, flesh, and mantle were shown to retain 65%, 25%, and 10% of the total toxin burden, respectively. All tissues reached levels above the safety limit recommended for human consumption, suggesting that evisceration alone, a practice widely used among local populations, is not enough to ensure seafood safety. In a second experiment, the toxin content in contaminated giant clams was followed at different time points (0, 2, 4, and 6 days post-exposure). Observations suggest that no toxin elimination is visible in *T. maxima* throughout 6 days of detoxification.

Keywords: giant clams; ex situ exposure to toxic algae; ciguatoxins; *Gambierdiscus polynesiensis*; anatomical distribution; toxin elimination; CBA-N2a; LC-MS/MS

Key Contribution: Giant clams were experimentally exposed to highly toxic cells of *G. polynesiensis*. Ciguatoxins were preferentially accumulated in viscera, although toxin contents monitored in two other tissues, i.e. mantle and flesh, also exceeded the recommended safety limit. No toxin elimination was observed over a period of 6 days.

---

1. Introduction

Pacific Island Countries and Territories (PICTs) communities are strongly dependent on marine products for their subsistence and income [1,2]. Thus, seafood contamination by marine biotoxins represents a serious impediment to the economy of PICTs populations. Although fish are the most widely consumed marine products in PICTs, several species of marine invertebrates, including giant clams, are also part of the diet of local populations. As an example, reported catches of giant clam meat reached tens of tons annually during the 1970s and 1980s in Tonga, Fiji, and French Polynesia [3]. While similar levels of exploitation were still reported in the early 2000s from small and isolated islands of French Polynesia [4], the local market of giant clam meat from the East Tuamotu and Australes archipelagos now exceeds 70 tons per year [5]. Giant clams grow in quite shallow depths, so are easily harvested by hand. Even though some people still consume it, the hepatopancreas is most often eviscerated before consumption in local communities, probably due to its visual appearance and bitter taste, but also as a precaution (in the same way as the liver of fish is typically removed).

Ciguatera fish poisoning (CFP) is a seafood-borne illness classically caused by the consumption of tropical coral reef fish contaminated with ciguatoxins (CTXs), polyether neurotoxins produced by dinoflagellates of the genus *Gambierdiscus* [6]. The clinical manifestations of the disease classically involve gastrointestinal disorders, cold allodynia, itching, paresthesia, asthenia, muscular pain, and dizziness, together with other neurological and cardio-vascular symptoms [7]. Although coral reef fish are usually regarded as the primary vectors of ciguatera, atypical ciguatera-like poisoning events following the consumption of giant clams (*Tridacna maxima*), marine gastropods (*Tectus niloticus*), and sea urchins (*Tripneustes gratilla*) are also occasionally reported from PICTs, most notably from French Polynesia, New Caledonia, Cook Islands, and the Republic of Vanuatu [8–14]. Besides the symptoms typical of CFP, atypical signs were also observed, such as the rapid onset of the disease (occurring within minutes), alteration of taste, burning sensation on the tongue and throat, as well as paralysis. In the case of poisoning events involving *T. maxima*, toxicological analyses using the receptor binding assay (RBA) and the neuroblastoma cell-based assay (CBA-N2a) confirmed the presence of liposoluble compounds acting on voltage-gated sodium channels (VGSCs). These toxins also displayed a mode of action and chemical properties very similar to those of CTXs [11,15,16]. Unfortunately, confirmation of the exact nature of these compounds could not be achieved from wild samples.

Recently, Roué et al. (2016) [17] provided confirmation that giant clams should be regarded as potential vectors of ciguatera, particularly in fishing areas frequently exposed to *Gambierdiscus* blooms. These authors showed that *T. maxima* fed highly toxic cells of *Gambierdiscus polynesiensis* were able to bio-accumulate algal CTXs, at concentrations well above the safety limit commonly recommended for human consumption in the Pacific region. However, since toxin analyses were conducted on extracts prepared from whole specimens, it was not possible to tell whether CTXs were differentially concentrated in the tissues of contaminated animals. Likewise, the study by Roué et al. (2016) [17] did not provide any information about the detoxification time for CTXs in giant clams. Indeed, such information has significant public health implications and may help identify practical measures to limit the risk of poisoning among consumers, e.g., by simply removing the contaminated tissues and/or issuing a quarantine period within distribution channels whenever necessary.

In this context, two additional series of ex situ exposures to toxic *Gambierdiscus* cells were conducted to address the following issues: (i) assess the differential tissue distribution of CTXs in contaminated individuals; and (ii) assess whether any elimination of CTXs is visible from contaminated giant clams over a short-term period of one week, assuming that these bivalves belong to the group of "rapid detoxifiers" [18]. To achieve this, giant clams were exposed to in vitro cultures of the highly toxic strain TB92—*G. polynesiensis* in two independent experiments. In the first experiment ("tissue distribution study"), the exposed specimens were dissected into three tissues: viscera, mantle and flesh (i.e., rest of the body tissues, including gills) which were further tested for their toxicity. In the second experiment ("detoxification study"), contaminated individuals were sacrificed at different time intervals (0, 2, 4, and 6 days) following their transfer into clean water, and tested for their toxicity. Detection of CTXs in samples was achieved using the CBA-N2a and liquid chromatography—tandem mass spectrometry (LC-MS/MS) analyses.

2. Results

2.1. Tissue Distribution Study

Giant clams (n = 3 per tank) were placed in four independent tanks and exposed to *G. polynesiensis* cells (tank No. 1, 2, and 3) or kept in clean seawater (control tank). After two days of exposure, giant clams were sacrificed and dissected into three tissues: viscera (consisting mainly of the hepatopancreas), mantle and flesh (i.e., rest of the body tissues, including gills). Then, each group of tissues was pooled for each tank before further analyses.

Extracts obtained from viscera, mantle and flesh of control giant clams showed no toxicity as confirmed by CBA-N2a and LC-MS/MS analyses (data not shown).

Conversely, all extracts corresponding to viscera (n = 3), mantle (n = 3) and flesh (n = 3) of exposed giant clams were found toxic using CBA-N2a. As shown in Figure 1, corresponding to the mean of the three tanks for each tissue, a sigmoidal dose-response curve was obtained in OV^+ conditions (i.e., in the presence of ouabain and veratridine) whereas no cytotoxic effects were observed on Neuro-2a cells in OV^- conditions (i.e., in absence of ouabain and veratridine), a response typical of the presence of CTXs in these samples. The mean EC_{50} values (n = 3 tanks) were 0.24 ± 0.02, 1.21 ± 0.29, and 1.65 ± 0.52 µg tissue/µL for viscera, flesh and mantle, respectively (Figure 1).

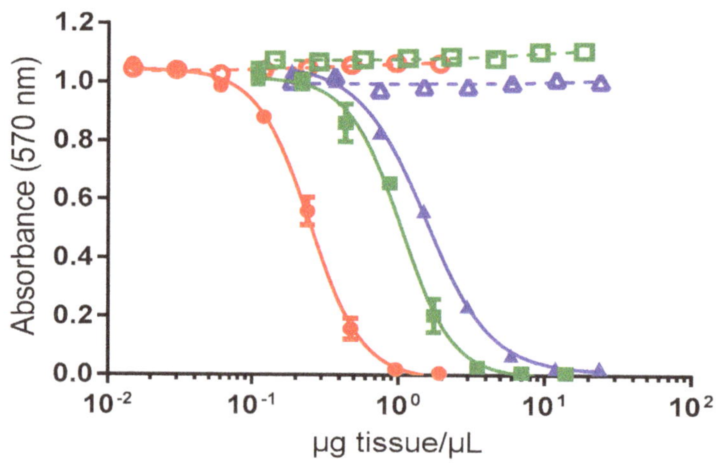

Figure 1. Dose-response curves of Neuro-2a cells in absence (open symbols) and presence (solid symbols) of ouabain and veratridine (OV^- and OV^+ conditions, respectively), when exposed to different concentrations of extracts obtained from viscera (○/●), flesh (□/■) and mantle (△/▲) of giant clams exposed to *G. polynesiensis* cells (TB92 strain). Data represent the mean ± SD of the three experimental tanks, each tissue tested in three independent CBA-N2a experiments (each run in triplicates).

Absolute toxicities of tissues (i.e., toxin contents in each tissue) from giant clams from each of the three tanks indicated that viscera was 4- to 6-fold more toxic than flesh and 4- to 9-fold more toxic than mantle (Table 1), with mean absolute toxicities (n = 3 tanks) of 13.0 ± 1.1, 2.7 ± 0.6, and 2.0 ± 0.6 ng P-CTX3C equiv./g tissue, respectively.

Regarding the relative weight contribution of each tissue to the whole body weight of giant clams, flesh was the most abundant tissue (around 50%), followed by mantle and viscera (around 25% each) (Table 1). The tissue distribution of toxins (as expressed per g of whole body) was then calculated as a function of absolute toxicity and relative weight contribution (Table 1). For each of the three tanks, despite the limited contribution of viscera to the total body mass of giant clams, the toxin distribution in this tissue was 2- to 3-fold and 5- to 8-fold higher than what is observed in the flesh and the mantle, respectively (Table 1). With a mean toxin distribution (n = 3 tanks) of 3.3 ± 0.3, 1.3 ± 0.3

and 0.5 ± 0.1 ng P-CTX3C equiv./g whole body for viscera, flesh and mantle, respectively (Figure 2a), the contribution of viscera to the whole toxin content averaged 65%, followed by the flesh and the mantle with 25% and 10%, respectively (Figure 2b).

Table 1. Absolute toxicities and tissue distribution of toxins in giant clams exposed to *G. polynesiensis* cells (TB92 strain) based on CBA-N2a data. In each tank, each group of tissues was pooled from three giant clams.

Tank	Tissue	Absolute Toxicity of Tissue [1]	Tissue Wet Weight [g] (Relative Weight Contribution, %)	Tissue Distribution of Toxins [2]	Tissue Contribution to the Total Body Toxin Content
1	Viscera	12.2 ± 0.4	36 (26%)	3.2 ± 0.1	59%
	Flesh	3.0 ± 0.2	75 (54%)	1.6 ± 0.1	31%
	Mantle	2.6 ± 0.2	28.5 (20%)	0.5 ± 0.0	10%
2	Viscera	13.1 ± 0.8	37.5 (25%)	3.3 ± 0.2	71%
	Flesh	2.0 ± 0.2	72 (48%)	0.9 ± 0.1	20%
	Mantle	1.4 ± 0.2	39.5 (27%)	0.4 ± 0.1	9%
3	Viscera	13.6 ± 1.4	33.5 (26%)	3.5 ± 0.4	64%
	Flesh	3.0 ± 0.1	58.5 (45%)	1.4 ± 0.0	25%
	Mantle	2.1 ± 0.3	38 (29%)	0.6 ± 0.1	11%

[1] The absolute toxicity of each tissue, expressed in ng P-CTX3C equiv./g tissue, was measured directly from the extracts of each tissue using the following formula: P-CTX3C EC_{50}/sample EC_{50}; [2] The tissue distribution of toxins, expressed in ng P-CTX3C equiv./g whole body, was calculated using the following formula: absolute toxicity × relative weight contribution.

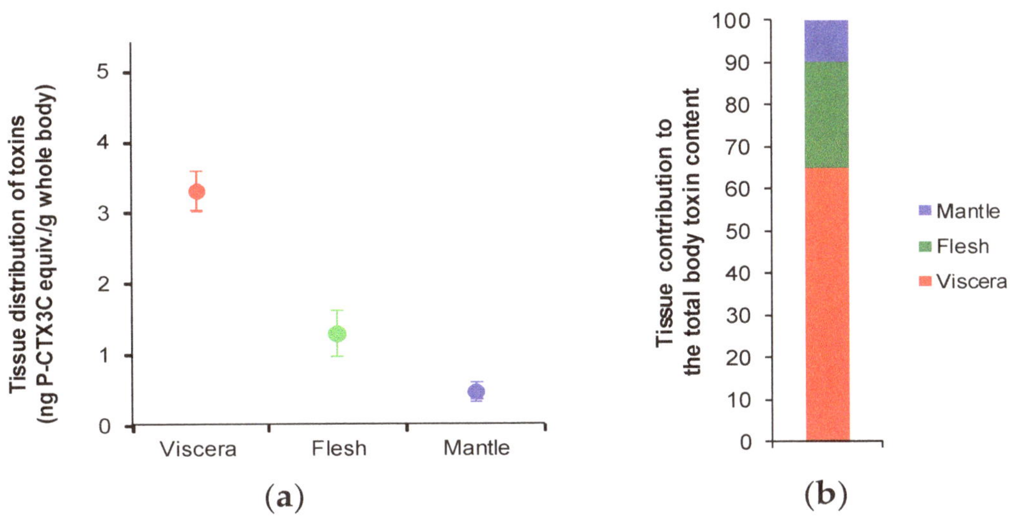

Figure 2. (**a**) Tissue distribution of toxins and (**b**) tissue contribution to the total body toxin content in giant clams exposed to *G. polynesiensis* cells (TB92 strain), as assessed by CBA-N2a data. Data represent the mean ± SD of the three experimental tanks, each tissue tested in three independent CBA-N2a experiments (each run in triplicates).

The presence of CTXs in toxic fractions was further confirmed by LC-MS/MS analyses, with P-CTX3B (RT 11.2 min) and P-CTX3C (RT 11.5 min) present in similar proportions (≈1:1) in all tissues (i.e., mantle, viscera or flesh) (Figure 3). The toxin content of viscera was higher than in the flesh and the mantle (where CTXs were detected at the limit of quantification) (Figure 3), confirming that viscera was the main storage tissue for CTXs compared to the other two tissue types. It should be noted that LC-MS/MS data in this study are mostly semi-quantitative as concentrations were low and no certified standards exist for any of the ciguatoxins.

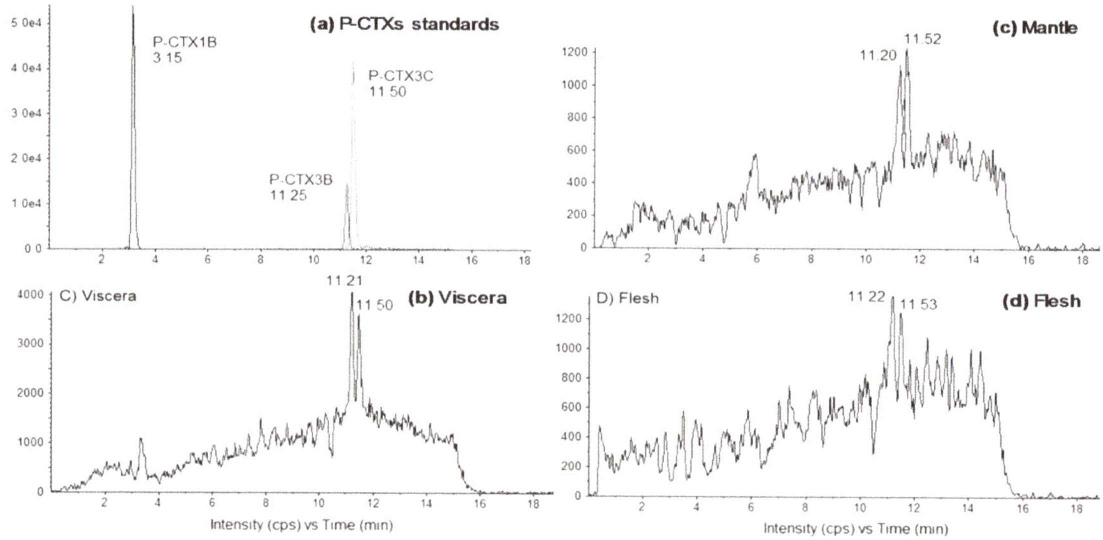

Figure 3. Liquid chromatography-tandem mass spectrometry (LC-MS/MS) chromatograms of (**a**) P-CTXs standards, and fractions obtained from (**b**) viscera, (**c**) mantle and (**d**) flesh of giant clams exposed to *G. polynesiensis* cells (TB92 strain). Chromatograms were acquired following the procedure described in Section 4.5, in positive multi-reaction monitoring mode. For (**d**), S/N < 3.

2.2. Detoxification Study

As in the first experiment, giant clams ($n = 3$ per tank) were placed in five independent tanks and exposed to *G. polynesiensis* cells ($n = 4$ tanks) or kept in clean seawater (control tank). After two days of exposure to *G. polynesiensis* cells, the seawater was renewed for the detoxification study. Each exposed tank was then randomly assigned a time point (0, 2, 4, and 6 days post-exposure) at which point the giant clams were sacrificed. After sacrifice, each whole animal was independently analyzed for its toxicity using CBA-N2a.

Fractions obtained from control animals showed no toxicity using CBA-N2a (data not shown). Conversely, all fractions obtained from contaminated giant clams after 0 ($n = 3$), 2 ($n = 3$), 4 ($n = 3$), and 6 ($n = 3$) days of detoxification were found toxic, with a sigmoidal dose-response curve typical of the presence of CTXs (data not shown). The mean toxin contents ($n = 3$ giant clams) were 1.8 $\pm$ 0.4, 2.5 $\pm$ 0.7, 2.0 $\pm$ 1.1, and 2.1 $\pm$ 0.7 ng P-CTX3C equiv./g whole body for 0, 2, 4, and 6 days of detoxification, respectively (Table 2). Thus, no significant elimination of toxins had occurred (ANOVA, $p > 0.05$) and whole giant clams contained a mean of 114 $\pm$ 54 ng P-CTX-3C equiv./giant clam all along the course of the detoxification period (Figure 4).

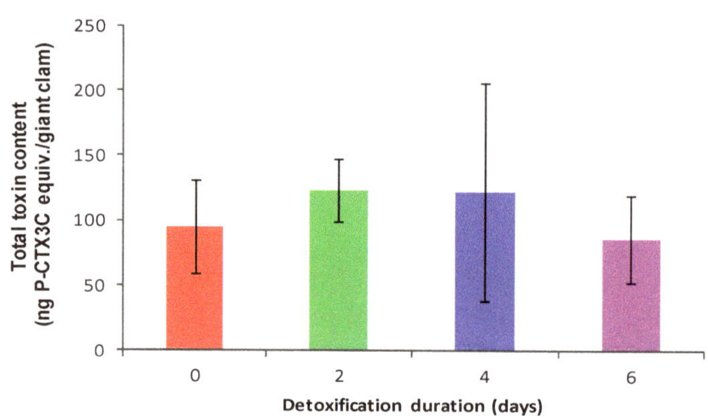

Figure 4. Mean total toxin burdens in contaminated giant clams, as assessed by CBA-N2a, at 0, 2, 4, and 6 days post-exposure. Data represent the mean $\pm$ SD of three individual giant clams, each of them tested in three independent CBA-N2a experiments (each run in triplicates).

Table 2. Toxin contents in contaminated giant clams after 0, 2, 4, and 6 days of detoxification, based on CBA-N2a data.

Days of Detoxification	Giant Clam Individual	Wet Weight [g]	Toxin Content [1]
0	1	39.9	1.4 ± 0.2
	2	57.1	1.9 ± 0.1
	3	60.8	2.2 ± 0.4
	Average of 1–3	53 ± 11	1.8 ± 0.4
2	1	39.0	3.3 ± 0.4
	2	40.8	2.3 ± 0.1
	3	67.2	1.9 ± 0.4
	Average of 1–3	49 ± 16	2.5 ± 0.7
4	1	71.0	3.3 ± 0.4
	2	54.3	1.6 ± 0.2
	3	57.6	1.1 ± 0.2
	Average of 1–3	61 ± 9	2.0 ± 1.1
6	1	46.6	1.4 ± 0.1
	2	43.9	2.8 ± 0.8
	3	32.1	2.0 ± 0.3
	Average of 1–3	41 ± 8	2.1 ± 0.7

[1] The toxin content, expressed in ng P-CTX3C equiv./g whole body, was measured directly from the extracts of each whole individual using the following formula: P-CTX3C EC_{50}/sample EC_{50}.

Each tank of three giant clams exposed to *G. polynesiensis* cells received a total toxin load of 17.25 μg P-CTX3C equiv. As shown in Figure 4, at the end of the exposure period (i.e., 0 days of detoxification), each giant clam had accumulated around 95 ng P-CTX3C equiv., consequently, it was estimated that each giant clam had retained only 0.6% of the total toxin load administered in each tank. This percentage stayed stable throughout the full detoxification experiment.

3. Discussion

3.1. Anatomical Distribution of CTXs in Toxic Giant Clams

Results of CBA-N2a analyses indicated that all the tissues (i.e., viscera, mantle, and rest of the flesh) of giant clams fed toxic *G. polynesiensis* cells, were able to retain CTXs. These findings confirmed previous observations on the ability of giant clams to bio-accumulate CTXs in their tissues [17]. In the present study, CTXs were preferentially concentrated in the viscera (mostly composed by the hepatopancreas), which was 2.5 and 6.7-fold more toxic than the flesh (i.e., the rest of body tissues, including gills) and mantle, respectively. This preferential accumulation of toxins in the viscera is consistent with a digestive uptake route of CTXs in this bivalve mollusk, as previously hypothesized by Roué et al. (2016) [17]. Indeed, following the filtration and ingestion of *G. polynesiensis* cells by *T. maxima*, the subsequent distribution of CTXs to non-visceral tissues could well explain the lower toxicities observed in the flesh and mantle. This differential accumulation of toxins in *T. maxima* tissues is coherent with the chemical nature of CTXs since lipophilic toxins are often concentrated in digestive glands; for instance, CTXs are known to be preferentially accumulated in the liver of fish [19,20]. Our results are also in agreement with previous studies conducted on other bivalve species, which concluded that in filter-feeding marine organisms exposed to harmful algal blooms, most of the toxin uptake accumulates in the viscera (most notably the digestive gland) despite the limited contribution of this tissue to the animal total body mass [18,21–30]. As an example, in the Japanese scallop *Patinopecten yessoensis* exposed to cultured *Dinophysis fortii* cells, dinophysistoxins (DTXs) and pectenotoxins (PTXs) were almost exclusively found in the digestive gland with only low levels being detected in the gill, mantle, gonad, and adductor muscle [27]. However, the preferential accumulation of toxins in non-digestive and non-visceral tissues has also been documented in a limited number of studies [28,31–33]. For example, the highest concentrations of okadaic acid-group (OA) toxins were detected in the adductor muscle of *M. chilensis* collected in areas where *Alexandrium catenella*, *Dinophysis acuminata* and *Dinophysis acuta* cells were observed [28]. Moreover, even if up to 90% of the azaspiracid (AZA) burden was found in digestive gland of mussels exposed to *Azadinium spinosum* cells, Jauffrais et al. (2012) [30] also showed that different algal densities or dissolved toxins may contribute to accumulation via different mechanisms, e.g., direct accumulation in gills or other tissues. Thus, the fact that up to approximately 40% of the total body burden was found in non-visceral tissues in this study may indicate that direct accumulation in mantle may also take place in giant clams.

In terms of public health implications, the distribution of toxins found in the viscera of toxic giant clams was 167-fold higher than the safety limit commonly recommended for human consumption, i.e., 0.01 ng P-CTX1B/g or 0.02 ng P-CTX3C/g [34,35]. Likewise, the flesh and mantle in toxic individuals also contained toxin levels well above this safety limit, i.e., 66 and 25-fold higher, respectively. Such observations clearly show that the common practice of removing the animal hepatopancreas before consumption of giant clams, will certainly contribute to reduce the amount of toxins ingested, but does not guarantee an effective protection of consumers as ca. 30–40% of the total toxin burden remains in non-visceral tissues. Similar conclusions apply to two bivalve mollusks, *V. antiqua* and *G. solida,* for which only the foot is consumed. Indeed, in their study, García et al. (2015) [28] showed that in addition to the digestive gland, foot also accumulates significant levels of toxins in these two species, and could thus be highly harmful to consumers. In contrast, the amount of DTXs detected in the adductor muscle of the Japanese scallops *P. yessoensis* was extremely low, confirming the validity of the practice of evisceration implemented in Japan since 1980 [27]. Still, poor dissection practices may also lead to contamination of non-visceral tissues as shown for domoic acid in scallops [36] and therefore, great care should be taken with evisceration as a means of toxin-reduction.

The TB92—*G. polynesiensis* strain is known to produce multiple P-CTX congeners, i.e., P-CTX3B, P-CTX3C, P-CTX3C/3B analogs, P-CTX4A, P-CTX4B, and M-seco-P-CTX4A [37], as confirmed by LC-MS/MS analyses performed on *Gambierdiscus* cultures used for exposure experiments [17]. Here, LC-MS/MS data indicated that P-CTX3B and P-CTX3C, present in almost similar proportions (assuming the same response factor in LC-MS/MS for P-CTX3B and P-CTX3C), were the only known analogs detected in all tissue samples, even though the potential presence of other CTX congeners at concentrations below the limit of detection of the technique cannot be completely ruled out. Interestingly, P-CTX3B and P-CTX3C were also among the major congeners detected in *Tectus niloticus* (Gastropod) and *Tripneustes gratilla* (Echinoid) specimens involved in CFP poisoning events in French Polynesia [13,14], suggesting that both congeners are preferentially bio-accumulated by marine invertebrates.

3.2. *Elimination of CTXs from T. maxima*

Results of CBA-N2a analyses showed that following the transfer of contaminated giant clams into clean seawater, the amount of toxins detected in *T. maxima* at 0, 2, 4, and 6 days post-exposure did not differ significantly. Moreover, giant clams were still not edible after one week in clean seawater (with a toxin content 105-fold higher than the safety limit commonly recommended for human consumption), suggesting that the time lime for toxin detoxification below the safety limit commonly recommended for human consumption in giant clams may take weeks or even months. These results are consistent with field studies conducted on other marine invertebrates species also involved in CFP poisoning events in French Polynesia, which indicated that *Tectus niloticus* (Gastropod) and *Tripneustes gratilla* (Echinoid) have a slow detoxification rate for CTXs. Indeed, based on CBA-N2a results, a 19-fold and 64-fold decrease in the overall ciguatoxicity of *Tectus niloticus* and *Tripneustes gratilla* samples, respectively, was observed over a two-year period, but toxin concentrations were still consistently above the safety limit recommended for human consumption [13,14]. Of note, the authors also underlined that these findings may be confounded by the additional accumulation of CTXs over the study period, even in the presence of low cell numbers of *Gambierdiscus*.

Elimination of toxins is mainly affected by their chemical properties. In general, lipophilic toxins are retained longer than the hydrophilic toxins. In bivalve mollusks, this elimination rate has been shown to also vary according to the species involved, taking from days up to several months, depending on initial concentration and environmental conditions [38]. Classically, bivalve species fall into two general categories in terms of their detoxification capacity: (i) rapid to moderate detoxifiers which eliminate toxins in a few days; and (ii) slow detoxifiers which need several months or even years to eliminate toxins [18,31,33,39–41]. For example, the blue mussel *Mytilus edulis* eliminate PSTs in less than 10 days whereas the butter clam *Saxidomus giganteus* need one to four months to depurate the same toxins [18]. In the same way, the clam *Perna viridis* showed a detoxification time for brevetoxins (PbTxs) significantly higher (above 4–5 months) than the oyster *Crassostrea virginica* and the clam *Mercenaria* after a bloom of *Karenia brevis* [41]. Finally, DA detoxification time has been shown to be species-specific and to have a wide-ranging variability, since most mussels depurate DA very quickly [39] whereas the razor clam *Siliqua patula* or the scallop *P. maximus* have much slower detoxification kinetics [31,40]. Results in the present study suggest that *T. maxima* falls into the slow detoxifiers' category since no significant toxin elimination was visible throughout one week of detoxification; however, further experiments with longer detoxification period will be necessary to firmly conclude this. Furthermore, as for other marine invertebrates [13,14], it would be interesting to study bioaccumulation and elimination rates of CTXs in wild giant clams.

At the end of the exposure period, each giant clam had retained only 0.6% of the toxin burden administered to each tank, corresponding to a total accumulation of 1.8% for each tank containing three giant clams, a result coherent with the study by Roué et al. (2016) [17] which showed an accumulation of around 3% of the toxins in the same experimental conditions. This low toxin uptake rate observed in giant clams is not surprising. For example, a study has shown that the absorption efficiency of DTX1 by the digestive gland of *P. yessoensis* was estimated at less than 3% of the total amount of DTX1 given to the scallops that were fed *D. fortii* cells [42]. Likewise, the bay scallops *Argopecten irradians* exposed to cultured cells of *Prorocentrum lima* showed a toxin assimilation efficiency in their tissues of less than 1% [43]. Interestingly, a biphasic detoxification kinetic is often observed in bivalves, consisting in an initial, more rapid, detoxification phase and a subsequent slower phase of toxin elimination [18,25,30,44,45]. It has been suggested that the initial rapid loss of toxin corresponds to clearance of unassimilated toxins from undigested cells or in dissolved form in the gut lumen, whereas the second slower phase represents the release of assimilated toxins either bound to the digestive gland or incorporated within other tissues [21,25,30,46,47]. In the present study, one hypothesis could be that giant clams could also have a biphasic detoxification kinetic, and that the rapid detoxification of unassimilated CTXs in contaminated animals actually occurred during the exposure period, and that these latter were in the slow phase of toxin elimination during the course of the detoxification study. Furthermore, another hypothesis could be that this low toxin uptake is in fact due to a low ingestion of *Gambierdiscus* cells by giant clams. In any case, further experiments with a longer exposure time may be necessary to better reflect conditions in the natural environment, and should be followed by a longer detoxification period and search for toxins in feces and system circulating-water as well as quantification of remaining *Gambierdiscus* cells in experimental tanks, in order to clarify actual accumulation and detoxification rates in giant clams.

4. Materials and Methods

4.1. Biological Material

4.1.1. Giant Clams

Giant clams (*T. maxima*) used in this study were purchased from an aquaculture farm in Tahiti (French Polynesia). For the tissue distribution study, giant clams (n = 12) were 15.3 ± 1.1 cm long, had a mean shell height of 8.4 ± 1.4 cm and provided a mean flesh wet weight of 60.6 ± 16.6 g. For the detoxification study, giant clams (n = 15) were 17.1 ± 1.6 cm long, had a mean shell height of 9.9 ± 1.4 cm and gave a mean flesh wet weight of 54.4 ± 14.2 g.

4.1.2. Cultures of *Gambierdiscus polynesiensis*

A highly toxic strain (TB92—*G. polynesiensis*) available from the algal collection of Institut Louis Malardé was used for the ex situ exposure experiments. TB92 cultures were obtained as previously described in Roué et al. (2016) [17]. Cultures were harvested in their late exponential/early stationary growth phase (i.e., 28-days post-inoculation) when cells exhibit highest CTX levels corresponding to an average toxicity of 5.83 $\pm$ 0.85 pg P-CTX3C equiv./cell.

4.2. *Ex Situ Exposure Experiments*

The experimental set-up was similar to the one previously described in Roué et al. (2016) [17]. Experiments were conducted in a closed environment, in tanks containing 20 L of seawater with a salinity value of 37. The temperature and the percentage of dissolved oxygen were stabilized at around 28.8 °C and 8 mg/L, respectively, and controlled daily. The light regime followed a 12:12 h (light:dark) photoperiod with an average irradiance of about 50 μmol photons/m/s of light (daylight fluorescent tubes). Each tank was equipped with a pump set at a flow rate of 200 L/h, in order to favor the suspension and/or dissemination of *Gambierdiscus* cells in the surrounding environment of giant clams. Three giant clams were placed in each experimental tank and acclimated to laboratory conditions during three days prior to exposure. Two independent series of experiments were conducted, in order to assess both the distribution of CTXs in three tissue types and the detoxification rate of CTXs from giant clams, respectively.

4.2.1. Tissue Distribution Study

In three tanks, a cell dose of 150,000 cells was administered 20 times to giant clams over a period of 48 h, to reach a total of 3×10^6 *Gambierdiscus* cells per tank, corresponding to an overall toxin load of 17.25 μg P-CTX3C equiv., which is representative of what can be found in natural blooms, as previously discussed in Roué et al. (2016) [17]. A fourth tank containing animals kept in the same laboratory conditions as exposed animals served as control. The water was not replaced in the course of the exposure period. No mortality was observed during the acclimation step, nor in the course of the exposure experiment.

At the end of the exposure period, each individual was collected and immediately sacrificed. The whole meat was removed from the shell and thoroughly rinsed twice in 0.5 L seawater to remove all trace of incubation water or intervalvar liquid potentially contaminated with *Gambierdiscus* cells and/or dissolved CTXs. Each giant clam sample was further dissected into three tissues: viscera (consisting mainly of the hepatopancreas), mantle and flesh (i.e., rest of the body tissues, including gills). Each group of tissue corresponding to the three giant clams from a same experimental tank was then pooled, ground and stored at −20 °C.

4.2.2. Detoxification Study

Four tanks received each a total cell load of 3×10^6 cells throughout a 48 h exposure period, corresponding to an overall toxin load of 17.25 μg P-CTX3C equiv., in the same conditions that for the tissue distribution study. A fifth tank containing animals maintained in the same laboratory conditions as exposed animals served as control. Following toxic exposure, seawater in each of the five tanks was entirely renewed by clean seawater in order to remove any remaining *Gambierdiscus* cells (data not available) and/or dissolved CTXs. In order, to assess whether any elimination of CTXs is visible from contaminated giant clams over a short-term period of one week, assuming that these bivalves belong to the group of "rapid detoxifiers" [18], each exposed tank was then randomly assigned a time point (0, 2, 4, and 6 days post-exposure) at which to sacrifice the three individuals from the respective tank. For each animal, the whole meat was removed from the shell and thoroughly rinsed twice in 0.5 L seawater. Each sample was then ground separately and stored at −20 °C.

4.3. Toxin Extration

Samples were extracted according to the procedure previously described in Roué et al. (2016) [17]. Briefly, 10 g of a given tissue (for tissue distribution study) or 10 g of the whole giant clam body (for detoxification study) were extracted in pure methanol (MeOH), followed by 50% aqueous MeOH. The resulting dried extract was partitioned between dichloromethane (CH_2Cl_2) and 60% aqueous MeOH. The CH_2Cl_2 phase, likely to contain CTXs, was dried under vacuum and defatted by a second solvent partition using cyclohexane and aqueous MeOH. The methanolic fraction was recovered and further purified using C_{18} Sep-Pak cartridges (360 mg, Waters®). The resulting 90% aqueous methanol fraction, likely to contain the majority of CTXs, was dried in a SpeedVac concentrator and stored at +4 °C until CBA-N2a and LC-MS/MS analyses.

4.4. Neuroblastoma Cell-Based Assay (CBA-N2a)

CBA-N2a analyses were conducted following the procedure previously described in Roué et al. (2016) [17] as adapted from Manger et al. (1993) [48]. All dry fractions were carefully weighed and re-suspended in methanol to reach a final concentration of 1 mg of dry extract/100 μL prior to CBA-N2a assays. Fractions were tested at a concentration range of 5–9524 pg/μL in eight distinct concentrations and three independent experiments (each run in triplicate). Absorbance data were fitted to a sigmoidal dose-response curve (variable slope) based on the four-parameter logistic model (4PL) allowing the calculation of EC_{50} values using Prism v6.0.7 software (GraphPad, San Diego, CA, USA). The half maximal effective concentration values (EC_{50}) ± standard deviations (SD) were expressed in g tissue equiv./μL or g whole body equiv./μL for tissue distribution and detoxification studies, respectively. Toxin contents (T) were then estimated using the following formula T = (P-CTX3C EC_{50}/sample EC_{50}) and expressed in ng P-CTX3C equiv./g tissue or ng P-CTX3C equiv./g whole body for tissue distribution and detoxification studies, respectively. According to Bricelj and Shumway (1998) [18], the contribution of a tissue to the total toxin body burden is a function of both its absolute toxicity and relative weight contribution. Thus, the tissue distribution of toxins, expressed in ng P-CTX3C equiv./g whole body, were calculated for each tissue (viscera, mantle, and flesh) using the following formula: absolute toxicity (expressed in ng P-CTX3C equiv./g tissue) × relative weight contribution (in %). The EC_{50} value obtained for P-CTX3C, the maximum concentration of dry extracts (MCE) to be tested in CBA-N2a for giant clam fractions and the limit of quantification (LOQ) for CBA-N2a in our assays conditions were also assessed and were 3.10 ± 0.76 fg/μL, 10,869 pg/μL, and 0.014 ng P-CTX3C equiv./g of giant clam, respectively.

4.5. Liquid Chromatography Coupled with Tandem Mass Spectrometry (LC-MS/MS)

The procedure for liquid chromatography coupled with tandem mass spectrometry (LC-MS/MS) analyses was adapted from Yogi et al. (2011) [49]. Analyses were performed using an LC system (UFLC XR Nexera, SHIMADZU, Kyoto, Japan) coupled to a hybrid triple quadrupole-linear ion trap API4000 QTRAP mass spectrometer (SCIEX, Redwood City, CA, USA) equipped with a TurboV® electrospray ionization source. A Zorbax Eclipse plus column (C_{18}, 1.8 μm, 50 mm × 2.1 mm, Agilent technologies, Santa Clara, CA, USA) was employed at 40 °C and P-CTXs were eluted at 400 μL/min with a linear gradient using water as eluent A and methanol as eluent B, both eluents containing 2 mM ammonium formate and 50 mM formic acid. The elution gradient ran from 78 to 88% B over 10 min and was held for 4 min before returning to initial conditions and re-equilibration during 5 min. Five microliters of sample were injected onto the column. The instrument control, data processing and analysis were conducted using Analyst software. The API4000 QTRAP was operated in positive mode using Multi Reaction Monitoring (MRM). The pseudomolecular ions $[M+NH_4]^+$ and $[M+H]^+$ were selected as precursor ions. The ions resulting in the successive losses of NH_4 and/or water molecules were selected as product ions (see Roué et al. (2016) [17]). The MRM experiments were established by using the following source settings: curtain gas set at 25, ion spray at 5500 V, a turbogas temperature of 300 °C, gas 1 set at 40 and gas 2 set at 60 psi with an entrance potential of 10 V. Calibration was carried out using P-CTX3C standard purchased from Wako chemical. In addition, three P-CTX standards: P-CTX1B, P-CTX3B, and P-CTX3C, obtained from the Institut Louis Malardé's bank of standards were injected to obtain a toxin profile for verification of retention times. The limit of detection (LOD) was statistically determined and found to be 0.66 ng P-CTX3C equiv./g of shellfish matrix.

Author Contributions: M.R., H.T.D. and M.C. conceived of and designed the study. A.U. performed the in vitro cultures of *Gambierdiscus* cells for exposure experiments. Mélanie Roué performed the tissue distribution and detoxification studies. Mélanie Roué performed chemical extractions prior to CBA-N2a and LC-MS/MS analyses. J.V. and M.S. conducted the CBA-N2a and LC-MS/MS analyses, respectively. M.R., H.T.D. and J.V. contributed to the CBA-N2a data analysis and M.S., P.H. and Z.A. to the LC-MS/MS data analysis. M.R. drafted the paper and coordinated its revisions while all authors contributed to the editing of the paper.

Acknowledgments: The work described in this paper was supported by funds from the countries of, France and French Polynesia (Contrat de Projet "Caractérisation écotoxicologique des lagons exploites" CELEX—No. 8192-2012/VP/DBP). The authors greatly acknowledge Nicolas Maihota (IRD), Philippe Cruchet (ILM) and Laura Dufranc (IRD/ILM) for their technical assistance.

References

1. Bell, J.D.; Kronen, M.; Vunisea, A.; Nash, W.J.; Keeble, G.; Demmke, A.; Pontifex, S.; Andréfouët, S. Planning the use of fish for food security in the Pacific. *Mar. Policy* **2009**, *33*, 64–76. [CrossRef]
2. Charlton, K.E.; Russell, J.; Gorman, E.; Hanich, Q.; Delisle, A.; Campbell, B.; Bell, J. Fish, food security and health in Pacific Island countries and territories: A systematic literature review. *BMC Public Health* **2016**, *16*. [CrossRef] [PubMed]
3. Van Wynsberge, S.; Andréfouët, S.; Gaertner-Mazouni, N.; Wabnitz, C.C.C.; Gilbert, A.; Remoissenet, G.; Payri, C.; Fauvelot, C. Drivers of density for the exploited giant clam *Tridacna maxima*: A meta-analysis. *Fish Fish.* **2016**, *17*, 567–584. [CrossRef]
4. Gilbert, A.; Andréfouët, S.; Yan, L.; Remoissenet, G. The giant clam *Tridacna maxima* communities of three French Polynesia islands: Comparison of their population sizes and structures at early stages of their exploitation. *ICES J. Mar. Sci.* **2006**, *63*, 1573–1589. [CrossRef]
5. Direction des Ressources Marines et Minières de Polynésie Française. Available online: http://www.peche.pf/spip.php?rubrique213 (accessed on 27 February 2018).
6. Dickey, R.W.; Plakas, S.M. Ciguatera: A public health perspective. *Toxicon* **2010**, *56*, 123–136. [CrossRef] [PubMed]
7. Friedman, M.A.; Fernandez, M.; Backer, L.C.; Dickey, R.W.; Bernstein, J.; Schrank, K.; Kibler, S.; Stephan, W.; Gribble, M.O.; Bienfang, P.; et al. An updated review of ciguatera fish poisoning: Clinical, epidemiological, environmental, and public health management. *Mar. Drugs* **2017**, *15*, 72. [CrossRef] [PubMed]
8. Bagnis, R. A propos de quelques cas d'intoxications par des mollusques du genre "bénitier" dans une île de la Société. *Bull. Soc. Pathol. Exot.* **1967**, *60*, 580–592.
9. Bagnis, R. Situation de l'endémicité ciguatérique aux îles Gambier. *Cahiers Pacifique* **1974**, *18*, 585–600.
10. Rongo, T.; van Woesik, R. Socioeconomic consequences of ciguatera poisoning in Rarotonga, southern Cook Islands. *Harmful Algae* **2012**, *20*, 92–100. [CrossRef]
11. Laurent, D.; Kerbrat, A.S.; Darius, H.T.; Rossi, F.; Yeeting, B.; Haddad, M.; Golubic, S.; Pauillac, S.; Chinain, M. Ciguatera Shellfish Poisoniong (CSP): A new ecotoxicological phenomenon from cyanobacteria to humans via giant clams. In *Food Chains: New Research*; Jensen, M.A., Muller, D.W., Eds.; Nova Science Publishers, Inc.: New York, NY, USA, 2012; pp. 1–44. ISBN 9781613243770.
12. Gatti, C.; Lonati, D.; Darius, H.T.; Zancan, A.; Roué, M.; Schicchi, A.; Locatelli, C.A.; Chinain, M. *Tectus niloticus* (*Tegulidae*, Gastropod) as a Novel Vector of Ciguatera Poisoning: Clinical Characterization and Follow-Up of a Mass Poisoning Event in Nuku Hiva Island (French Polynesia). *Toxins* **2018**, *10*, 102. [CrossRef] [PubMed]
13. Darius, H.T.; Roué, M.; Sibat, M.; Viallon, J.; Gatti, C.M.I.; Vandersea, M.W.; Tester, P.A.; Litaker, R.W.; Amzil, Z.; Hess, P.; et al. *Tectus niloticus* (*Tegulidae*, Gastropod) as a Novel Vector of Ciguatera Poisoning: Detection of Pacific Ciguatoxins in Toxic Samples from Nuku Hiva Island (French Polynesia). *Toxins* **2018**, *10*, 2. [CrossRef] [PubMed]

14. Darius, H.T.; Roué, M.; Sibat, M.; Viallon, J.; Gatti, C.; Vandersea, M.; Tester, P.A.; Litaker, R.W.; Amzil, Z.; Hess, P.; et al. Toxicological investigations on the sea urchin *Tripneustes gratilla* (*Toxopneustidae*, Echinoid) from Anaho Bay (Nuku Hiva, French Polynesia): Evidence for the presence of Pacific ciguatoxins. *Mar. Drugs* **2018**, *16*, 122. [CrossRef] [PubMed]
15. Laurent, D.; Kerbrat, A.S.; Darius, H.T.; Girard, E.; Golubic, S.; Benoit, E.; Sauviat, M.P.; Chinain, M.; Molgo, J.; Pauillac, S. Are cyanobacteria involved in Ciguatera Fish Poisoning-like outbreaks in New Caledonia? *Harmful Algae* **2008**, *7*, 827–838. [CrossRef]
16. Pawlowiez, R.; Darius, H.T.; Cruchet, P.; Rossi, F.; Caillaud, A.; Laurent, D.; Chinain, M. Evaluation of seafood toxicity in the Australes archipelago (French Polynesia) using the neuroblastoma cell-based assay. *Food Addit. Contam. Part A Chem. Anal. Control. Expo. Risk Assess.* **2013**, *30*, 567–586. [CrossRef] [PubMed]
17. Roué, M.; Darius, H.T.; Picot, S.; Ung, A.; Viallon, J.; Gaertner-Mazouni, N.; Sibat, M.; Amzil, Z.; Chinain, M. Evidence of the bioaccumulation of ciguatoxins in giant clams (*Tridacna maxima*) exposed to *Gambierdiscus* spp. cells. *Harmful Algae* **2016**, *57*, 78–87. [CrossRef]
18. Bricelj, V.M.; Shumway, S.E. Paralytic Shellfish Toxins in Bivalve Molluscs: Occurrence, Transfer Kinetics, and Biotransformation. *Rev. Fish. Sci.* **1998**, *6*, 315–383. [CrossRef]
19. Lewis, R.J.; Sellin, M.; Street, R.; Holmes, M.H.; Gillepsie, N.C. Excretion of ciguatoxin from moray eels (Muraenidae) of the central Pacific. In *Proceedings of the Third International Conference on Ciguatera Fish Poisoning*; Tosteson, T.R., Ed.; Polysciences Publications: Quebec, QC, Canada, 1992; pp. 131–143.
20. Chateau-Degat, M.L.; Huin-Blondey, M.O.; Chinain, M.; Darius, T.; Legrand, A.M.; Ngoc, L.N.; Laudon, F.; Chansin, R.; Dewailly, E. Prevalence of chronic symptoms of ciguatera disease in French Polynesian adults. *Am. J. Trop. Med. Hyg.* **2007**, *77*, 842–846. [PubMed]
21. Chen, C.Y.; Chou, H.N. Accumulation and depuration of paralytic shellfish poisoning toxins by purple clam *Hiatula rostrata* Lighttoot. *Toxicon* **2001**, *39*, 1029–1034. [CrossRef]
22. Blanco, J.; Acosta, C.P.; Bermúdez de la Puente, M.; Salgado, C. Depuration and anatomical distribution of the amnesic shellfish poisoning (ASP) toxin domoic acid in the king scallop *Pecten maximus*. *Aquat. Toxicol.* **2002**, *60*, 111–121. [CrossRef]
23. Kwong, R.W.M.; Wang, W.X.; Lam, P.K.S.; Yu, P.K.N. The uptake, distribution and elimination of paralytic shellfish toxins in mussels and fish exposed to toxic dinoflagellates. *Aquat. Toxicol.* **2006**, *80*, 82–91. [CrossRef] [PubMed]
24. Blanco, J.; Mariño, C.; Martín, H.; Acosta, C.P. Anatomical distribution of diarrhetic shellfish poisoning (DSP) toxins in the mussel *Mytilus galloprovincialis*. *Toxicon* **2007**, *50*, 1011–1018. [CrossRef] [PubMed]
25. Lassus, P.; Amzil, Z.; Baron, R.; Séchet, V.; Barillé, L.; Abadie, E.; Bardouil, M.; Sibat, M.; Truquet, P.; Bérard, J.B.; et al. Modelling the accumulation of PSP toxins in Thau Lagoon oysters (*Crassostrea gigas*) from trials using mixed cultures of *Alexandrium catenella* and *Thalassiosira weissflogii*. *Aquat. Living Resour.* **2007**, *20*, 59–67. [CrossRef]
26. Murray, S.A.; O'Connor, W.A.; Alvin, A.; Mihali, T.K.; Kalaitzis, J.; Neilan, B.A. Differential accumulation of paralytic shellfish toxins from *Alexandrium minutum* in the pearl oyster, *Pinctada imbricata*. *Toxicon* **2009**, *54*, 217–223. [CrossRef] [PubMed]
27. Matsushima, R.; Uchida, H.; Nagai, S.; Watanabe, R.; Kamio, M.; Nagai, H.; Kaneniwa, M.; Suzuki, T. Assimilation, accumulation, and metabolism of dinophysistoxins (DTXs) and pectenotoxins (PTXs) in the several tissues of Japanese scallop *Patinopecten yessoensis*. *Toxins* **2015**, *7*, 5141–5154. [CrossRef] [PubMed]
28. García, C.; Pérez, F.; Contreras, C.; Figueroa, D.; Barriga, A.; López-Rivera, A.; Araneda, O.F.; Contreras, H.R. Saxitoxins and okadaic acid group: Accumulation and distribution in invertebrate marine vectors from Southern Chile. *Food Addit. Contam. Part A Chem. Anal. Control. Expo. Risk Assess.* **2015**, *32*, 984–1002. [CrossRef] [PubMed]
29. Kameneva, P.A.; Krasheninina, E.A.; Slobodskova, V.V.; Kukla, S.P.; Orlova, T.Y. Accumulation and tissue distribution of dinophysitoxin-1 and dinophysitoxin-3 in the mussel *Crenomytilus grayanus* feeding on the benthic dinoflagellate *Prorocentrum foraminosum*. *Mar. Drugs* **2017**, *15*, 330. [CrossRef] [PubMed]
30. Jauffrais, T.; Marcaillou, C.; Herrenknecht, C.; Truquet, P.; Séchet, V.; Nicolau, E.; Tillmann, U.; Hess, P. Azaspiracid accumulation, detoxification and biotransformation in blue mussels (*Mytilus edulis*) experimentally fed *Azadinium spinosum*. *Toxicon* **2012**, *60*, 582–595. [CrossRef] [PubMed]
31. Drum, A.S.; Siebens, T.L.; Crecelius, E.A.; Elston, R.A. Domoic acid in the Pacific razor clam *Siliqua patula* (Dixon, 1789). *J. Shellfish Res.* **1993**, *12*, 443–450.

32. Wekell, J.C.; Gauglitz, E.J.; Bamett, H.J.; Hatfield, C.L.; Simons, D.; Ayres, D. Occurrence of domoic acid in washington state razor clams (*Siliqua patula*) during 1991–1993. *Nat. Toxins* **1994**, *2*, 197–205. [CrossRef] [PubMed]
33. Sagou, R.; Amanhir, R.; Taleb, H.; Vale, P.; Blaghen, M.; Loutfi, M. Comparative study on differential accumulation of PSP toxins between cockle (*Acanthocardia tuberculatum*) and sweet clam (*Callista chione*). *Toxicon* **2005**, *46*, 612–618. [CrossRef] [PubMed]
34. European Food Safety Authority. Scientific Opinion on marine biotoxins in shellfish-Emerging toxins: Ciguatoxin group. *EFSA J.* **2010**, *8*, 1627. [CrossRef]
35. Hossen, V.; Soliño, L.; Leroy, P.; David, E.; Velge, P.; Dragacci, S.; Krys, S.; Flores Quintana, H.; Diogène, J. Contribution to the risk characterization of ciguatoxins: LOAEL estimated from eight ciguatera fish poisoning events in Guadeloupe (French West Indies). *Environ. Res.* **2015**, *143*, 100–108. [CrossRef] [PubMed]
36. Anonymous. *Domoic Acid in the King Scallop, Pecten Maximus*; Unpublished Report Prepared for the EU ASP Working Group; UK National Reference Laboratory for Marine Biotoxins: Belfast, UK, 2001.
37. Chinain, M.; Darius, H.T.; Ung, A.; Cruchet, P.; Wang, Z.; Ponton, D.; Laurent, D.; Pauillac, S. Growth and toxin production in the ciguatera-causing dinoflagellate *Gambierdiscus polynesiensis* (Dinophyceae) in culture. *Toxicon* **2010**, *56*, 739–750. [CrossRef] [PubMed]
38. Lee, R.; Lovatelli, A.; Ababouch, L. *Bivalve Depuration: Fundamental and Practical Aspects*; FAO Fisheries Technical Paper No 511; Food and Agriculture Organization of the United Nations: Rome, Italy, 2008; ISBN 9789251060063.
39. Blanco, J.; Bermúdez De La Puente, M.; Arévalo, F.; Salgado, C.; Moroño, Á. Depuration of mussels (*Mytilus galloprovincialis*) contaminated with domoic acid. *Aquat. Living Resour.* **2002**, *15*, 53–60. [CrossRef]
40. Blanco, J.; Perez-Acosta, C.; Mariño, C.; Muñiz, S.; Martín, H.; Moroño, A.; Correa, J.; Arévalo, F.; Salgado, C. Depuration of domoic acid from different body compartments of the king scallop *Pecten maximus* grown in raft culture and natural bed. *Aquat. Living Resour.* **2006**, *19*, 257–265. [CrossRef]
41. McFarland, K.; Jean, F.; Soudant, P.; Volety, A.K. Uptake and elimination of brevetoxin in the invasive green mussel, *Perna viridis*, during natural *Karenia brevis* blooms in southwest Florida. *Toxicon* **2015**, *97*, 46–52. [CrossRef] [PubMed]
42. Suzuki, T.; Ota, H.; Yamasaki, M. Direct evidence of transformation of dinophysistoxin-1 to 7-*O*-acyl-dinophysistoxin-1 (dinophysistoxin-3) in the scallop *Patinopecten yessoensis*. *Toxicon* **1999**, *37*, 187–198. [CrossRef]
43. Bauder, A.G.; Cembella, A.D.; Bricelj, V.M.; Quilliam, M.A. Uptake and fate of diarrhetic shellfish poisoning toxins from the dinoflagellate *Prorocentrum lima* in the bay scallop *Argopecten irradians*. *Mar. Ecol. Prog. Ser.* **2001**, *213*, 39–52. [CrossRef]
44. Plakas, S.M.; Wang, Z.; El Said, K.R.; Jester, E.L.E.; Granade, H.R.; Flewelling, L.; Scott, P.; Dickey, R.W. Brevetoxin metabolism and elimination in the Eastern oyster (*Crassostrea virginica*) after controlled exposures to *Karenia brevis*. *Toxicon* **2004**, *44*, 677–685. [CrossRef] [PubMed]
45. Medhioub, W.; Guéguen, M.; Lassus, P.; Bardouil, M.; Truquet, P.; Sibat, M.; Medhioub, N.; Soudant, P.; Kraiem, M.; Zouher, A. Detoxification enhancement in the gymnodimine-contaminated grooved carpet shell, *Ruditapes decussatus* (Linné). *Harmful Algae* **2010**, *9*, 200–207. [CrossRef]
46. Silvert, W.L.; Cembella, A.D. Dynamic modelling of phycotoxin kinetics in the blue mussel, *Mytilus edulis*, with implications for other marine invertebrates. *Can. J. Fish. Aquat. Sci.* **1995**, *52*, 521–531. [CrossRef]
47. Yu, K.N.; Kwong, R.W.M.; Wang, W.-X.; Lam, P.K.S. Biokinetics of paralytic shellfish toxins in the green-lipped mussel, *Perna viridis*. *Mar. Pollut. Bull.* **2007**, *54*, 1068–1071. [CrossRef] [PubMed]
48. Manger, R.L.; Leja, L.S.; Lee, S.Y.; Hungerford, J.M.; Wekell, M.M. Tetrazolium-based cell bioassay for neurotoxins active on voltage-sensitive sodium channels: Semiautomated assay for saxitoxins, brevetoxins, and ciguatoxins. *Anal. Biochem.* **1993**, *214*, 190–194. [CrossRef] [PubMed]
49. Yogi, K.; Oshiro, N.; Inafuku, Y.; Hirama, M.; Yasumoto, T. Detailed LC-MS/MS Analysis of Ciguatoxins Revealing Distinct Regional and Species Characteristics in Fish and Causative Alga from the Pacific. *Anal. Chem.* **2011**, *83*, 8886–8891. [CrossRef] [PubMed]

11

Analytical and Sample Preparation Techniques for the Determination of Food Colorants in Food Matrices

Konstantina Ntrallou [1], Helen Gika [2,3] and Emmanouil Tsochatzis [1,3,*]

1 Department of Chemical Engineering, Aristotle University of Thessaloniki, 54124 Thessaloniki, Greece; kondrallou@gmail.com
2 Laboratory of Forensic Medicine & Toxicology, Department of Medicine, Aristotle University of Thessaloniki, 54124 Thessaloniki, Greece; gkikae@auth.gr
3 BIOMIC AUTH Center for Interdisciplinary Research of the Aristotle University of Thessaloniki, Innovation Area of Thessaloniki, 57001 Thermi, Greece
* Correspondence: tsochatzism@gmail.com

Abstract: Color additives are widely used by the food industry to enhance the appearance, as well as the nutritional properties of a food product. However, some of these substances may pose a potential risk to human health, especially if they are consumed excessively and are regulated, giving great importance to their determination. Several matrix-dependent methods have been developed and applied to determine food colorants, by employing different analytical techniques along with appropriate sample preparation protocols. Major techniques applied for their determination are chromatography with spectophotometricdetectors and spectrophotometry, while sample preparation procedures greatly depend on the food matrix. In this review these methods are presented, covering the advancements of existing methodologies applied over the last decade.

Keywords: food colorants (synthetic, natural); food matrices; instrumental analysis; sample preparation

---

1. Introduction

Codex Alimentarius gives a definition for food additives as "any substance that its intentional addition of which to a food aiming for a technological (including organoleptic) purpose in the manufacture, processing, preparation treatment, packing, packaging, transport or holding of such food results, or may be reasonably expected to result, in it or its by-products becoming a component of the food or otherwise affecting the characteristics of such foods" [1,2]. Carocho et al. highlighted that the definition given by the Codex Alimentarius does not include the term contaminants or substances added to food for maintaining or improving nutritional qualities [2].

In food technology, food colorants, of several types, are chemical substances that are added to food matrices, to enhance or sustain the sensory characteristics of the food product, which may be affected or lost during processing or storage, and in order to retain the desired color appearance [3–5]. These are classified based on several criteria: firstly, based on their origin in nature, nature-identical or, if synthetic, whether they are organic or inorganic. Another classification could be based on their solubility (e.g., soluble or insoluble) or covering ability (e.g., transparent or opaque), though an overlap may exist among one or more of these classifications. The most common and widely used classification is based on the distinction between soluble and insoluble color additives (colorants or pigments), which can be further categorized as natural or synthetic [4].

In addition, as described by Martins et al., there were several food additives that had been used extensively in the past but are no longer allowed, due to existing evidence of their side effects, toxicity in the medium- and long-term, as well as a high frequency of potential health incidents [6]. It is also important to note that, apart from synthetic food colorants, certain commercial additives of plant or animal origin have also been suspended [3,6–8].

It is clear that the analysis of trace amounts of food colorants is essential with the proper analytical techniques applied, with high specificity and selectivity. Ni et al. has reported that there is increasing interest in the monitoring of the concentration of synthetic food colorants in various products [9].

The analytical methods and sample preparation protocols presented hereafter cover the main techniques that have been applied over the last decade (2008 onwards).

2. Natural Food Colorants

Natural additives have been used since ancient times. In certain cases, they were used for the preservation of foodstuffs. Nowadays, most consumers seem to be in favor of the use of the natural, as opposed to the synthetic ones, which are considered by the food industry to be more efficient. In the meantime, there is also considerable interest in the overall reduction of food colorants to food products [4,5,10]. The classification of naturally derived colorants can become very complex because of the wide variety of innate properties of the coloring substances. They can be derived from a variety of sources in nature, and therefore, natural colorants also exhibit a wide variety of chemical compositions that affect properties, solubilities, and stabilities differently, and they can have different sources as plant-origin or animal-origin [10].

As reported by Carocho et al., there are benefits linked with the use of natural additives over their respective synthetic ones, which in certain cases present a greater potency over the synthetic ones. The latter in most cases present a single effect on the foodstuff in question. Nevertheless, natural additives are often produced using different methods, i.e., extraction from plants or produced by microorganisms, although there is a tendency to consider them safer than their respective synthetic additive. In general, toxicity is a factor that must be thoroughly assessed and evaluated, to ensure health and safety [2,5,10].

Synthetic colorants have a large span of application and are proportionally lower in cost, than their respective natural substances. However, natural colorants are gradually replacing the synthetic ones as they tend to be considered safer, while presenting higher color specificity, no side effects or related toxicity, and conferring health improving effects and functional benefits to the food itself [6,11,12]. A good example for this beneficial effect is the class of yeast-derived natural pigments (e.g., monascin; a yellow natural pigment). These present certain features, apart from food coloring, such as biological activity, reported potential anti-cancer, anti-inflammatory, anti-diabetic, and anti-cholesterolemic effects [6,13,14].

As reported by Martins et al., numerous references highlighted the effective and/or selective use of food colorants. Therefore, for the approved food colorants with an "E" code, individual Acceptable Daily Intakes (ADI) have been approved and established, expressed mostly as mass fractions (i.e., mg/kg per body weight (b.w), which can be used for specific purposes (i.e., colorants) in specific food products (i.e., biscuits, chocolates, cheeses etc.) [6].

Commonly, naturally occurring food colorants can be allocated in different sub-categories, namely anthocyanins, carotenoids, beet colorants, and phenolic compounds. In addition, annatto, carminic acid, and some curcuminoids have been studied, particularly curcumin. Finally, other colorants remain to be assessed and evaluated in order to be authorized with an "E" code.

Anthocyanins are a widely studied natural food colorants group, mainly obtained from flowers, fruits, leaves, and even whole plants with a color range that goes from red to purple and blue. Commercial anthocyanins, such as cyanidin 3-glucoside, pelargonidin 3-glucoside, and peonidin 3-glucoside have been used effectively [2,4,6].

Carotenoids are another cluster of naturally derived colorants with a renowned technological effect. They present coloring attributes along with certain bioactive as well as antioxidant properties and are being used extensively in the food industry as natural preservatives [4,6,7,10,15] apart from food colorants [7]. Their main source is extracts from plant roots, flowers, and leaves, as well as from algae, yeasts, and aquatic animals. This category mainly includes Lutein, astaxanthin, and lycopene [2,6], the most widely used carotenoids used with others such as crocin and crocetin, still under investigation [4–6].

Red-purple colorants derived from beets and beetroot (Beta vulgaris L.) root is the principal source of these natural colorants but also fruit of *Hylocereus polyrhizus* (Weber) Britton and Rose, *Opuntia ficus-indica* [L.] Miller, *Opuntia stricta* (Haw). Haw and *Rivina humilis* L. are also rich in these colorant substances, namely, the betacyanins and betalains, which are the most frequently studied and already authorized (E162). They are being used in various food products such as burgers, desserts, ice creams, jams, jellies, soups, sauces, sweets, drinks, dairy products, and yogurts [2,4–6].

Other natural food colorants are considered the phenolic compounds, where flavanones, flavones (4′,5,7-trihydroxyflavones), and flavonols (fisetin, myricetin, myricitin, quercetin, and rutin) have been studied. As reported by Martins et al., currently only the commercially available products are being used (i.e., myricetin and myricitrin from *Myrica cerifera* L. roots). Phenolic compounds do not yet have an approved "E" code nor an ADI value [6] with many still being studied and examined since their safety, stability, and spectrum of activity still remain unclear [6,16].

Another category of natural food colorant is the curcuminoids with the most widely known and used colorant in this group being curcumin (E100), usually isolated from Curcuma longa L. rhizomes.

Other natural used colorants are the annatto (E160b) group, as well as bixin and norbixin which are extracts from *Bixa orellana* L. seeds [2,4–6]. In addition, carminic acid (E120) with a yellow to red-orange food color is already largely used, either naturally occurring or of synthetic origin with an ADI of 5 mg/kg b.w [6] or crocin. Nevertheless, there are other food colorants under investigation, including c-phycocyanin (blue pigment isolated from Arthrospira platensis) and c-phycoerythrin (red-orange pigment from blue-green algae). Other naturally occurring pigments, which are commercially available, are being studied, such as geniposide, monascorubrin, and purple corn color [4–6].

3. Synthetic Food Colorants

Based on increasing demand, mainly from the consumer, for products that are more visually attractive, several synthetic food colorants have been developed for use in food production, to increase certain quality and organoleptic characteristics. However, it is reported that over time, most of the synthetic food colorants were excluded due to repeated side effects as well as to their short- and/or long-term toxicity and eventually to potential carcinogenic effects [3,6,11].

Thus, a change in consumer expectations has been reported, which is largely in favor of the natural colorants [6,17].

Apart from this, also from a regulatory point of view, there is increasing attention and interest related to the risk assessment of these colorants used in food products (i.e., azo-dyes). In case of the azo-dyes, a limiting factor for their use is their potential carcinogenicity, which occurs after their reduction to carcinogenic metabolites into the intestine [3,18,19]. These metabolites are produced in the human body, though their toxic effect depends on the ingested amount of the target substance/colorant [3,18,20]. However, it is reported that regular evaluation and assessment of potential toxicity of food colorants by regulatory authorities is necessary [3,18,21].

4. Toxicological Aspects and Regulatory Framework

Based on various scientific findings, several toxicity effects, have been reported including behavioral effects on children, effects on the respiratory system, connection with allergies, development of attention deficit hyperactivity disorder (ADHD) in children, or neuro-developmental effects at the No-Adverse Effect Limit levels [3,18,21]. In any case, further investigation to assess the potential associated risks of these compounds is needed [3–9,11,14,18].

Several groups have indicated the toxic effect of some of groups of these substances. As an example, Mpountoukas et al. have tested the food colorants amaranth, erythrosine, and tartrazine by in vitro experiments, and they concluded there was an in vitro toxic effect on human lymphocytes as they bound to DNA [22]. Many other studies have shown the chemical property of synthetic colorants, namely, Tartrazine [23], azorubine [17,24,25], Allura Red [17,26,27] Sunset Yellow, Quinoline Yellow [17], and Patent Blue [28], to bind to human serum albumin (HSA). Masone and Chanforan compared binding affinities of artificial colorants to human serum albumin (HSA), exhibiting more affinity to HSA than to their natural equivalents' colorants and interacting with its functions. The results supported the hypothesis of their potential risk to human health [17]. Finally, there are dyes, which are rather inexpensive, and which have been used in the food industry, such as Sudan I–IV, which are classified as both a toxic and carcinogenic [24–31]. In Figure 1, basic structures of colorants used in the food industry some of them with toxicological concern (Sudan I–IV) are presented.

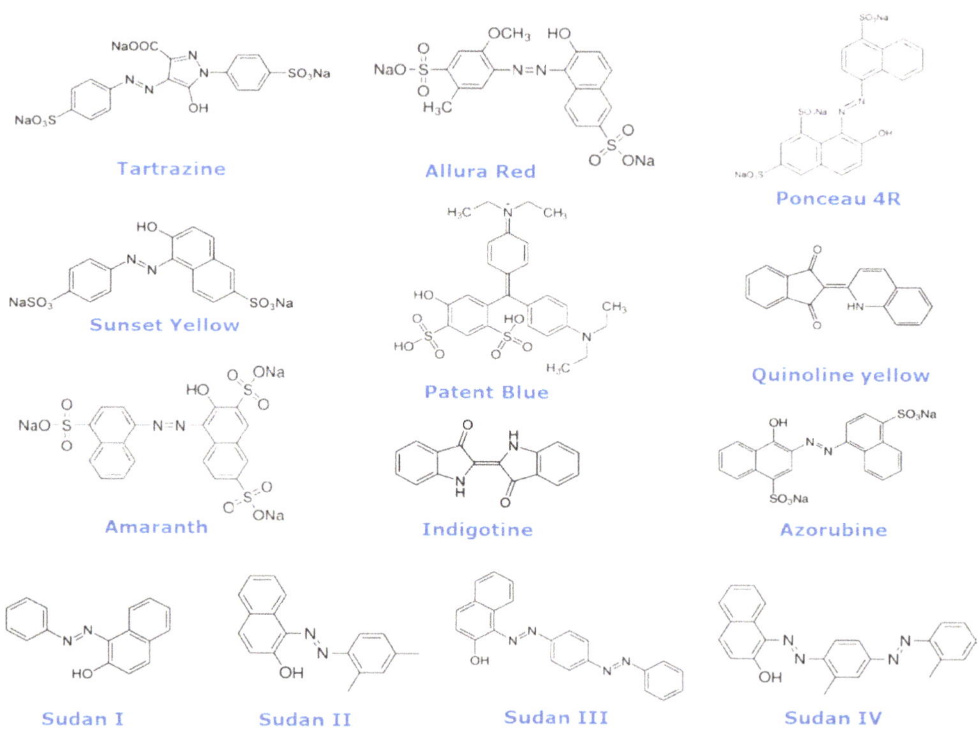

Figure 1. Chemical structures of selected regulated food colorants.

The main regulatory authorities, EFSA in Europe and the US Food and Drug Administration (FDA) in the United States, are responsible for the evaluation and assessment of food products to enhance and promote health safety [2,4,5]. The European Union, set a re-evaluation program of food additives, including food colorants, to be performed by EFSA by 2020, based on the EU Regulation 257/2010. This re-evaluation program was set in order to assess the safety of all authorized food additives in the European Union before 20 January 2009 [32].

The regulatory framework in Europe, in brief, contains the authorization procedure in Regulation (EU) No. 1331/2008, the rules on food additives with a list of approved color additives and their conditions of use in Regulations (EU) No 1333/2008 and 1129/2011, the specifications for food additives in Regulation (EU) 231/2012, and finally for labelling in Regulations (EU) No. 1169/2011 and 1333/2008. Respectively, in the United States, the color additives are included in Title 21 CFR Part 70, listing food additives (exempt from certification, including specifications and conditions of use) in Title 21 CFR Part 73, and certification of donor additives in Title 21 CFR Part 80 [4,5,10,33].

However, despite the existence of different regulatory frameworks, the overall approach follows similar steps, which are based on well-established risk assessment procedures [33].

Authorization for the use of food colorants in the production of food products is subject to a number of toxicity tests, in order to define and evaluate acute, sub-chronic and chronic toxicity, hepatotoxicity, carcinogenicity, mutagenicity, teratotoxicity, genotoxicity, reproductive toxicity, accumulation in the body, bioenergy effects, and immunotoxicity [3–9,11,14,18].

5. Analytical Methodologies for the Determination of Food Colorants

5.1. Analytical Techniques in the Use of Natural Food Colorants Determinations

The available bibliography concerning the methods of analysis for the natural colorants is limited, compared to that for the synthetic ones, and it is exclusively oriented to their determination in the different naturally deriving products.

All the relative information concerning analytical methods for natural colorants, including tested matrices, analytical technology, type of detection and settings, analytical columns if used, elution parameters, mobile phases, injection volumes, and analytical figures of merit (LOD, LOQ), have been reviewed and are summarized in Table 1.

It can be concluded from Table 1 that evaluation of methods' performance criteria was not within the aims of the above-mentioned reports, as they were focusing in activity, bioavailability, processing impact, and adulteration. Thus, no analytical figures of merit are reported in these papers.

From Table 1 and Figure 2 it could be perceived that the predominant technique is HPLC combined with spectrophotometric (UV-Vis) or Diode Array (DAD) detectors, followed by HPLC by MS/MS. Spectrophotometric UV-Vis methods seem also to be preferred by the researchers in this field as they show low instrument cost and do not involve expert skill. However, it should be considered that the individual features of the spectra obtained for single colors are highly dependent on the pH-adjustment of the solution or the mobile phase, using proper acid or alkali. The pH adjustment certainly affects maximum absorption wavelength, where shifts and intensities based on the different pH can be observed. Although sample preparation is much less demanding in comparison to the LC methods, these techniques present a significant disadvantage, which is the lack of ability to analyze simultaneously a bigger number of food colorants.

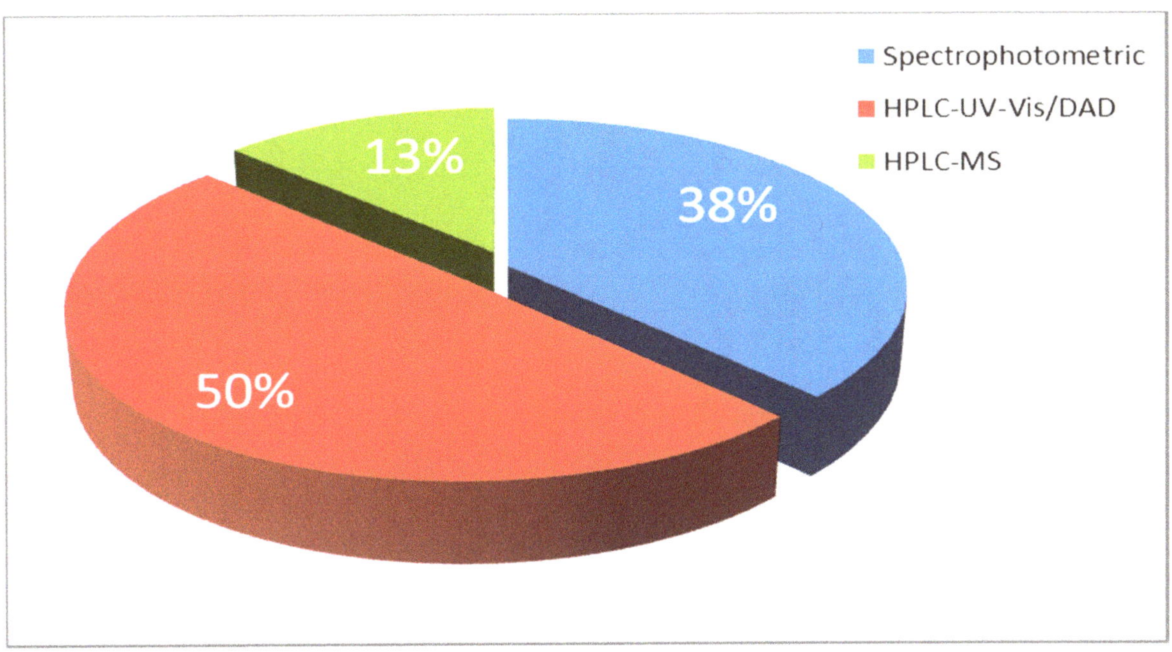

Figure 2. Distribution of techniques used for the analysis of natural food colorants.

Table 1. Methods for the analysis of natural food colorants in various food products.

Food Colorant	Food Matrix	Analytical Technique	Detection	Detection Settings (i.e., λ, Ionisation)	Column	Elution	Mobile Phase	Inj. Volume	Figures of Merit (LOD, LOQ, Linear Range)	Ref.
3-Deoxy-anthocyanidins	*Sorghum bicolor* (L.) Moench seeds	High Pressure Liquid Chromato-graphy (HPLC)	Diode Array Detection (DAD)	485 nm	Luna C18 column (150 × 4.6 mm, 5 mm)	Gradient	4% HCOOH in H_2O (*v/v*) (Solvent A) and acetonitrile (Solvent B)	20 µL	n/a	[34]
Anthocyanin-derived extracts	*Acacia decurrens* Willd. Bark	Spectrophotometric analysis	UV-Vis	400–800 nm	n/a	n/a	n/a		n/a	[35]
	-*Tulipa gesneriana* L.	Spectrophotometric analysis	UV-Vis	765 nm	n/a	n/a	n/a	n/a	n/a	[36]
Cyanidin 3-glucoside	-*Pistacia lentiscus* L. fruits; -*Santalum album* L. fruits	HPLC	DAD	520 nm, 440 nm, 310 nm and 280 nm	SS Wakosil C18 (150 × 4.6 mm, 5 µm)	Gradient	0.1% trifluoroacetic acid (TFA) in H_2O (solvent A) and 0.1% TFA in acetonitrile (Solvent B)	20 µL	n/a	[37]
Cyanidin 3-glucoside	-*Pistacia lentiscus* L. fruits; -*Santalum album* L. fruits	HPLC	ESI-MS		SS Wakosil C18 (150 × 4.6 mm, 5 µm)	Gradient	0.1% TFA in H_2O (solvent A) and 0.1% TFA in acetonitrile (Solvent B	20 µL	n/a	[37]
Betacyanins	-Hylocereus polyrhizus	HPLC	MS	ESI (+)	AQUA C18-reversed phase column, 5 µm	Gradient	(A) 2% (*v/v*) CH_3COOH in H_2O and (B)0.5% CH_3COOH in H_2O/acetonitrile (50/50, *v/v*)		n/a	[38]
Betalains	*Beta vulgaris* L. roots	HPLC	UV-Vis	538 nm; 480 nm	Lichrocart 250 × 4 RP-18 (5 µm)	Gradient	H_2O (A) and acetonitrile (B).	20 µL	n/a	[39]
Betalains	*Opuntia ficus-indica* [L.]	HPLC	UV-Vis	245 nm	Luna C18(2) column (250 × 4.6 mm, 5 µm)	Isocratic	20 mM KH_2PO_4/ Acetonitrile 95:5 *v/v*	20 µL	n/a	[40]
Betalains	*Opuntia ficus-indica* [L.]	Spectrophotometric analysis	UV-Vis	λ = 536 nm	n/a	n/a	n/a	n/a	n/a	[41]
Betalains	*Rivina humilis* L. fruits, juice	Spectrophotometric analysis	UV-Vis	λ = 535 nm	n/a	n/a	n/a	n/a	n/a	[42]
α-carotene	*Daucus carota* L. roots	HPLC	UV-Vis	450 nm	Supelcosil LC-18 column (15 cm × 4.6 cm, 5 µm)	Isocratic	Methanol/10% (*v/v*) Acetonitrile: H_2O	50 µL	n/a	[43]
Lutein	Commercial/Milk	HPLC	UV-Vis	450 nm	RP C30 YMC (250 × 4.6 mm, 5 µm)	Isocratic	ethanol, tert-butyl-methyl-ether (MTBE) as the mobile phase	50 µL	n/a	[44]

Table 1. *Cont.*

Food Colorant	Food Matrix	Analytical Technique	Detection	Detection Settings (i.e., λ, Ionisation)	Column	Elution	Mobile Phase	Inj. Volume	Figures of Merit (LOD, LOQ, Linear Range)	Ref.
Lutein	Hawaii "T. erecta", Carmen "T. patula".	HPLC	DAD	450 nm	Waters-Spherisorb column SC-04 (125 × 4.0 mm, ODS2, 3.0 µm)	Gradient	(A) Acetonitrile–methanol (9:1 *v*/*v*): (B) Ethyl acetate	100 µL	n/a	[45]
Astaxanthin	Microalgae and yeasts	HPLC	DAD	470 nm	Chiralcel OD-RH column (5 µm, 150 mm × 4.6 mm)		(A) Acetonitrile and (B) phosphoric acid (3.5 mM)	n/a	n/a	[46]
Crocetin, Crocin	Grape seed, monascus, gardenia, and red radish	Spectrophotometric analysis	UV-Vis	438 nm; 462 nm	n/a	n/a	n/a	n/a	n/a	[47]
Monascus red pigments	Beetroot red and paprika extract	High Resolution Mass Spectrometry	HPLC-QTOF-MS	ESI (+)	Kinetex c18 column (2.6 µm, 50 mm × 4.6 mm)	Gradient	(A) Acetonitrile; (B) H_2O; (C) aqueous HCOOH 1% *v*/*v*	n/a	n/a	[48]

5.2. Sample Preparation for Natural Colorant Analysis

Several sample preparation protocols are reported in the literature by applying various techniques. The applied protocol is strongly dependent by the type and nature of the food sample. Below in Table 2, a short description of the sample preparation protocols is given, along with their application for the clean-up of food samples, for the quantification of natural food colorants. A hydrolysis step with a deprotonation step (ethanol, HCl solution) is being reported depending on the food matrix, including dilution methods and SFE with supercritical CO_2.

Table 2. Sample preparation techniques for the analysis of natural food colorants in food products.

Food Colorant	Extraction/Sample Preparation	Ref.
3-Deoxyanthocyanidins	Ground sample, with 1% HCl in methanol, centrifugation hydrolysis;	[34]
Anthocyanin-derived extracts	Comparison of different extraction methods (ultrasonic and natural extraction) vs. magnetic stirring	[35]
	Extraction with ethanol: H_2O (1:1 *v/v*) acidified with 0.01% HCl	[36]
Cyanidin 3-glucoside	Extraction with 0.1% HCl (*v/v*) in methanol, combination of the extracts, evaporation, and dissolution	[37]
Betacyanins	Mixing with water, filtration, addition of ethanol (precipitation of pectic substances and proteins)	[38]
Betalains	Sample dissolution in ethanol, agitated and homogenized	[39]
Betalains	Filtration of water extract (no pH adjustment)	[40]
Betalains	Lyophilization and macerated with PBS (pH 5.0) in 1:5 *w/w* ratio, followed by spray-drying	[41]
Betalains	Dilution of the juice; filtration; addition of Se^{4+}, Zn^{2+}, and Cu^{2+}	[42]
α-carotene	Comparison between simple extraction and Supercritical Fluid extraction (CO_2); Simple extraction: Hexane/acetone; SFE: SC-CO_2 (SFE)	[43]
Lutein	Sample dilution in 95% ethanol and extraction with acetone and petroleum ether. Evaporation and reconstitution	[44]
Lutein	Extraction with organic solvent (isopropanol), centrifugation and supernatant extracted with hexane	[45]
Astaxanthin	Extraction with ethyl acetate, filtration	[46]
Crocetin, Crocin	Dilution in DMSO	[47]

5.3. Analytical Techniques in the Use of Synthetic Food Colorants Determinations

The need to determine synthetic colorants in food matrices originating from their known toxicity, renders the analytical task even more challenging as food matrices are ordinarily very complex. Various analytical techniques are used to determine synthetic food colorants in food samples, including spectrophotometry, thin layer chromatography, capillary electrophoresis, high performance liquid chromatography and mass spectrometry (MS).

Certain chemical properties and characteristics of the substances/colorants that influence their separation, such as hydrophilicity/hydrophobicity, existence of acidic or alkaline groups should to be taken into account. Using a Reversed Phase (RP) liquid chromatography separation, more polar compounds are eluting first followed by the less polar. However, their chromatographic separation is normally performed at neutral pH (ca. 7), and thus, any presence of acidic or alkaline groups could affect the elution sequence.

Ordinarily, organic solvents such as methanol, acetonitrile, or their mixture are used for analysis by HPLC. The addition of acetonitrile improves significantly chromatographic peaks' shape (i.e., asymmetry). Nevertheless, the addition of an inorganic electrolyte as a chemical modifier to the mobile phase can be considered as important in order to advance the separation of all the ionizable species [12,28,37,49].

Food colorants are compounds that absorb exceedingly in the visible region. Thus, spectrophotometry is sufficient and appropriate for their quantitative analysis. It is generally preferred as a quite straightforward technique, with respective low instrumental cost (i.e., compared to MS/MS). However, in several cases, its main drawback is the lack of specificity, as in case of mixtures of absorbing species. A solution to overcome the problem of specificity is the application of mass spectrometry (MS). In this case, all spectral interventions or interferences, presented on UV–Vis/DAD detectors, are overpassed. High analytical sensitivity could succeed, even in more difficult food matrices, though after proper clean-up. In addition, tandem MS technique could provide structural information based on the molecular mass/ion and the respective fragmentation pattern. Regarding the ionization mode, in most cases, for synthetic colorants, the electro spray ionization (ESI) is preferred because synthetic food colorants are polar molecules, and their ionization efficiency depends on the existence of matrix interferences, present in sample or in the mobile phase. In general, negative mode (ESI-) is more effective, though in other non-regulated substances (i.e., Sudan I-IV) the positive ionization is preferred. During the MS/MS analysis, chemical modifiers (i.e., $HCOONH_4$ or CH_3COONH_4) are added to the mobile phases, in order to improve and facilitate the better ionization of each target analyte.

Capillary electrophoresis follows in frequency of use the HPLC-DAD/UV-Vis or MS/MS techniques, applied for the quantification of food colorants. These methods present good separation of both small and large molecules, using high voltages. Other reported techniques are FIA (Flow Injection Analysis) and TLC (Thin Layer Chromatography). These could be considered as relatively simple analytical techniques, even for quantification, though in certain cases they could lack specificity and could be affected by matrix interferences.

For synthetic food colorants, all the respective references containing details about the tested matrices, analytical techniques, detection and settings, analytical columns if used, elution, mobile phases, injection volumes, and figures of merit (LOD, LOQ) are presented below in Table 3.

As it could be extrapolated from Table 3, a significant number of LC-MS, LC-MS/MS or LC-UV/Vis methods are available, which are dedicated to simultaneous detection of either a significant or limited number of artificial colorants (whether authorized or delisted), even including illegal Sudan-type dyes. In addition, to Table 3, Figure 3 gives the percentage distribution of the analytical techniques, regarding the analysis of synthetic food colorants. It could be easily concluded that HPLC/U(H)PLC is the most frequently applied technique, followed by capillary electrophoresis and enzyme-linked immunosorbent assay (ELISA) as well as other residual methods. In the case of ELISA, it needs to be highlighted that it cannot be applied for a group of substances/food colorants but only for standalone substances, for which the monoclonal antibodies have been developed.

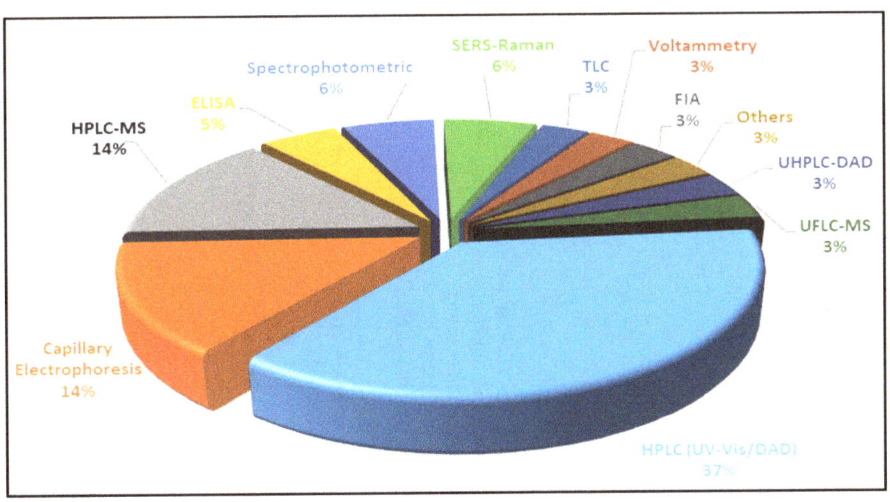

Figure 3. Distribution of techniques for the analysis of synthetic food colorants.

Table 3. Analytical techniques for the determination of natural food colorants in food samples.

Food Colorant	MATRIX	Analytical Technique	Detection	Column	Elution	Mobile Phase	Inj. Volume	Figures of Merit (LOD, LOQ)	Ref.
Brilliant blue	Liquid foods	CE	UV (λ = 220 nm) 36 cm capillary; Separation voltage (8 kV, −8 kV);	Fused-silica cappilaries of 375 od μm and 75 μm i.d	-	30 mM PBS buffer (pH6), with 0.9 mg/mL dASNPs and 2 mM β-cyclodextrin (CD)	Electro kinetic injection	LOD = 0.36 mg/L LOQ = 0.63 mg/L	[2]
Amaranth, ponceau 4R, sunset yellow, tartrazine, brilliant blue		spectrophotometric kinetic method	UV-Vis $\lambda_{Prussian\ blue}$ = 760 nm	n/a	n/a	n/a	n/a	LOD = 0.2–6.0 mg/L	[9]
Sudan I	Non-alcoholic drinks, sweets, jellies	Enzyme-linked Immuno-sorbent assay (ELISA)	n/a	n/a	n/a	n/a	n/a	LOD = 0.07 ng/mL	[29]
Sudan I	Non-alcoholic drinks, jellies	HPLC	UV- Vis (478 nm)	C_{18} (250 × 4.6 mm, 5.0 μm)	Gradient	methanol/2% CH_3COOH	20 μL	LOD = 0.14 ng/mL	[29]
Tartrazine, quinoline Yellow, sunset yellow, Carmoisine, Amaranth, ponceau 4R, Erythrosine, Red 2G, allura Red AC, Patent Blue V, Indigo Carmine, brilliant blue, Green S	Beverages, dairy powders, jellies, candies, condiments, icings, syrups, extracts	HPLC	DAD Various wavelengths	Discovery C18 (250 mm × 4.6 mm 5 μm)	Gradient	CH_3COONH_4 0.13 M (pH = 7.5; NaOH)/methanol: acetonitrile 80:20 v/v	20 μL	LOD = 1.87–22.1 μg/L	[49]
Tartrazine, sunset yellow, brilliant Blue, acid red	Powder	SERS-Raman	confocal microscope Raman spectrometer system	n/a	n/a	n/a	n/a	LOD = 10^{-7} M	[50]
Allura red, sunset yellow, tartrazine	Soft drinks	HPLC	DAD	n/a	Gradient	methanol (HPLC grade) and NaH_2PO_4/ Na_2HPO_4 buffer (0.10 M, pH = 7.0).	n/a	LOD = 0.06–0.30 μg/mL	[51]
Allura red, sunset yellow, tartrazine	Soft drinks	HLA-Go	-	-	-	-	-	-	[51]
Azorubine, amaranth, cochineal red A, red 2G, allura red, azocarmine B (AZO B), azocarmine G (AZO G), ponceau 2R, ponceau 6R, tartrazine, sunset yellow, quinoline yellow, orange II, metanil yellow (MY), patent blue V, indigo carmine, brilliant blue	Solid food/liquid beverages	HPLC	DAD DAD, $\lambda_{quant.}$: −620 nm (blue); −515 nm (red); −420 and 480 nm (yellow).	C8 (150 × 4.6 mm, 3 μm)	Gradient	Acetonitrile/sodium acetate (pH = 7)	20 μL	5–300 mg/kg (solid food samples) 5–100 mg/L (drinks)	[52]

Table 3. *Cont.*

Food Colorant	MATRIX	Analytical Technique	Detection	Column	Elution	Mobile Phase	Inj. Volume	Figures of Merit (LOD, LOQ)	Ref.
Brilliant blue, Tartrazine, amaranth, carmine, sunset yellow, allura red, erythrosine	Wine and soft drinks	UFLC	ESI (−)-MS/MS	ODS II (100 mm × 2.0 mm; 2.2 µm)	Gradient	A: Acetonitrile: CH_3COONH_4 5.0 mM/ (B) H_2O: CH_3COONH_4 5.0 mM	5 µL	LOD = 0.45–1.51 µg/L LOQ = 1.51–5.00 µg/L	[53]
Brilliant blue, tartrazine, amaranth, sunset yellow	Wine and soft drinks	TLC-UV-Vis	UV-Vis	TLC-PET 20 × 20 silica gel	n/a	8 mL 2-propanol and 3 mL NH_4OH	5 µL (standards) and 30 µL sample	n/a	[54]
Allura red, sunset yellow, tartrazine	Solid food/liquid beverages	Spectrophoto-metric BLLS/RBL	Absorbance spectra-pH data Spectral measurements (300–600 nm) at different pH	n.a	n.a	n.a	n.a	LOD = 0.54 mg/L	[55]
Sunset yellow	Beverage	HPLC	ESI (−)-MS	C18-ether column (150 mm × 4.6 mm, 5 µm)	Isocratic	(A) 63% aqueous solution 20 mM CH_3COONH_4: (B) 37% methanol	20 µL	n/a	[56]
Carmoisine, sunset yellow	Beverage	HPLC	ESI (−)-MS	C18 column (250 mm × 2 mm, 4 µm)	Isocratic	(A) Methanol and (B) 10 mM $HCOONH_4$ (45:55, *v*/*v*)	20 µL	LOD = 10–12 µg/L	[57]
Allura red	Beverage	HPLC	ESI (−)-MS	HSS-T3 column (2.1 mm 100 mm, 1.8 µm)	Gradient	A: H_2O: CH_3COONH_4 1.0 mM/ (B) Methanol: CH_3COONH_4 1.0 mM	20 µL	n/a	[58,59]
Brilliant blue, tartrazine, allura red, amaranth, Azorubine, patent Blue V, ponceau 4R	Various food products	HPLC	DAD Various wavelengths	Xterra RP18 column (250 × 4.6 mm, 5 µm)	Gradient	A) 0.1 M CH_3COONH_4 in water and (B) 0.1 M CH_3COONH_4 in methanol	20 µL	LOD = 0.02–1.49 mg/L	[60]
Brilliant blue, indigo carmine, allura red, carminic acid, ponceau 4R, sunset yellow, tartrazine	Dairy powders, color beverages, jellies, candies, condiments, icings, syrups,	CE	UV (200 nm) Condiiton with 1 M NaOH, H_2O electrode polarity (25kV)	-	-	Running buffer of pH 10 (20 mM NaOH solution to 15 mM disodium tetraborate (borax) to 20 mM NaOH, until the desired pH	Large-volume injection	LODs 0.05–0.40 µg/mL	[61]
Brilliant blue, indigo carmine, allura red, carminic acid, ponceau 4R, sunset yellow, tartrazine, fast green FCF	Liquid foods	CE	UV (λ = 200 nm) Condiiton with 1 M NaOH, H_2O electrode polarity (25kV)	-		Running buffer of pH 10 (20 mM NaOH to 15 mM disodium tetraborate (borax) to 20 mM NaOH, until desired pH	Large-volume injection	LOD = 0.002–0.026 µg/mL	[62]

Table 3. *Cont.*

Food Colorant	MATRIX	Analytical Technique	Detection	Column	Elution	Mobile Phase	Inj. Volume	Figures of Merit (LOD, LOQ)	Ref.
Sunset yellow, carmoisine, amaranth, ponceau 4R, erythrosine, red 2G, allura red	Soft drinks	HPLC	DAD Various wavelengths	Symmetry C18 (Waters, Milford, USA) column (150 mm × 4.6 mm, 5 µm)	Gradient	CH_3COONH_4 buffer (1% *w/v*) (0.13 M) (pH: 7.5) by addition of 0.1 M aq. NH_3 (solvent A), methanol (solvent B) and acetonitrile (solvent C)	n/a	0.5–1.4 µg/mL	[63]
Tartrazine, quinoline yellow, sunset yellow, carmoisine, ponceau 4R, allura red, indigo carmine, brilliant blue	Various foods and medicines	HPLC	DAD Various wavelengths	C18 column (250 mm × 4.6 mm, 5 µm)	Isocratic	A) Triton X-100 (0.25%, *v/v*) and (B) 50 mmol/L PBS (pH 7)	20 µL	LOD = 0.05–0.44 µg/mL LOQ = 0.05–1.12 µg/mL	[64]
Tartrazine, sunset yellow, azorubine, amaranth, cochineal red, red 2G, allura red AC, Brilliant Black BN, brown FK, Brown HT Patent Blue V, brilliant Blue FCF, Green S	Fish roe	HPLC	DAD Various wavelengths	Xterra RP18 column (250 × 4.6 mm, 5 µm)	Gradient	(a) 100 mmol/L CH_3COONa buffer (pH 7.0) and (b) Acetonitrile	20 µL	LOD = 0.02–1.49 mg/L	[65]
Brilliant blue, Indigo carmine, allura red, erythrosine, ponceau 4R, sunset yellow, Lemon yellow	Protein-rich samples	HPLC	DAD Various wavelengths	RP-C18 Column	Gradient	Methanol- 20 mM of CH_3COONH_4	20 µL	LOD = 0.1–0.4 mg/kg	[66]
Brilliant blue, tartrazine, sunset yellow, amaranth, carminic acid, acid red, allura red	Meat products	UHPLC	PDA Various wavelengths	BEH C18 (100 × 2.1 mm, 1.7 µm	Gradient	Acetonitrile/ CH_3COONH_4	5 µL	LOD = 0.01 mg/kg LOQ = 0.05 mg/kg	[67]
Carminic acid, sunset yellow, tartrazine	Non-alcoholic drinks, sweets, jellies	Capillary Electrophoresis (CE)	UV (λ = 220 nm) 36 cm capillary; Separation voltage (8 kV, −8 kV);	Fused-silica capillaries of 375 od µm and 75 µm i.d	-	30 mM PBS buffer (pH6), with 0.9 mg/mL dASNPs and 2 mM β-cyclodextrin (CD)	Electrokinetic injection	LOD = 0.03 -0.072 mg/L LOQ = 0.16–0.31 mg/L	[68]
Amaranth, Ponceau 4R, Sunset yellow, tartrazine, Sudan I-IV	Soft drinks/solid samples	HPLC-ESI (+)-MS	ESI (+)-MS	Spherigel C18 (250 × 4.6 mm, 5 µm)	gradient	aq. methanol 0.1% HCOOH/aqueous methanol 20 mM CH_3COONH_4/1% CH_3COOH	20 µL	LOD = 2.0–3.5 ng LOQ = 5.4–10.5 ng	[69]
Brilliant blue, allura red, amaranth, Erythrosine, ponceau 4R, sunset yellow, tartrazine	Soft drinks and processed meats	HPLC	DAD Various wavelengths	Inertsil ODS-SP column (250 × 4.6 mm, 5 µm)	Gradient	CH_3COONH_4 (0.1 M, pH = 7.2)—methanol-acetonitrile (9:1 *v/v*)	20 µL	LOD: 0.005 µg/mL LOQ: 0.018 µg/mL	[70]

Table 3. *Cont.*

Food Colorant	MATRIX	Analytical Technique	Detection	Column	Elution	Mobile Phase	Inj. Volume	Figures of Merit (LOD, LOQ)	Ref.
Brilliant blue, sunset yellow, tartrazine	Dairy powders, beverages, jellies, candies, syrups, extracts	Flow injection (FIA)	Amperometric detection (boron-doped diamondelectrode)	-	-	$E_{det.}2 = -450$ mV (100 ms duration) vs. Ag/AgCl (3.0 M KCl).		LOD = 0.8–3.5 μM	[71]
Brilliant blue, sunset yellow, tartrazine	Beverages, jellies, candies, condiments, icings, syrups	Differential pulse voltammetry (DPV)	Cathodically pretreated boron-doped diamond (BDD) elctrode	-	-	30 Hz; amplitude(a), 40 mV; for DPV scan rate (v), 0 mVs	-	LOD = 13.1–143 nM	[72]
Erythrosine, carmoisine, amaranth, ponceau 4R, Red 3G	Syrups	Capillary electrophoresis (CE)	Laser-induced fluorescence detection Various $\lambda_{exc.}/\lambda_{emis.}$: (Nd:YAG laser with wavelength 532 nm and power 5 mW)	Fused silica capillary with I.D. 50 μm, O.D. 360 μm, length 30 cm	n/a	V = +17 kV (intensity of electrical field was 460 Vcm^{-1}).	n/a	LOD = 0.2–0.4 μg/mL	[73]
Tartrazine, sunset yellow, azorrubine, bordeaux S, ponceau 4R, erytrosine, red no 40, patent blue V, indigo carmin, brilliant blue	Alcoholic beverages	CE	UV/vis PBS 10 mM with sodium dodecyl sulfate 10 mM, pH 11, and +25 kV of voltage	Fused silica capillary (73 cm)	n/a	Phosphate buffer solution of 10 mM with sodium dodecyl sulfate 10 mM, pH 11, and +25 kV voltage	n/a	LOD = 0.4–2.5 μg/mL LOQ = μg/mL	[74]
Tartazine, Amaranth, Sunset yellow, allura red, Lutein, lycopene, β-carotene	Various foodstuff	HPLC	DAD Various wavelengths	C18 column (250 mm × 4.6 mm, 5 μm)	Gradient	1% CH_3COONH_4, methanol and acetone;	20 μL	LOD = 0.2–50 ng/mL	[75]
Tartrazine, Quinoline yellow, Sunset Yellow, Carmoisine, Brilliant Blue	Solid foods	spectrophotometric method	UV-Vis Various wavelengths	n/a	n/a	n/a	n/a	LOQ = 1–5 μg/mL	[76]
Allura red	Liquid foods	spectrophotometric method	UV-Vis (506 nm)	n/a	n/a	n/a	n/a	LOD = 2.35 μg/L	[77]

Table 3. *Cont.*

Food Colorant	MATRIX	Analytical Technique	Detection	Column	Elution	Mobile Phase	Inj. Volume	Figures of Merit (LOD, LOQ)	Ref.
Tartrazine, New red, Amaranth, Ponceau 4R, Sunset yellow, Allura red, Acid red, Brilliant Blue, Acid red, Erythrosine, Acid orange, Basic flavine O, Basic orange, Siperse blue 106, Crystal violet, Leucine malachite green, Leucine crystal violet	Meat	UHPLC	DAD (200–800 nm)	C18 column (2.1 mm × 50 mm, 1.7 µm)	Gradient	(A) 20 mM CH_3COONH_4–0.02% acetic acid (pH 5) and (B) acetonitrile	2 µL	LOD = 0.96–2.16 µg/kg LOQ = 1.61–7.19 µg/kg	[78]
New red, Amaranth, Camine, Sunset yellow, Acid Red G, Allura red, Acid Scarlett GR, Erythrosine, Rhodamine B, Sudan I, Para red, Sudan II, Sudan III, Sudan red 7B, Sudab IV, Sudan Orange G	Hotpot condiment	HPLC	DAD Various wavelengths	C18 column (4.6 mm × 250 mm, 5 µm)	Gradient	(A) Methanol; (B) 0.01 M PBS (pH = 7.5)	20 µL	LOD = 0.001–0.00 3 mg/kg	[79]
Allura Red, Ponceau 4R	Granulated drinks	UV–visible spectrophotometer	UV–visible spectrophotometer (ZCDS)	n/a	n/a	n/a	n/a	LOD = 0.059–0.102 µg/mL LOQ = 0.198–0.341 µg/mL	[80]
Brilliant Blue, Sunset Yellow, Tartrazine	Non-alcoholic drinks, sweets, jellies	HPLC	UV-Vis (630 nm, 480 nm, 430 nm)	Modified C18 column (250 × 4.6 mm, 5 µm) with a 0.25% (*v*/*v*) Triton X-100 aq. solution at pH 7	Isocratic	0.25 mL of Triton X-100 (Sigma) up to 100 mL with 50 mmol l–1 phosphate buffer solution at pH 7	20 µL	LOD = 0.143–0.080 mg/L	[81]
Tartrazine, Amaranth, Sunset Yellow, Allura red, Ponceau 4R, Erythrosine	Soft drinks, sugar and gelatin based confectionery	HPLC-UV	UV 430 nm, 510 nm	C18 column (250 mm × 4.6 mm, 5 µm	Gradient	(A) 0.1 mol/L ammonium acetate aqueous solution (pH 7.5, adjusted with 10 mol/L NaOH-methanol–acetonitrile (30:70, *v*/*v*)	20 µL	LOD = 0.015–0.32 ng/mL	[82]
Allurea red, Amaranth, Erythrosine, Ponceau 4R, Sunset Yellow	Beverages, alcoholic drinks and fish foe	SERS-Raman	Radiation of 514.5 nm from an air-cooled argon ion laser was used for SERS excitation	n/a	n/a	n/a	n/a	LOD = 10^{-7}–10^{-5} M	[83]
Sunset yellow	beverage, dried bean curd, braised pork	ELISA	n/a	n/a	n/a	n/a		LOD = 25 pg mL^{-1}	[84]

Table 3. *Cont.*

Food Colorant	MATRIX	Analytical Technique	Detection	Column	Elution	Mobile Phase	Inj. Volume	Figures of Merit (LOD, LOQ)	Ref.
(40 food colorants) Ponceau 6R, Tartrazine, Fast yellow AB, Amaranth, Indigotine, Naphthol yellow S, Chrysoine, Ponceau 4R, Sunset yellow FCF, Red 10B, Orange G, Acid violet 7, Brilliant black PN, Allura red AC, Yellow 2G, Red 2G, Uranine, Fast red E, Green S, Ponceau 2R, Azorubine, Orange I, Quinoline yellow, Martius yellow, Ponceau SX, Ponceau 3R, Fast green FCF, Eosine, Brilliant blue FCF, Orange II, Orange RN, Acid blue 1, Erythrosine, Amido black 10B, Acid red 52, Patent blue V, Acid green 9, Phloxine B, Benzyl violet 4B, Rose bengal	Drinks, syrups and candies	HPLC	HPLC–DAD	C18 column (50 mm × 4.6 mm, 1.8 μm)	Gradient	(A) 0.1 mol/L of CH_3COONH_4 pH 6.7 and (B) was Methanol–Acetonitrile (50:50, *v*/*v*)	5 μL	LOD = 0.03–0.1 μg/g	[85]
Ponceau 4R, Sunset Yellow, Allura Red, Azophloxine, Ponceauxylidine, Erythrosine, Orange II	Animal feed and meat	LC	ESI (−)-MS	C18 column (2.1 mm × 150 mm, 5 μm)	Gradient	(A) 20 mmol/L CH_3COONH_4: Acetonitrile		LOD = 0.02–21.83 ng/mL	[86]
New Coccine, Indigo Carmine, Erythrosine, Tartrazine, Sunset Yellow FCF, Fast Green FCF, Brilliant Blue FCF, Allura Red AC, Amaranth, Dimethyl Yellow, Fast Garnet GBC, Para Red, Sudan I, Sudan II, Sudan III, Sudan IV, Sudan Orange G, Sudan Red 7B, Sudan Red B, Sudan Red G	Chili powders; commercial syrup preserved fruits	LC	ESI (−) and ESI (+) MS/MS	Acclaim Polar Advantage C16 (3 mm, 4.6 × 150 mm)	Gradient	(A) Acetonitrile and (B) 20 mM CH_3COONH_4 −1.0% CH_3COOH		LOQ = 0.005–1 mg/kg	[87]
Multi-class (53 food colorants)	Spices	UHPLC	QTOF-MS (sequential window acquisition of all theoretical fragment-ion spectra-SWATH)	Acquity UPLC BEH C18 column (2.1 × 100 mm, 1.7 μm,)	Gradient	(A) Acetonitrile and (B) 10 mM CH_3COONH_4 pH = 6.7	n/a	n/a	[88]
Multi-class (34 water soluble synthetic food colorants)	Beverages, syrup, chewing gum	HPLC	DAD-IT-TOF/MS (λ = 200–700 nm)	Atlantis™ dC18 (4.6 mm × 250 mm, 5 μm)	Gradient	(A)20 mM $HCOONH_4$ buffer; (B) methanol/acetonitrile (1:1 *v*/*v*)	20 μL	LODs = 0.009–0.102 μg/mL; LOQs = 0.045–0.203 μg/mL	[89]

The applied analytical techniques are followed by proper detection approaches. In this framework, simple detector UV-Vis/DAD is mostly applied, followed by MS/MS detectors, UV-Vis spectrometry, and electrochemical detection. The UV-Vis/DAD detection wavelengths depend on the analyte color (i.e., blue, yellow, red) set in any case in the maximum absorbance.

Regarding the MS, listed and EU-approved food colorants could be analyzed in the negative ionization, while for other substances (i.e., Sudan I-IV) positive ionization is applied.

From observation among the available methods of analysis (Table 3 and Figure 3), it could be concluded that traditional TLC methods require a significant sample preparation step and a time-consuming analytical procedure. On the other hand, the HPLC methods need longer analysis time, compared to the respective LC-MS/MS methods, in order to obtain good separation for the same number of analytes [87–89].

As reported recently by Periat et al., full-scan screening methods using HR-MS (High Resolution Mass Spectrometry) have proven to be an alternative to triple quadrupole methods as they could maximize the number of control and analyzed target colorants. Main advantages of the HR-MS can be the reduced sample preparation and the combined targeted analysis with untargeted screening of food colorants with high MS resolving power. Quadropole Time-of-Flight (QTOF) used by Li et al. and by Periat et al. for the detection and identification of coloring compounds in spices provided not only mass accuracy but also MS/MS spectra information and thus increased selectivity. A drawback of the approach could be the high cost of the instrumentation [85,86]. As reported by Li et al., HR-MS accurate mass measurements can detect a large number of target analytes, avoiding isobaric interferences in complex samples [89]. A combination of an ESI (or APCI) ionization with an anion trap analyzer linked to a TOF mass analyzer (ESI/APCI-IT-TOF/MS) provides simultaneously multi tandem MS (up to MS^2) with respective mass accuracy. Currently, there is an increasing interest on the fragmentation mechanism of synthetic food dyes; use of ESI-IT-TOF/MS^n in positive as well as in negative ionization modes [87–89] has been increased.

5.4. Sample Preparation for the Determination of Synthetic Colorants in Foods

Currently, there is no generally accepted/standard method for synthetic colorant extraction in laboratories. Nevertheless, most extraction procedures follow a common approach, which normally involves firstly the release of desired analytes from their matrices, followed afterwards by removal of extraneous matter/interferences by applying an efficient extraction protocol (i.e., solid–liquid or liquid–liquid extraction) [90].

The applied sample preparation protocols are strongly dependent on the type and nature of the food sample. A short description of the sample preparation protocols, along with their application to the clean-up of food samples, for the analysis of synthetic food colorants is given in Table 4.

Membrane filtration involves the permeation of the analyte through a thin layer of material. Explicitly, in case of beverages, when filtration is involved, a degassing step needs to be done in advance, in order to remove CO_2 [90].

Solid phase extraction (SPE) is one of the most commonly used techniques in determination of food colorants, presenting certain advantages such as simplicity. Polyamide resin used for SPE cleanup retains polar compounds with chemical groups that can be protonated. In acidic pH, during SPE, the colorants are adsorbed to the polyamide stationary phase mainly by Van der Waals interactions. Other hydrophilic substances can mask SPE interaction sites by reducing their binding power for the colorants and consequently reducing the capacity of the cartridges. Some substances, such as amaranth, are strongly retained by SPE cartridges, and the ammonia solution used for elution could be insufficient for its release (low recoveries).

Table 4. Sample preparation techniques for the determination of synthetic food colorants in food samples.

Food Colorant	MATRIX	Extraction/Sample Preparation	Ref.
Amaranth, Ponceau 4R, Sunset Yellow, Tartrazine and Brilliant Blue		Reduction of iron (III) in sodium acetate/hydrochloric acid solution (pH 1.71) followed by a chromogenic reaction with potassium hexacyanoferrate (III) to form the Prussian blue species	[9]
Sudan I	Non-alcoholic drinks, sweets, jellies	Extraction, sonication, centrifugation, filtration. Sample extracts or liquid samples were filtrated	[29]
Tartrazine, quinoline Yellow, Sunset Yellow, Carmoisine, Amaranth, Ponceau 4R, Erythrosine, Red 2G, Allura Red AC, Patent Blue V, Indigo Carmine, Brilliant Blue FCF, Green S	Dairy powders, color beverages, jellies, candies, condiments, icings, syrups, extracts	Beverages: degassed by stirring Solid: dissolved in water, ultrasonication, filtration	[49]
Tartrazine, Sunset yellow, Brilliant Blue, Acid red	Powder	Fabrication of flower-like silver nanostructures by adding 10 mL of ultrapure water, 2 mL of PVP solution (1%) and 0.2 mL of silver nitrate solution (1 mol/L)	[50]
Allura Red, Sunset Yellow, and Tartrazine	Soft drinks	Food solutions were prepared with dilution with methanol–water mixture (*v/v*, 50/50)	[51]
Azorubine, amaranth, cochineal red A, red 2G, allura red, azocarmine B (AZO B), azocarmine G (AZO G), ponceau 2R, ponceau 6R, tartrazine, sunset yellow, quinoline yellow, orange II, metanil yellow (MY), patent blue V, indigo carmine and brilliant blue FCF.	Solid food/liquid beverages	4 g solid sample +20 mL ethanol-H_2O (1:1 *v/v*), ultrasound and shaking, centrifugation, separation, and SPE in polyamide (PA) cartridge Beverages: degas (ultrasound), diluted 1.1 with H_2O, centrifuge	[52]
Brilliant Blue FCF, Tartrazine, Amaranth, Carmine, Sunset yellow, Allura red, Erythrosine	Wine and soft drinks	Magnetic dispersive solid-phase extraction (M-dSPE):	[53]
Brilliant Blue FCF, Tartrazine, Amaranth, Sunset yellow,	Wine and soft drinks	Degassing followed by SPE with Sep-Pack C18 and elution with 2-propanol	[54]
Allura red, sunset yellow, tartrazine	Solid food/liquid beverages	Sample dilution (0.5–2.0 g) in 100 mL H_2O	[55]
Sunset yellow	Beverage	No extraction	[56]
Carmoisine, sunset yellow	Beverage	Samples were diluted with water and filtered through 0.2 μm polypropylene membrane	[57]
Allura red	Beverage	filtered through 0.45 μm nylon filter and diluted 1:20 (*v/v*) in ultrapure water	[58,59]
Brilliant blue, tartrazine, allura red Amaranth, Azorubine, Patent Blue V, Ponceau 4R	Various food products	Beverages: sample sonicated, addition of aq. NH_3, filtration; Solid: homogenization, addition of aq. NH_3, sonication, centrifugation	[60]
Brilliant Blue FCF, Indigo carmine, Allura red, carminic acid, Ponceau 4R, Sunset yellow, tartrazine	Dairy powders, color beverages, jellies, candies, condiments, icings, syrups, extracts	Flavored milk samples: diluted with ethanol (1:1 *v/v*), SPE with PA cartridge	[61]
Brilliant Blue FCF, Indigo carmine, Allura red, carminic acid, Ponceau 4R, Sunset yellow, tartrazine, fast green FCF	Liquid foods	Beverages: degas (ultrasound) and directly to CE Milk: diluted with ethanol (1:1 *v/v*), SPE with PA cartridge Jelly: blended with ethanol: H_2O 1:1 *v/v* at 65 °C × 4 h + SPE (Polyamide cartridge)	[62]
Sunset yellow, Carmoisine, Amaranth, Ponceau 4R, Erythrosine., Red 2G, Allura red	Soft drinks	Extraction stage, followed by sonification, centrifugation, and concentration step + clean up via SPE on polyamide (PA) cartridges	[63]
Tartrazine, Quinoline Yellow, Sunset Yellow, Carmoisine, Ponceau, 4R, Allura Red, Indigo Carmine, Brilliant Blue	Various foods and medicines	Homogenization, dissolution, filtration	[64]
Tartrazine, Sunset Yellow FCF, Azorubine Amaranth Cochineal Red, Red 2G), Allura Red AC, Brilliant Black BN, Brown FK and Brown HT, Patent Blue V, Brilliant Blue FCF, and Green S	Fish roe	Extraction with aq. NH_3, centrifugation, pH adjustment, addition of PA sorbent and extraction with methanol	[65]
Brilliant blue, Indigo carmine, allura red, erythrosine, ponceau 4R, sunset yellow, Lemon yellow	Protein-rich samples	Purification/deproteinization with chitosan	[66]

Table 4. *Cont.*

Food Colorant	MATRIX	Extraction/Sample Preparation	Ref.
Brilliant blue, tartrazine, sunset yellow, amaranth, carmininic acid, acid red, allura red	Meat products	ASE (static) with ethanol-H_2O_NH_3 75:24:1 *v/v/v*, 85oC for 10 min	[67]
Carminic acid, sunset yellow, tartrazine, brilliant blue.	Non-alcoholic drinks, sweets, jellies	d-SPME with diamino-moiety functionalized silica nanoparticles (dASNPs) and β-cyclodextrin (β-CD) and pseudo-stationary phases (PSPs). Optimization of pH (2.5), sorbent, amount of dASNPs; ionic strength; extraction time and mode; desorption time; interferences (no interferences identified)	[68]
Amaranth, Ponceau 4R, Sunset yellow, tartrazine, Sudan I-IV	Soft drinks/solid samples	Liquid: filtration, degassing Solid: homogenization, extraction with DMSO, sanitation, centrifugation, and filtration	[69]
Brilliant blue FCF, Allura red, Amaranth, Erythrosine, Ponceau 4R, Sunset Yellow, Tartrazine	Soft drinks and processed meats		[70]
Brilliant blue, sunset yellow, tartrazine	Dairy powders, color beverages, jellies, candies, condiments, icings, syrups, extracts	Homogenization, addition in 0.1 M H_2SO_4 (gelatin dissolution), ultrasonic and dilution with supporting electrolyte	[71]
Brilliant blue, sunset yellow, tartrazine	Dairy powders, color beverages, jellies, candies, condiments, icings, syrups, extracts	-	[72]
Erythrosine, carmoisine, amaranth, ponceau 4R, Red 3G	Syrups	Dilution to PBS (20 mM, pH 11) as a background electrolyte (BGE) in the ratio of 1:10 or 1:2 for CE–LIF	[73]
Tartrazine, sunset yellow, azorrubine, bordeaux S, ponceau 4R, erytrosine, red no 40, patent blue V, indigo carmine, brilliant blue FCF	Alcoholic beverages	degassed by mechanical agitation and filtered	[74]
Tartazine, Amaranth, Sunset yellow, Allura red, Lutein, Lycopene, β-carotene	Various foodstuff	Ultrasound-assisted solvent extraction: immersion to methanol, sonication, centrifugation, extraction with acetone, evaporation. Liquid: 0.5 mL + 1 mL methanol; Solid: 0.2 g + 1 mL methanol	[75]
Tartrazine, Quinoline yellow, Sunset Yellow, Carmoisine, Brilliant Blue	Solid foods	Solid: a portion of food was diluted in H_2O, centrifuges and diluted with equal volume of CH_3COOH 3 M Liquid: a portion of was diluted in a mixture of NH_3: ethanol 2:73%, mix, and centrifuged.	[76]
Allura red	Liquid foods	Sample filtered, pH adjusted (4.0), extraction with Acetonitrile by SPE (MCI GEL CHP20P resin)	[77]
Tartrazine, New red, Amaranth, Ponceau 4R, Sunset yellow, Allura red, Acid red, Brilliant Blue, Acid red, Erythrosine, Acid orange, Basic flavine O, Basic orange, Siperse blue 106, Crystal violet, Leucine malachite green, Leucine crystal violet	Meat	Microwave assisted extraction: sample with methanol/H_2O (95:5 *v/v*) followed by SPE (C18 column), evaporation to dryness and reconstitution with methanol	[78]
New red, Amaranth, Carmine, Sunset yellow, Acid Red G, Allura red, Acid Scarlett GR, Erythrosine, Rhodamine B, Sudan I, Para red, Sudan II, Sudan III, Sudan red 7B, Sudab IV, Sudan Orange G	Hotpot condiment	Direct solvent extraction: sample with solvent (acetone-methanol), vortex, centrifugation, evaporation, pH adjustment	[79]
Allura Red, Ponceau 4R	Granulated drinks	Powdered sample dissolved in distilled water,	[80]
Brilliant Blue, Sunset Yellow, Tartrazine	Non-alcoholic drinks, sweets, jellies	Solid: dilution in H_2O and filtration	[81]
Tartrazine, Amaranth, Sunset Yellow, Allura red, Ponceau 4R, Erythrosine	Soft drinks, sugar and gelatin based confectionery	Ionic liquid dispersive liquid phase microextraction. Sample + ionic liquid [1-Octyl-3-methylimidazolium tetrafluoroborate ($[C_8MIM][BF_4]$)] and methanol addition.	[82]

Table 4. *Cont.*

Food Colorant	MATRIX	Extraction/Sample Preparation	Ref.
Allura red, Amaranth, Erythrosine, Ponceau 4R, Sunset Yellow	Beverages, alcoholic drinks and fish foe	Synthesis of the G/Ag nanoparticle composite	[83]
Sunset yellow	Beverage, dried bean curd, braised pork		[84]
(40 food colorants) Ponceau 6R, Tartrazine, Fast yellow AB, Amaranth, Indigotine, Naphthol yellow S, Chrysoine, Ponceau 4R, Sunset yellow FCF, Red 10B, Orange G, Acid violet 7, Brilliant black PN, Allura red AC, Yellow 2G, Red 2G, Uranine, Fast red E, Green S, Ponceau 2R, Azorubine, Orange I, Quinoline yellow, Martius yellow, Ponceau SX, Ponceau 3R, Fast green FCF, Eosine, Brilliant blue FCF, Orange II, Orange RN, Acid blue 1, Erythrosine, Amido black 10B, Acid red 52, Patent blue V, Acid green 9, Phloxine B, Benzyl violet 4B, Rose bengal.	Drinks, syrups, and candies	Drinks: degas, evaporation Solid: grind, mixing with solvent and SPE with PA column with 1% NH_3/ethanol solution	[85]
Ponceau 4R, Sunset Yellow, Allura Red, Azophloxine, Ponceauxylidine, Erythrosine Orange II	Animal feed and meat	Homogenization, extraction with ethanol:NH_3:H_2O (80:1:19 *v*/*v*), evaporation, reconstitution	[86]
New Coccine, Indigo Carmine, Erythrosine, Tartrazine, Sunset Yellow FCF, Fast Green FCF, Brilliant Blue FCF, Allura Red AC, Amaranth, Dimethyl Yellow, Fast Garnet GBC, Para Red, Sudan I, Sudan II, Sudan III, Sudan IV, Sudan Orange G, Sudan Red 7B, Sudan Red B, Sudan Red G	Chili powders; commercial syrup preserved fruits	Homogenization and extraction with Acetonitrile twice	[87]
Multi-class (53 food colorants)	Spices	Extraction with H_2O/methanol/acetonitrile/THF, 9:1:5:5, *v*/*v*/*v*/*v*)	[88]
Multi-class (34 water soluble synthetic food colorants)	Beverages, syrup, chewing gum	Beverages: degassing, pH adjustment, and dilution. Syrup: dilution, sonication, and pH adjustment. Chewing gum: washing with water, pH adjustment.	[89]

Dispersive solid phase extraction (d-SPE) analysis is a simple sample preparation methodology that is suitable for a wide variety of food and agricultural products, as is also QuEChERS, introduced for pesticides from Anastassiades et al. [91]. In case of synthetic colorants, a modified QuEChERS method has been reported (magnetic-dSPE) using cross-linking magnetic polymer (NH_2-LDC-MP) containing less hydrophilic amino groups and more lipophilic styrene monomer for cleaning up the synthetic food colorants from wine and soft drinks [53].

Liquid–liquid extraction (LLE) deals with the separation of substances based on their relative solubility in two different immiscible liquids. Common solvents for the extraction of synthetic food colorants from food matrices are water, ethanol, methanol, isopropyl alcohol, ammoniac ethanol, ethyl acetate, ammonia, cyclohexane, and tetra-n-butyl ammonium phosphate. Wu et al. has also reported an extraction method based on Ionic liquid dispersive liquid phase microextraction using the ionic liquid (1-Octyl-3-methylimidazolium tetrafluoroborate ((C_8MIM)(BF_4))) [81].

In the literature, a limited number of protocols exists dealing with other types of extraction methods for synthetic food colorants, such as MAE and Ultrasound Assisted Extraction (UAE). These kinds of extractions require special instrumentation and most probably can be beneficial for a laboratory, as extractions with organic solvents are characterized by consumption of high volumes of solvents, are time consuming, and in some cases have low recoveries [90].

6. Conclusions

The use of food colorants in the production of foods leads to the need for the development of accurate, precise, sensitive, and selective analytical methods for their analysis and quantification. Certain interest in the impacts of food colorants is being reported worldwide. There is a plethora of analytical research works that deal with the analytical challenge of the analysis and quantification of either natural or synthetic food colorants. The research community gives more attention to the appropriate analysis, in sufficient concentration or mass fraction levels, mostly to synthetic food colorants rather than natural ones.

Analytical methodologies have much more to offer in this direction and, as it could be concluded from synthetic colorants, HPLC is the most frequently used followed by capillary electrophoresis. In terms of detection methods, the simple UV-Vis/DAD is the predominant one followed by tandem MS. The analytical techniques and sample preparation methodologies presented cover the existing methodologies mainly applied during the last decade.

Regarding sample preparation, this is highly sample dependent. It could involve the application of different extraction techniques, such as membrane filtration, liquid–liquid and solid phase extraction techniques, for cleaning-up the highly complex matrix of food products. Sample preparation is of great importance and must be carefully developed, in order to avoid or eliminate existing matrix interferences aiming to the development of simple, selective, and precise methods of extraction.

In the case of simple liquid samples, dilution and injection are preferred, though in other cases such as high protein content foods, specific steps need to be followed for sufficient sample clean-up.

Author Contributions: Conceptualization, E.T., H.G., and K.N.; methodology, K.N.; formal analysis, E.T. and K.N.; investigation, K.N.; resources, E.T. and K.N.; data curation, E.T.; writing—original draft preparation, K.N.; writing—review and editing, H.G. and E.T.; visualization, E.T.; supervision, E.T.; project administration, E.T. All authors have read and agreed to the published version of the manuscript.

Acknowledgments: We acknowledge support of this work by the project "FoodOmicsGR Comprehensive Characterisation of Foods" (MIS 5029057) which is implemented under the Action "Reinforcement of the Research and Innovation Infrastructure", funded by the Operational Programme Competitiveness, Entrepreneurship and Innovation (NSRF2014-2020) and co-financed by Greece and the European Union (European Regional Development Fund).

References

1. Codex Alimentarius. Available online: http://www.codexalimentarius.org/standards/gsfa/ (accessed on 18 September 2018).
2. Carocho, M.; Morales, P.; Ferreira, I.C.F.R. Natural food additives: Quo vadis? *Trans Food Sci. Technol.* **2015**, *45*, 284–295. [CrossRef]
3. Amchova, P.; Kotolova, H.; Ruda-Kucerova, J. Health safety issues of synthetic food colorants. *Reg. Toxic. Pharm.* **2015**, *73*, 914–922. [CrossRef]
4. Aberamound, A. A Review Article on Edible Pigments Properties and Sources as Natural Biocolorants in Foodstuff and Food Industry. *World J. Dairy Food Sci.* **2011**, *6*, 71–78.
5. Burrows, A.J.D. Palette of Our Palates: A Brief History of Food Coloring and Its Regulation. *Compreh. Rev. Food Sci. Food Saf.* **2009**, *9*, 394–408. [CrossRef]
6. Martins, N.; Roriz, C.L.; Morales, P.; Barros, L.; Ferreira, I.C. Food colorants: Challenges, opportunities and current desires of agro-industries to en-sure consumer expectations and regulatory practices. *Trends Food Sci. Technol.* **2016**, *52*, 1–15. [CrossRef]
7. Rodriguez-Amaya, D.B. Natural food pigments and colorants. *Curr. Opin. Food Sci.* **2016**, *7*, 20–26. [CrossRef]
8. Tumolo, T.; Lanfer-Marquez, U.M. Copper chlorophyllin: A food colorant with bioactive properties? *Food Res. Int.* **2012**, *46*, 451–459. [CrossRef]
9. Ni, Y.; Wang, Y.; Kokot, S. Simultaneous kinetic spectrophotometric analysis of five synthetic food colorants with the aid of chemometrics. *Talanta* **2009**, *78*, 432–441. [CrossRef]
10. Sigurdson, G.T.; Tang, P.; Giusti, M.M. Natural colorants: food colorants from natural sources. *Annu. Rev. Food Sci. Technol.* **2017**, *8*, 261–280. [CrossRef]
11. Carocho, M.; Barreiro, M.F.; Morales, P.; Ferreira, I.C. Adding molecules to food, pros and cons: A review on synthetic and natural food additives. *Compr. Rev. Food Sci. Food Saf.* **2014**, *13*, 377–399. [CrossRef]

12. Dias, M.I.; Ferreira, I.C.; Barreiro, M.F. Microencapsulation of bioactives for food applications. *Food Funct.* **2015**, *6*, 1035–1052. [CrossRef] [PubMed]
13. Patakova, P. Monascus secondary metabolites: Production and biological activity. *J. Ind. Microbiol. Biotechnol.* **2013**, *40*, 169–181. [CrossRef] [PubMed]
14. Wang, C.; Chen, D.; Chen, M.; Wang, Y.; Li, Z.; Li, F. Stimulatory effects of blue light on the growth, monascin and ankaflavin production in Monascus. *Biotechnol. Lett.* **2015**, *37*, 1043–1048. [CrossRef]
15. Dias, M.G.; Camoes, M.F.G.; Oliveira, L. Carotenoids in traditional Portuguese fruits and vegetables. *Food Chem.* **2009**, *113*, 808–815. [CrossRef]
16. Carocho, M.; Ferreira, I.C. A review on antioxidants, prooxidants and related controversy: Natural and synthetic compounds, screening and analysis methodologies and future perspectives. *Food Chem. Toxicol.* **2013**, *51*, 15–25. [CrossRef]
17. Masone, D.; Chanforan, C. Study on the interaction of artificial and natural food colorants with human serum albumin: A computational point of view. *Comput. Biol. Chem.* **2015**, *56*, 152–158. [CrossRef]
18. Oplatowska-Stachowiak, M.; Elliottt, C.T. Food colors: Existing and emerging food safety concerns. *Crit. Rev. Food Sci. Nutr.* **2017**, *57*, 524–548. [CrossRef]
19. Feng, J.; Cerniglia, C.E.; Chen, H. Toxicological significance of azo dye metabolism by human intestinal microbiota. *Front. Biosci.* **2012**, *4*, 568–586. [CrossRef]
20. Golka, K.; Kopps, S.; Myslak, Z.W. Carcinogenicity of azo colorants: Influence of solubility and bioavailability. *Toxicol. Lett.* **2004**, *151*, 203–210. [CrossRef]
21. Vojdani, A.; Vojdani, C. Immune reactivity to food coloring. *Altern. Ther. Health Med.* **2015**, *21*, 52–62.
22. Mpountoukas, P.; Pantazaki, A.; Kostareli, E.; Christodoulou, P.; Kareli, D.; Poliliou, S.; Mourelatos, C.; Lambropoulou, V.; Lialiaris, T. Cytogenetic evaluation and DNA interaction studies of the food colorants amaranth, erythrosine and tartrazine. *Food Chem. Toxic.* **2010**, *48*, 2934–2944. [CrossRef]
23. Pan, X.; Qin, P.; Liu, R.; Wang, J. Characterizing the Interaction between tartra-zine and two serum albumins by a hybrid spectroscopic approach. *J. Agric. Food Chem.* **2011**, *59*, 6650–6656. [CrossRef] [PubMed]
24. Basu, A.; Kumar, G.S. Study on the interaction of the toxic food additive Carmoisine with serum albumins: A microcalorimetric investigation. *J. Hazard Mater.* **2014**, *273*, 200–206. [CrossRef] [PubMed]
25. Datta, S.; Mahapatra, N.; Halder, M. pH-insensitive electrostatic interaction of carmoisine with two serum proteins: A possible caution on its uses in food and pharmaceutical industry. *J. Photochem. Photobiol. B* **2013**, *124*, 50–62. [CrossRef] [PubMed]
26. Wang, L.; Zhang, G.; Wang, Y. Binding properties of food colorant Allura Red with human serum albumin in vitro. *Mol. Biol. Rep.* **2014**, *41*, 3381–3391. [CrossRef]
27. Wu, D.; Yan, J.; Wang, J.; Wang, Q.; Li, H. Characterisation of interaction between food colourant A Allura red AC and human serum albumin: Multi-spectroscopic analyses and docking simulations. *Food Chem.* **2015**, *170*, 423–429. [CrossRef]
28. Tellier, F.; Steibel, J.; Chabrier, R.; Ble, F.X.; Tubaldo, H.; Rasata, R.; Chambron, J.; Duportail, G.; Simon, H.; Rodier, J.F.; et al. Sentinel lymph nodes fluorescence detection and imaging using Patent Blue V bound to human serum albumin. *Biomed. Opt. Exp.* **2012**, *3*, 2306–2316. [CrossRef]
29. Wang, Y.; Wei, D.; Yang, H.; Yang, Y.; Xing, W.; Li, Y. Development of a highly sensitive and specific monoclonal antibody-based enzyme-linked immunosorbent assay (ELISA) for detection of Sudan I in food samples. *Talanta* **2009**, *77*, 1783–1789. [CrossRef]
30. Moller, P.; Wallin, H. Genotoxic hazards of azo pigments and other colorants related to 1-phenylazo-2-hydroxynaphthalene. *Mutat. Res.* **2000**, *462*, 13–30. [CrossRef]
31. Stiborova, M.; Martinek, V.; Rydlova, H.; Hodek, P.; Frei, E. Sudan I is a Potential Carcinogen for Humans: Evidence for Its Metabolic Activation and Detoxication by Human Recombinant Cytochrome P450 1A1 and Liver Microsomes. *Cancer Res.* **2002**, *62*, 5678–5684.
32. European Food Safety Authority. Available online: http://www.efsa.europa.eu (accessed on 15 September 2018).
33. Lehto, S.; Buchweitz, M.; Klimm, A.; Straßburger, R.; Bechtold, C.; Ulberth, F. Comparison of food colour regulations in the EU and the US: A review of cur-rent provisions. *Food Addit. Contam.* **2017**, *34*, 335–355. [CrossRef] [PubMed]
34. Dykes, L.; Rooney, W.L.; Rooney, L.W. Evaluation of phenolics and antioxidant activity of black sorghum hybrids. *J. Cereal Sci.* **2013**, *58*, 278–283. [CrossRef]

35. Sivakumar, V.; Vijaeeswarri, J.; Anna, J.L. Effective natural dye extraction from different plant materials using ultrasound. *Ind. Crop. Prod.* **2011**, *33*, 116–122. [CrossRef]
36. Sagdic, O.; Ekici, L.; Ozturk, I.; Tekinay, T.; Polat, B.; Tastemur, B.; Senturk, B. Cytotoxic and bioactive properties of different color tulip flowers and degradation kinetic of tulip flower anthocyanins. *Food Chem. Toxicol.* **2013**, *58*, 432–439. [CrossRef] [PubMed]
37. Longo, L.; Scardino, A.; Vasapollo, G. Identification and quantification of anthocyanins in the berries of *Pistacia lentiscus* L., *Phillyrea latifolia* L. and *Rubia peregrina* L. *Innov. Food Sci. Emerg. Technol.* **2007**, *8*, 360–364. [CrossRef]
38. Stintzing, F.C.; Schieber, A.; Carle, R. Betacyanins in fruits from red-purple pitaya, Hylocereus polyrhizus (Weber) Britton & Rose. *Food Chem.* **2002**, *77*, 101–106.
39. Ravichandran, K.; Saw, N.M.M.T.; Mohdaly, A.A.; Gabr, A.M.; Kastell, A.; Riedel, H.; Cai, Z.-Z.; Knorr, D.; Smetanska, I. Impact of processing of red beet on betalain content and antioxidant activity. *Food Res. Int.* **2013**, *50*, 670–675. [CrossRef]
40. Cassano, A.; Conidi, C.; Drioli, E. Physico-chemical parameters of cactus pear (Opuntia ficus-indica) juice clarified by microfiltration and ultrafiltration processes. *Desalination* **2010**, *250*, 1101–1104. [CrossRef]
41. Otalora, M.C.; Carriazo, J.G.; Iturriaga, L.; Nazareno, M.A.; Osorio, C. Micro-encapsulation of betalains obtained from cactus fruit (Opuntia ficus- indica) by spray drying using cactus cladode mucilage and maltodextrin as encapsulating agents. *Food Chem.* **2015**, *187*, 174–181. [CrossRef]
42. Khan, M.I.; Giridhar, P. Enhanced chemical stability, chromatic properties and regeneration of betalains in Rivina humilis L. berry juice. *LWT—Food Sci. Technol.* **2014**, *58*, 649–657. [CrossRef]
43. Sun, M.; Temelli, F. Supercritical carbon dioxide extraction of carotenoids from carrot using canola oil as a continuous co-solvent. *J. Supercrit. Fluids* **2006**, *37*, 397–408. [CrossRef]
44. Sobral, D.; Costa, R.G.B.; Machado, G.M.; Paula, J.C.J.; Teodoro, V.A.M.; Nunes, N.M.; Pinto, M.S. Can lutein replace annatto in the manufacture of Prato cheese? *LWT—Food Sci. Technol.* **2016**, *68*, 349–355. [CrossRef]
45. Khalil, M.; Raila, J.; Ali, M.; Islam, K.M.S.; Schenk, R.; Krause, J.-P.; Schweigert, F.H.; Rawel, H. Stability and bioavailability of lutein ester supplements from Tagetes flower prepared under food processing conditions. *J. Funct. Foods* **2012**, *4*, 602–610. [CrossRef]
46. Grewe, C.; Menge, S.; Griehl, C. Enantioselective separation of all-E- astaxanthin and its determination in microbial sources. *J. Chromatogr. A* **2007**, *1166*, 97–100. [CrossRef] [PubMed]
47. Wada, M.; Kido, H.; Ohyama, K.; Ichibangase, T.; Kishikawa, N.; Ohba, Y.; Nakashima, M.N.; Kuroda, N.; Nakashima, K. Chemiluminescent screening of quenching effects of natural colorants against reactive oxygen species: Evaluation of grape seed, monascus, gardenia and red radish extracts as multi-functional food additives. *Food Chem.* **2007**, *101*, 980–986. [CrossRef]
48. Thalhamer, B.; Buchberger, W. Adulteration of beetroot red and paprika extract based food colorant with Monascus red pigments and their detection by HPLC-QTof MS analyses. *Food Control* **2019**, *105*, 58–63. [CrossRef]
49. Minioti, K.S.; Sakellariou, C.F.; Thomaidis, N.S. Determination of 13 synthetic food colorants in water-soluble foods by reversed-phase high- performance liquid chromatography coupled with diode-array detector. *Anal. Chim. Acta* **2007**, *583*, 103–110. [CrossRef]
50. Ai, Y.-J.; Wu, Y.-X.; Dong, Q.-M.; Li, J.-B.; Xu, B.-J.; Yu, Z.; Ni, D. Rapid qualitative and quantitative determination of food colorants by both T Raman spectra and Surface-enhanced Raman Scattering (SERS). *Food Chem.* **2018**, *241*, 427–433. [CrossRef]
51. Al-Degs, Y.S. Determination of three dyes in commercial soft drinks using HLA/GO and liquid chromatography. *Food Chem.* **2009**, *117*, 485–490. [CrossRef]
52. Bonan, S.; Fedrizzi, G.; Menotta, S.; Elisabetta, C. Simultaneous determination of synthetic dyes in foodstuffs and beverages by high-performance liquid chromatography coupled with diode-array detector. *Dyes Pigment.* **2013**, *99*, 36–40. [CrossRef]
53. Chen, X.H.; Zhao, Y.G.; Shen, H.Y.; Zhou, L.X.; Pan, S.D.; Jin, M.C. Fast determination of seven synthetic pigments from wine and soft drinks using magnetic dispersive solid-phase extraction followed by liquid chromatography–tandem mass spectrometry. *J. Chromatogr. A* **2014**, *1346*, 123–128. [CrossRef]
54. De Andrade, F.I.; Guedes, M.I.F.; Vieira, Í.G.; Mendes, F.N.P.; Rodrigues, P.A.S.; Maia, C.S.C.; Ávila, M.M.M.; Ribeiro, L.d.M. Determination of synthetic food dyes in commercial soft drinks by TLC and ion-pair HPLC. *Food Chem.* **2014**, *157*, 193–198. [CrossRef] [PubMed]

55. El-Sheikh, A.H.; Al-Degs, Y.S. Spectrophotometric determination of food dyes in soft drinks by second order multivariate calibration of the absorbance spectra-pH data matrices. *Dyes Pigments* **2013**, *97*, 330–339. [CrossRef]
56. Gosetti, F.; Gianotti, V.; Polati, S.; Gennaro, M.C. HPLC–MS degradation study of E110 Sunset Yellow FCF in a commercial beverage. *J. Chromatogr. A* **2005**, *1090*, 107–115. [CrossRef] [PubMed]
57. Gosetti, F.; Frascarolo, P.; Mazzucco, E.; Gianotti, V.; Bottaro, M.; Gennaro, M.C. Photodegradation of E110 and E122 dyes in a commercial aperitif: A high performance liquid chromatography–diode array–tandem mass spectrometry study. *J. Chromatogr. A* **2008**, *1202*, 58–63. [CrossRef]
58. Gosetti, F.; Chiuminatto, U.; Mazzucco, E.; Calabrese, G.; Gennaro, M.C.; Marengo, E. Identification of photodegradation products of Allura Red AC (E129) in a beverage by ultra-high performance liquid chromatography–quadrupole- time-of-flight mass spectrometry. *Anal. Chim. Acta* **2012**, *746*, 84–89. [CrossRef]
59. Gosetti, F.; Chiuminatto, U.; Mazzucco, E.; Calabrese, G.; Gennaro, M.C.; Marengo, E. Non-target screening of Allura Red AC photodegradation products in a beverage through ultra-high performance liquid chromatography coupled with hybrid triple quadrupole/linear ion trap mass spectrometry. *Food Chem.* **2013**, *136*, 617–623. [CrossRef]
60. Harp, B.P.; Miranda-Bermudez, E.; Barrows, J.N. Determination of seven certified color additives in food products using liquid chromatography. *J. Agric. Food Chem.* **2013**, *61*, 3726–3736. [CrossRef]
61. Huang, H.-Y.; Shih, Y.-C.; Chen, Y.-C. Determining eight colorants in milk beverages by capillary electrophoresis. *J. Chromatogr. A* **2002**, *959*, 317–325. [CrossRef]
62. Huang, H.-Y.; Chiu, C.-W.; Sue, S.-L.; Cheng, C.-F. Analysis of food colorants by capillary electrophoresis with large- volume sample stacking. *J. Chromatogr. A* **2003**, *995*, 29–36. [CrossRef]
63. Karanikolopoulos, G.; Gerakis, A.; Papadopoulou, K.; Mastrantoni, I. Determination of synthetic food colorants in fish products by an HPLC-DAD method. *Food Chem.* **2015**, *177*, 197–203. [CrossRef] [PubMed]
64. Khanavi, M.; Hajimahmoodi, M.; Ranjbar, A.M.; Oveisi, M.R.; Ardekani, M.R.S.; Mogaddam, G. Development of a green chromatographic method for simultaneous determination of food colorants. *Food Anal. Methods* **2012**, *5*, 408–415. [CrossRef]
65. Kirschbaum, J.; Krause, C.; Brückner, H. Liquid chromatographic quantification of synthetic colorants in fish roe and caviar. *Eur. Food Res. Technol.* **2006**, *222*, 572–579. [CrossRef]
66. Kong, C.; Fodjo, E.K.; Li, D.; Cai, Y.; Huang, D.; Wang, Y.; Shen, X. Chitosan-based ad-sorption and freeze deproteinization: Improved extraction and purifica-tion of synthetic colorants from protein-rich food samples. *Food Chem.* **2015**, *188*, 240–247. [CrossRef]
67. Liao, Q.G.; Li, W.H.; Luo, L.G. Applicability of accelerated solvent extraction for synthetic colorants analysis in meat products with ultrahigh performance liquid chromatography–photodiode array detection. *Anal. Chim. Acta* **2012**, *716*, 128–132. [CrossRef]
68. Liu, F.-J.; Liu, C.-T.; Li, W.; Tang, A.-N. Dispersive solid-phase micro- extraction and capillary electrophoresis separation of food colorants in beverages using diamino moiety functionalized silica nanoparticles as both extractant and pseudostationary phase. *Talanta* **2015**, *132*, 366–372. [CrossRef]
69. Ma, M.; Luo, X.; Chen, B.; Su, S.; Yao, S. Simultaneous determination of water-soluble and fat-soluble synthetic colorants in foodstuff by high- perfo-mance liquid chromatography-diode array detection-electrospray mass spectrometry. *J. Chromatogr. A* **2006**, *1103*, 170–176. [CrossRef]
70. Ma, K.; Yang, Y.N.; Jiang, X.X.; Zhao, M.; Cai, Y.Q. Simultaneous determination of 20 food additives by high performance liquid chromatography with photo-diode array detector. *Chin. Chem. Lett.* **2012**, *23*, 492–495. [CrossRef]
71. Medeiros, R.A.; Lourencao, B.C.; Rocha-Filho, R.C.; Fatibello-Filho, O. Flow injection simultaneous determination of synthetic colorants in food using multiple pulse amperometric detection with a boron-doped diamond electrode. *Talanta* **2012**, *99*, 883–889. [CrossRef]
72. Medeiros, R.A.; Lourencao, B.C.; Rocha-Filho, R.C.; Fatibello-Filho, O. Simultaneous voltammetric determination of synthetic colorants in food using a cathodically pretreated boron-doped diamond electrode. *Talanta* **2012**, *97*, 291–297. [CrossRef]
73. Ryvolova, M.; Taborský, P.; Vrabel, P.; Krasenský, P.; Preisler, J. Sensitive determination of erythrosine and other red food colorants using capillary electrophoresis with laser-induced fluorescence detection. *J. Chromatogr. A* **2007**, *1141*, 206–211. [CrossRef] [PubMed]

74. Prado, M.A.; Boas, L.F.V.; Bronze, M.R.; Godoy, H.T. Validation of methodology for simultaneous determination of synthetic dyes in alcoholic beverages by capillary electrophoresis. *J. Chromatogr. A* **2006**, *1136*, 231–236. [CrossRef] [PubMed]
75. Shen, Y.; Zhang, X.; Prinyawiwatkul, W.; Xu, Z. Simultaneous determination of red and yellow artificial food colourants and carotenoid pigments in food products. *Food Chem.* **2014**, *157*, 553–558. [CrossRef] [PubMed]
76. Sorouraddin, M.H.; Rostami, A.; Saadati, M.A. simple and portable multi-colour light emitting diode based photocolourimeter for the analysis of mixtures of five common food dyes. *Food Chem.* **2011**, *127*, 308–313. [CrossRef]
77. Soylak, M.; Unsal, Y.E.; Tuzen, M. Spectrophotometric determination of trace levels of allura red in water samples after separation and pre-concentration. *Food Chem. Toxicol.* **2011**, *49*, 1183–1187. [CrossRef]
78. Sun, H.; Sun, N.; Li, H.; Zhang, J.; Yang, Y. Development of multiresidue analysis for 21 synthetic colorants in meat by microwave-assisted extraction–solid-phase extraction–reversed-phase ultrahigh performance liquid chromatography. *Food Anal. Methods* **2013**, *6*, 1291–1299. [CrossRef]
79. Tang, B.; Xi, C.; Zou, Y.; Wang, G.; Li, X.; Zhang, L.; Chen, D.; Zhang, J. Simultaneous determination of 16 synthetic colorants in hotpot condiment by high performance liquid chromatography. *J. Chromatogr. B* **2014**, *960*, 87–91. [CrossRef]
80. Turak, F.; Ozgur, M.U. Simultaneous determination of allura red and ponceau 4R in drinks with the use of four derivative spectrophotometric methods and comparison with high-performance liquid chromatography. *J. AOAC Int.* **2013**, *96*, 1377–1386. [CrossRef]
81. Vidotti, E.C.; Costa, W.F.; Oliveira, C.C. Development of a green chromatographic method for determination of colorants in food samples. *Talanta* **2006**, *68*, 516–521. [CrossRef]
82. Wu, H.; Guo, J.B.; Du, L.M.; Tian, H.; Hao, C.X.; Wang, Z.F.; Wang, J.Y. A rapid shaking-based ionic liquid dispersive liquid phase micro extraction for the simultaneous determination of six synthetic food colourants in soft drinks, sugar-and gelatin-based confectionery by high-performance liquid chromatography. *Food Chem.* **2013**, *141*, 182–186. [CrossRef]
83. Xie, Y.; Li, Y.; Niu, L.; Wang, H.; Qian, H.; Yao, W. A novel surface-enhanced Raman scattering sensor to detect prohibited colorants in food by graphene/silver nanocomposite. *Talanta* **2012**, *100*, 32–37. [CrossRef] [PubMed]
84. Xing, Y.; Meng, M.; Xue, H.; Zhang, T.; Yin, Y.; Xi, R. Development of a polyclonal antibody-based enzyme-linked immunosorbent assay (ELISA) for detection of sunset yellow FCF in food samples. *Talanta* **2012**, *99*, 125–131. [CrossRef] [PubMed]
85. Yoshioka, N.; Ichihashi, K. Determination of 40 synthetic food colors in drinks and candies by high-performance liquid chromatography using a short column with photodiode array detection. *Talanta* **2008**, *74*, 1408–1413. [CrossRef] [PubMed]
86. Zou, T.; He, P.; Yasen, A.; Li, Z. Determination of seven synthetic dyes in animal feeds and meat by high performance liquid chromatography with diode array and tandem mass detectors. *Food Chem.* **2013**, *138*, 1742–1748. [CrossRef] [PubMed]
87. Tsai, C.-F.; Kuo, C.-H.; Shih, D.Y.-C. Determination of 20 synthetic dyes in chili powders and syrup-preserved fruits by liquid chromatography/tandem mass spectrometry. *J. Food Drug Anal.* **2015**, *23*, 453–462. [CrossRef] [PubMed]
88. Périat, A.; Bieri, S.; Mottier, N. SWATH-MS screening strategy for the determination of food dyes in spices by UHPLC-HRMS. *Food Chem. X* **2019**, *1*, 100009. [CrossRef]
89. Li, X.Q.; Zhang, Q.H.; Ma, K.; Li, H.M.; Guo, Z. Identification and determination of 34 water-soluble synthetic dyes in foodstuff by high performance liquid chromatography–diode array detection–ion trap time-of-flight tandem mass spectrometry. *Food Chem.* **2015**, *182*, 316–326. [CrossRef]
90. Yamjala, K.; Nainar, M.S.; Ramisetti, N.R. Methods for the analysis of azo dyes employed in food industry—A review. *Food Chem.* **2016**, *192*, 813–824. [CrossRef]
91. Anastassiades, M.; Lehotay, S.J.; Stajnbaher, D.; Schenck, F.J. Fast and easy multiresidue method employing acetonitrile extraction/partitioning and "dispersive solid-phase extraction" for the determination of pesticide residues in produce. *J. AOAC Int.* **2003**, *86*, 412–431.

12

Harmful Elements (Al, Cd, Cr, Ni, and Pb) in Wild Berries and Fruits Collected in Croatia

Michaela Zeiner [1,2,*] and Iva Juranović Cindrić [3]

1 Man-Environment-Technology Research Centre, School of Science and Technology, Örebro University, Gymnastikgatan 1, 70182 Örebro, Sweden
2 Division of Analytical Chemistry, Department of Chemistry, BOKU—University of Natural Resources and Life Sciences, Muthgasse 18, 1190 Vienna, Austria
3 Department of Chemistry, Faculty of Science, University of Zagreb, Horvatovac 102a, 10000 Zagreb, Croatia; ijuranovic@chem.pmf.hr
* Correspondence: michaela.zeiner@oru.se

Abstract: Fruits and vegetables are considered a beneficial contribution to the human diet. Especially, berries contain a great deal of bioactive compounds, such as anthocyanins, organic acids, tannins, phenols, and antioxidants. Apart from organic substances, inorganic nutrients are also present in fruits. Some metals and metalloids are essential for humans, whilst others may exhibit harmful effects. Wild grown berries, collected in so-called unpolluted areas, are considered to be free of any potentially toxic ingredients. However, due to transmission processes pollutants can also reach remote areas and, furthermore, metal uptake from the soil via roots has to be taken into account. Thus, the presented study focused on the determination of Al, Cd, Cr, Ni, and Pb in lingonberries, blueberries, and rose hips collected in a non-polluted area in Croatia. Neither Cd nor Cr could be found in any sample. Ni levels were mainly up to 25 mg/kg, in a comparable range to the literature data. No health threat is to be expected by eating these fruits and berries regarding Cd, Cr, and Ni. Rose hips, however, contain Pb beyond the stipulated limit in fruits, and also Al is present at a high level (8 mg/g).

Keywords: blueberries; lingonberries; rose hips; aluminium; cadmium; chromium; nickel; lead; provisional tolerable intake

---

1. Introduction

Berries have been collected for nutritional purposes for a long time. They can be either consumed directly as fresh or dried fruits or in various processed forms, such as jams, syrups, infusions, juices, and jellies, or as ingredient of yoghurts or ice-creams. Diet in the Western world includes vegetables and fruits, as well as other plant parts, such as roots, leaves, stems, and seeds, from more than 40 botanical families [1]. Many demonstrate a presence of phytochemicals in fruit and vegetables, which contribute to good human health by influencing metabolic processes [1], e.g., free radical scavenging, stimulating the immune system, inducing apoptosis, and detoxification. Including fruits and vegetables in the daily diet is recommended in dietary guidelines worldwide. However, the consumers' awareness of nutraceutical components is increasing. Thus, ongoing scientific interest in berries, especially wild grown ones, is needed [2]. Regarding public interest, it was estimated in the 1990s, that one family gathers approx. 20–30 kg of berries per year [3]. Even if gathering berries and fruits by oneself is nowadays still carried out, it has started to change from self-collecting in the woods to the usage of pick-your-own farms, which offer a wide range of berries, such as strawberries, raspberries, or blackberries, and processed products [4,5].

Berry fruits, in general, contain a wide range of micro- and macronutrients, such as fibres, minerals, vitamins, and folate, but their biological properties have been attributed mainly to hydrophilic phenolic-type phytochemicals. Usually not only one bioactive compound determines the positive effects of a fruit, but the presence of multiple phytochemicals exhibiting complementary, synergistic, and/or additive effects results in the wide spectrum of health promoting effects [1,2]. Thus, berry research has been focused mainly on organic composition [6–11].

Regarding inorganic analytes, investigations dealing with nutrients in different berry types, e.g., major and minor essential elements, have been published for decades [12–17]. Using old data, such as from 1944 [12] or 1974 [13], it has to be taken into account that the analytical methodology, as well as the environmental influences, have changed, reducing their significance for current evaluations. In addition to the nutritional value, the content of minerals present also influences food quality parameters, like appearance, taste, texture, stability, and even flavonoid production [18]. Conversely, ripeness affects the mineral composition, and some nutrients have been found to decrease with maturity [15], whereas the level of the harmful element arsenic increased [16]. Furthermore potentially harmful metals and metalloids have to be taken into account. Berries collected in highly contaminated areas have been analysed for this purpose [3,19,20]. Additionally, in remote areas heavy metal pollution may occur due to transmission processes. Thus, food safety is of concern [21]. Ten chemicals have been classified to be of major public health concern, including four metals/metalloids (arsenic, cadmium, lead, and mercury) [22]. In addition to the top-ten, other metals may exhibit harmful effects to humans through food intake, e.g., nickel [23,24]. Therefore, maximum allowed limits have been established for harmful elements, and food on the market should not contain certain metals in levels that might result in exceeding the respective allowable intakes. PTWI (provisional tolerable weekly intake), PTDI (provisional tolerable daily intake), or PTMI (provisional tolerable monthly intake) levels have been published by the FAO/WHO [25].

The aim of the present investigation was to determine the content of five elements: aluminium (Al), cadmium (Cd), chromium (Cr), nickel (Ni), and lead (Pb) in wild berries and fruits, since they are considered to be potentially toxic contaminants. Regarding the methodology, previously optimised digestion and measurement procedures were applied, i.e., acidic microwave-assisted digestion followed by inductively-coupled plasma atomic emission spectrometry (ICP-AES) [26]. Closed-vessel microwave digestion as a sample preparation method has been approved for the determination of 31 elements in foodstuffs [27].

2. Materials and Methods

2.1. Chemicals and Glass/Plastic Ware

Reagents for sample and CRM digestion, i.e., hydrogen peroxide (H_2O_2) and supra pure nitric acid (HNO_3) were purchased from Sigma (Munich, Germany). The standard stock solution used for preparation of calibration standards (ICP Multielement Standard IV) was obtained from Merck (Darmstadt, Germany). Ultra-pure water, resistivity ≥ 18 MΩ·cm, was prepared by in-house equipment. Quality control measurements were based on the certified reference material strawberry leaves (CRM: LGC7162) from LGC Standards (Middlesex, UK). All glass- and plasticware used for sample storage and preparation were pre-cleaned with diluted nitric acid.

2.2. Samples and Sample Preparation

Wild berries and fruits, namely blueberries (*Vaccinium myrtillus*), lingonberries (*Vaccinium vitis-idaea*), and rosehip (*Rosa canina*), were collected during summer and fall 2010 when optimally ripe in a non-polluted rural area of Croatia (Slunj: N, 45.07°; E, 15.36°; A: ~280 m). The sampling site covered approx. 200 m^2. For each fruit type five plants, randomly distributed in the sampling area, were sampled by collecting 5 to 15 fruits. Immediately after picking, the fruits were frozen and kept at −20 °C prior to further work-up. After thawing the fruits were rinsed with ultra-pure water. Rose hips were separated into flesh (mesocarp) and seeds. All fruits were dried at 105 °C for 24 h. Using a metal-free mortar, the samples were then ground and homogenised, resulting in a pooled sample for each plant (i.e., five subsamples for each fruit type) In order to destroy the organic matrix, all samples underwent a previously-optimised acidic microwave assisted digestion [26], whereby approx. 0.25 g to 0.5 g (weighed to the nearest 0.1 mg) were treated with 5 mL nitric acid (c = 7 mol/L), applying the following three-step digestion procedure: (1) 150 °C/10 min, (2) 160 °C/10 min, and (3) 190 °C/20 min. The obtained clear solutions were then brought to 10.0 mL with ultrapure water.

2.3. Apparatus and Measurements

Microwave-assisted digestion of the fruits, CRM, and blanks were done in a MWS-2 Microwave System Speedwave instrument (Berghof Laborprodukte GmbH, Eningen, Germany). The subsequent quantitative determination of metals and metalloids was carried out using a Prodigy High Dispersive ICP-AES spectrometer (Teledyne Leeman, Hudson, NH, USA), using a simultaneous mode, the optimal instrumental conditions are listed in Table 1. The emission lines selected along with the respective characteristics of the analytical method are given in Table 2.

Table 1. Operating conditions of the Prodigy High Dispersive ICP-AES.

Parameter	Settings
Spectrometer	High resolution Echelle polychromator Large format programmable array detector (L-PAD)
RF-Generator	40 MHz "free-running"
Output power	1.1 kW
Argon flows	Coolant: 18 L min^{-1} Auxiliary: 0.8 L min^{-1} Nebuliser: 1.0 L min^{-1}
Peristaltic pump	1.0 mL min^{-1}
Nebuliser	Pneumatic (glass concentric)
Spray chamber	Glass cyclonic
Plasma viewing	Axial
Sample uptake delay	30 s

External standards in the concentration range from 0.0500 mg/L to 5.00 mg/L were used to calibrate all analytes. The standard solutions were prepared by diluting a multi-element standard stock solution (1000 mg/L) with 1% w/w HNO_3. This diluted nitric acid is also the medium of the blank solution, whose intensity was subtracted from all sample values for blank correction. In case of sample concentrations outside the calibration range, appropriate dilutions to the sample were performed with 1% w/w HNO_3.

2.4. Calculations

All solutions, i.e., fruit and berry digests, blanks, and CRM digest were measured thrice. The blank-corrected values (mass concentrations) were converted into contents in mg/kg dried plant material considering dilution factor, final volume, and mass of dry matter. Finally, means and standard deviations were calculated for all samples. In order to see statistically significant differences between the three fruits analysed, ANOVA test was carried out, based on $p < 0.05$. All calculations were performed using Microsoft Excel 2010 and 2013.

2.5. Characterisation of the Analytical Method

In order validate the analysis, parameters of trueness, precision, and day-to-day repeatability were determined from strawberry leaves which were chosen as a certified reference material for plant matrix. The samples were analysed after calibration on two different days. The recoveries were obtained using the following formula:

$$\text{recovery}_x \text{in } \% = \frac{\text{content}_{x,\ \text{found}} \text{ in mg/kg}}{\text{content}_{x,\ \text{certified}} \text{ in mg/kg}} \times 100$$

Based on Bouman's procedure [28], the limits of detection (LOD; 3σ) and quantification (LOQ; 10σ) were calculated.

3. Results

3.1. Analytical Method

The limits of detection (LOD) for the dried fruits, based on a digested mass of 0.25 g and a final volume of 10.0 mL, are below 0.3 mg/kg for all analytes except for Al, whose LOD is 2.9 mg/kg. In addition to the LOD values, the recoveries obtained by analysing the CRM (ranging from 85% to 113%) are given in Table 2. These figures of merit are in the range of reported ones for similar studies [26,29]. The coefficients of determination (R^2) of the calibration curves are all beyond 0.9995. The precision presented as relative standard deviation (RSD) for triplicate measurements is below 1.2% for all analytes. The RSD obtained for the day-to-day-repeatability is not higher than 2.5%.

Table 2. Characteristics of the analytical method.

Analyte	Wavelength (nm)	LOD in Digest Solution (mg/L)	LOD in Dried Fruits (mg/kg)	Recovery (%)
Aluminum	308.215	0.074	2.9	110
Cadmium	214.441	0.00071	0.028	113
Chromium	206.149	0.00045	0.018	101
Nickel	231.604	0.0038	0.15	85
Lead	220.353	0.0070	0.28	101

3.2. Elemental Content of Wild Berries and Fruits

The results obtained for the three kinds of berries and fruits for the five metals analysed are summarised in Table 3. In all samples the Cd, as well as Cr, levels were below the respective LODs. For the other analytes (Al, Ni, Pb) the minimum, mean, and maximum values are given. Statistically significant differences in the element content were found for Al and Ni, the p-values are 2.6×10^{-14} and 0.0023, respectively. Conversely, the Pb contents do not differ statistically significantly between the three sample types analysed ($p = 0.24$)

Table 3. Minimum–mean–maximum metal content in fruit material (mg/kg d.w. [1]), n = 5.

Metal	Lingonberries	Rose Hip	Blueberries
Aluminum	34.9–42.5–63.9	7527–8242–8836	1093–1248–1463
Cadmium	<0.028	<0.028	<0.028
Chromium	<0.018	<0.018	<0.018
Nickel	1.81–2.49–12.9	10.6–11.3–23.5	21.0–24.8–56.2
Lead	0.542–0.601–9.28	3.00–3.34–15.3	1.19–1.66–2.42

[1] d.w. = dry weight.

4. Discussion

4.1. Analytical Method

The above described figures of merit for the analytical method applied are in the range for the determination of trace elements in biological samples. LODs in a similar range were also obtained by other working groups [17,29]. Limit values for Cd and Pb in fruits and vegetables are stipulated (see Table 4). For both metals the calculated LOD is lower than the maximum allowed content. The latter are given for fresh weight. Since the water content of fruits and berries is around 90% [14], the elemental content referred to fresh weight is 10 times lower than referred to dry weight. Based on the characteristic data found, the analytical method chosen has been proven to be appropriate for the given analytical task.

Table 4. Limits for metals in food.

Metal		Limit in Food		Intake Limit [1]	Lit.
Aluminum				1.00 mg/kg bw/week (PTWI)	[25]
Cadmium		Fruits and vegetables	0.050 mg/kg f.w.		[30]
				0.025 mg/kg bw/month (PTMI)	[22]
				0.001 mg/kg bw/day (Rfd)	[31]
Chromium	insoluble Cr(III)-salts			1.500 mg/kg bw/day (Rfd)	[32]
	Cr(VI)			0.003 mg/kg bw/day (Rfd)	[33]
Nickel	soluble salts			0.020 mg/kg bw/day (Rfd)	[34]
				0.020 mg/kg bw/day (Trv)	[23]
Lead		Fruit, excluding cranberries, currants, elderberries and strawberry tree fruit	0.10 mg/kg f.w.		[35]
		cranberries, currants, elderberries and strawberry tree fruit	0.20 mg/kg f.w.		[35]
				PTI value withdrawn 2010	[22]

[1] f.w. = fresh weight; bw = body weight; PTWI = Provisional tolerable weekly intake; PTMI = Provisional tolerable monthly intake; Rfd = Reference oral dose; Trv = Toxicity reference value.

4.2. Elemental Content of Wild Berries and Fruits

Based on the five elements determined, all three fruit types differ statistically significantly from each other regarding Al and Ni content. In the following all elements analysed are discussed in detail.

Aluminium has no known physiological role in the human body. Until the 1970s it was considered to be innocuous, but starting in the 1980s toxic effects to animals, plants, and humans have been reported [36]. Al rarely shows acute toxicity, but chronic intoxications have to be considered, especially due to the fact that Al accumulates with age [37,38]. Diet is regarded as being a significant contributor to the body burden of aluminium [39], and its weekly intake is limited by the FAO/WHO [25]. The Al content found in the fruits and berries analysed was highest in rose hips, followed by blueberries and, finally, lingonberries, the mean values being 8242 mg/kg, 1248 mg/kg, and 42.5 mg/kg, respectively. The data differ from each other in a wide range, even since all samples were collected in the same area having a similar environmental background. Thus, it can be clearly seen that the uptake differs between the plants. Nile and Park do not give any value for Al in berries in their review on berry composition, they just mention that berries are rich in Al [10]. Rose hips used for the preparation of infusions collected in Turkey show a mean Al content of 157 mg/kg [40], whereby it is not reported if the entire fruit or only the flesh (as in the present study) was analysed. However, comparable Al contents were found in strawberries from Pakistan (Lahore region), whereby the values decreased during ripeness from 740 mg/kg to 230 mg/kg d.w. [15]. The PTWI given by the FAO/WHO limits the weekly Al dietary intake with 1 mg/kg bw. Thus, an 80 kg-person should not consume more than 100 g of rose hips per week. For blueberries and lingonberries more than 600 g, and even 2 kg, are to be eaten resp. in order to reach the maximum, considering no other Al sources. Chinese blueberries contained 41 mg/kg [41], much lower than the results from this study, but in the same order of magnitude than the data obtained for lingonberries. This fact can be explained by the impact of bedrock composition on Al content in plants. A potential health threat might only be caused by rose hip intake. This is a drawback, since rose hips have recently been defined as a functional food based on bioactive ingredients, as well as due to their content of essential elements, such as Ca, Mg, K, S, Si, Se, Mn, and Fe [42].

Cadmium, being a highly toxic metal, is one of ten chemicals classified by the WHO to be a major public health concern [22]. It occurs naturally in soil, but it is also a pollutant in the environment due to anthropogenic impact. Since it is easily taken up and accumulated by plants and crops through the root systems, it may easily enter food [43], especially berries and fruits. In all samples, including blueberries, lingonberries, as well as rose-hips, no Cd was detected, meaning that the contents are below the LOD of 0.028 mg/kg d.w. Commission Regulation (EU) No. 2015/1005 limits Cd in fruits and vegetables to 0.050 mg/kg f.w., corresponding to 0.50 mg/kg d.w. [30]. Thus, all fruits analysed are supposed to be safe for humans regarding this element. The PTMI is 0.025 mg/kg bw, which would be 2 mg per month for an 80 kg person. Considering Cd present in the berries at the LOD level, this corresponds to more than 700 kg of fresh blueberries in one month. An investigation of blackberries, elderberries, autumn olives, and candleberry-myrtles revealed that they also contained Cd only below LOD [17]. Whereas no Cd was found in Azorean blueberry, levels between LOD and limit of quantitation (LOQ) were registered for Madeiran blueberry [17]. A Russian study reports Cd in blue- and cranberries in the range from 0.03 mg/kg to 0.06 mg/kg d.w. [3]. Strawberries collected in Serbia did not contain Cd at detectable concentrations either [29]. Blueberries collected in China were reported to contain 0.034 mg/kg Cd [41], slightly above the LOD of the current study. Rose hips from Turkey have been reported to contain quite a high amount of Cd, namely 0.81 mg/kg d.w. [44]. A working group from Bangladesh found Cd contents ranging from 0.012 mg/kg and 0.216 mg/kg in leafy and non-leafy vegetables [43].

Chromium, as Cr(III), plays a crucial role in human health and is, thus, defined as an essential trace element [45]. Special attention has to be drawn to its interaction in glucose tolerance and Cr supplementation in persons with diabetes, hypoglycaemia, and obesity [46]. In addition to the positive effects of Cr(III), Cr(VI) is a toxic, carcinogenic substance, this fact being reflected in the 500-times lower reference oral dose for the latter species [32,33] compared to the former species. No limit for Cr in foods is given in the Codex Alimentarius Commission, or by Australia, New Zealand, Japan, the United States, and Taiwan [41]. Even if Cr(III) is reported to be mainly found in fruits, vegetables, and grain products [47,48], no Cr could be detected in all samples of this investigation (<0.018 mg/kg). Data for Cr in blueberries are given by Hua et al., who found 0.77 mg/kg [41]. Additionally, in fresh strawberries the Cr content is quite low, ranging from 0.01 mg/kg up to 0.03 mg/kg [29]. Wild berries from Portugal, analysed by Llorent-Martínez and colleagues, found Cr in blackberries, elderberries, autumn olives, and candleberry-myrtles up to 0.6 mg/kg fresh weight [17]. Cranberries from Russia have been reported to contain Cr from 0.02 mg/kg up to 1.5 mg/kg, and blueberries 0.03 to 0.06 mg/kg d.w. [3]. Literature data for rose hips are given by Duran and colleagues, who found 0.80 mg/kg d.w. Cr [44].

Nickel is naturally occurring in soils ranging from 4 to 80 mg/kg in the USA [49] and, for agricultural soil in Velika Gorica, Croatia (N 45.7173°–E 16.0571°; A 107 m), a mean content of 55 mg/kg was found [50]. The main exposure route to nickel is via food intake, approx. 0.100 to 0.300 mg/day are taken up by per adult [23,49]. Apart from contact dermatitis, genotoxicity, haematotoxicity, teratogenicity, immunotoxicity, and carcinogenicity have been identified as harmful effects of Ni [24]. Thus, its intake is limited to 0.020 mg/kg bw per day [23,34]. In the present study the highest Ni content was found in blueberries (25 mg/kg), followed by rose hips (11 mg/kg) and lingonberries (2.5 mg/kg). The daily consumption of 100 g fresh blueberries would lead to an intake of 0.250 mg, being approx. 16% of the allowed limit for an 80 kg person. Thus, no harmful effects are to be expected by the berries analysed. Cranberries from polluted areas (close to a Ni-Cu-smelter) showed elevated levels (up to 97 mg/kg d.w.), whereas the background contents were in the same range as the results from the current study, i.e., 2 mg/kg to 9 mg/kg [3]. No high Ni accumulation in lingonberries collected in the vicinity of a chromium mine was reported by a Finnish research group [51]. Blueberries from the same contaminated area did not accumulate Ni that much, with contents being in the range from 4 mg/kg to 11 mg/kg for all sampling sites [3]. A Chinese working group found 2.2 mg/kg d.w. Ni in blueberries [41]. Wild berries from Portugal contain Ni in the same order of magnitude, namely from 0.6 mg/kg to 1.6 mg/kg [17], as our results for lingonberries. Additionally, results for strawberries are in the same range [29]. Regarding rose hips, the literature data are available for Turkish fruits, with the mean content being 6.5 mg/kg d.w. [44].

Lead is well known as a toxic element, exhibiting developmental neurotoxicity in young children and cardiovascular effects and nephrotoxicity in adults [35]. Thus, the Pb content in fruits and berries is limited to 0.10 mg/kg f.w. and 0.20 mg/kg f.w., respectively [35], whereas the PTI value was withdrawn in 2010 [22]. The highest maximum and mean level for Pb were found in rose hips (15 mg/kg; 3.3 mg/kg, resp.), exceeding the limit value in foods, which would correspond to approx. 1 mg/kg d.w. This result is comparable to that form Turkish rose hips with 10 mg/kg d.w. [44]. Blueberries' Pb content showed the smallest range, compared to rose hips and lingonberries. Only one sample exceeds the limit in berries (2.4 mg/kg > 2 mg/kg). The mean value of Pb in blueberries (1.66 mg/kg) is similar to the data from Barcan et al., who reported contents ranging from 0.7 mg/kg up to 1.5 mg/kg [3]. Conversely, the Chinese working group found less, their average being 0.135 mg/kg d.w. [41]. Lingonberries contained the least Pb of all fruits analysed, except for one sample (9.3 mg/kg) all results are below 1 mg/kg, thus being within the allowed range for berries. Even less Pb was found in lingonberries from Finland, whose contents ranged from 0.0006 mg/kg to 0.0011 mg/kg f.w. [51], corresponding to approx. 0.006 mg/kg to 0.011 mg/kg dried berries. Wild berries gathered in Portugal all had Pb levels below the LOD (0.012 mg/kg f.w.) [17].

5. Conclusions

Wild berries and fruits collected in a remote and, thus, considered unpolluted area in Croatia have been analysed for potentially toxic elements, revealing that no health threat is to be expected by eating these fruits and berries regarding Cd, Cr, and Ni. Due to high Al and Pb contents in rose hips, their intake should not exceed 100 g per week for an adult, especially in the case of chronic renal failure. Only in cases of high consumption of blueberries are harmful effects by Al and Pb to be expected. Lingonberries were found to have the lowest contents of all metals investigated.

Author Contributions: M.Z. and I.J.C. conceived and designed the experiments; I.J.C. carried out the digestions and measurements; M.Z. and I.J.C. analysed the data; M.Z. wrote the paper; and M.Z. revised the paper.

References

1. Lampe, J.W. Health effects of vegetables and fruit: Assessing mechanisms of action in human experimental studies. *Am. J. Clin. Nutr.* **1999**, *70*, 475S–490S. [CrossRef] [PubMed]
2. Seeram, N.P. Berry Fruits: Compositional Elements, Biochemical Activities, and the Impact of Their Intake on Human Health, Performance, and Disease. *J. Agric. Food Chem.* **2008**, *56*, 627–629. [CrossRef] [PubMed]
3. Barcan, V.S.H.; Kovnatsky, E.F.; Smetannikova, M.S. Absorption of heavy metals in wild berries and edible mushrooms in an area affected by smelter emissions. *Water Air Soil Pollut.* **1998**, *103*, 173–195. [CrossRef]
4. Willkommen im Erdbeerland. Available online: www.erdbeerland.at (accessed on 18 April 2018).
5. Biobeerengarten Hummel. Available online: www.biobeerengarten.at/ (accessed on 18 April 2018).
6. Faria, A.; Oliveira, J.; Neves, P.; Gameiro, P.; Santos-Buelga, C.; de Freitas, V.; Mateus, N. Antioxidant Properties of Prepared Blueberry (*Vaccinium myrtillus*) Extracts. *J. Agric. Food Chem.* **2005**, *53*, 6896–6902. [CrossRef] [PubMed]

7. Jimenez-Garcia, S.N.; Guevara-Gonzalez, R.G.; Miranda-Lopez, R.; Feregrino-Perez, A.A.; Torres-Pacheco, I.; Vazquez-Cruz, M.A. Functional properties and quality characteristics of bioactive compounds in berries: Biochemistry, biotechnology, and genomics—A review. *Food Res. Int.* **2013**, *54*, 1195–1207. [CrossRef]
8. Pantelidis, G.E.; Vasilakakis, M.; Manganaris, G.A.; Diamantidis, G. Antioxidant capacity, phenol, anthocyanin and ascorbic acid contents in raspberries, blackberries, red currants, gooseberries and Cornelian cherries. *Food Chem.* **2007**, *102*, 777–783. [CrossRef]
9. Marinova, D.; Ribarova, F.; Atanassova, M. Total phenolics and total flavonoids in Bulgarian fruits and vegetables. *J. Chem. Technol. Metall.* **2005**, *40*, 255–260.
10. Nile, S.H.; Park, S.W. Edible berries: Bioactive components and their effect on human health—A review. *Nutrition* **2014**, *30*, 134–144. [CrossRef] [PubMed]
11. Yi, W.; Fischer, J.; Krewer, G.; Akoh, C.C. Phenolic Compounds from Blueberries Can Inhibit Colon Cancer Cell Proliferation and Induce Apoptosis. *J. Agric. Food Chem.* **2005**, *53*, 7320–7329. [CrossRef] [PubMed]
12. Chandler, F.B. *Composition and Uses of Blueberries*; University of Maine, Maine Agricultural Experiment Station: Orono, ME, USA, 1944; Volume 428, pp. 1–39.
13. Koivistoinen, P.; Ahlström, A.; Varo, P.; Nissinen, H. Mineral Element Composition of Finnish Vegetables, Fruits, and Berries. *Acta Agric. Scand.* **1974**, *24*, 131–134. [CrossRef]
14. De Souza, V.R.; Pereira, P.A.P.; da Silva, T.A.T.; de Oliveira Lima, L.C.; Pio, R.; Queiroz, F. Determination of the bioactive compounds, antioxidant activity and chemical composition of Brazilian blackberry, red raspberry, strawberry, blueberry and sweet cherry fruits. *Food Chem.* **2014**, *156*, 362–368. [CrossRef] [PubMed]
15. Mahmood, T.; Anwar, F.; Iqbal, T.; Bhatti, I.A.; Ashraf, M. Mineral composition of strawberry, mulberry and cherry fruits at different ripening stages as analysed by inductively coupled plasma—Optical emission spectroscopy. *J. Plant Nutr.* **2012**, *35*, 111–122. [CrossRef]
16. Zeiner, M.; Juranović Cindrić, I.; Majić, B.; Stingeder, G. Study of the Accumulation of Toxic and Essential Ultra-Trace Elements in Fruits of *Sorbus domestica* L. *Int. J. Environ. Res. Public Health* **2017**, *14*, 341–348. [CrossRef] [PubMed]
17. Llorent-Martínez, E.J.; Spínola, V.; Castilho, P.C. Evaluation of the inorganic content of six underused wild berries from Portugal: Potential new sources of essential minerals. *J. Food Compos. Anal.* **2017**, *59*, 153–160. [CrossRef]
18. Konieczynski, P.; Arceusz, A.; Wesolowski, M. Relationships between flavonoids and selected elements in infusions of medicinal herbs. *Open Chem.* **2015**, *13*, 68–74. [CrossRef]
19. Von Hoffen, L.P.; Säumel, I. Orchards for edible cities: Cadmium and lead content in nuts, berries, pome and stone fruits harvested within the inner city neighbourhoods in Berlin, Germany. *Ecotoxicol. Environ. Saf.* **2014**, *101*, 233–239. [CrossRef] [PubMed]
20. Rodushkin, I.; Ödman, F.; Holmström, H. Multi-element analysis of wild berries from northern Sweden by ICP techniques. *Sci. Total Environ.* **1999**, *231*, 53–65. [CrossRef]
21. D'Mello, J.P.F. *Food Safety: Contaminants and Toxins*; CABI Publishing: Wallingford, UK, 2003; ISBN 0 85199 607 8.
22. World Health Organization (WHO). International Programme on Chemical Safety—Health Impacts of Chemicals, Ten Chemicals of Major Public Health Concern. 2010. Available online: http://www.who.int/ipcs/assessment/public_health/chemicals_phc/en/ (accessed on 23 April 2018).
23. Haber, L.T.; Bates, H.K.; Allen, B.C.; Vincent, M.J.; Oller, A.R. Derivation of an oral toxicity reference value for nickel. *Regul. Toxicol. Pharmacol.* **2017**, *87*, S1–S18. [CrossRef] [PubMed]
24. Zdrojewicz, Z.; Popowicz, E.; Winiarski, J. Nickel—Role in human organism and toxic effects. *Pol. Med. J.* **2016**, *242*, 115–118. (in Polish).
25. FAO/WHO: Food Standards Programme Codex Committee on Contaminants in Foods, Fifth Session. In Proceedings of the Codex Committee on Food Additives and Contaminants, The Hague, The Netherlands, 21–25 March 2011.
26. Juranović Cindrić, I.; Zeiner, M.; Mihajlov-Konanov, D.; Stingeder, G. Inorganic Macro- and Micronutrients in "Superberries" Black Chokeberries (*Aronia melanocarpa*) and Related Teas. *Int. J. Environ. Res. Public Health* **2017**, *14*, 539–548. [CrossRef] [PubMed]
27. Chevallier, E.; Chekri, R.; Zinck, J.; Guérin, T.; Noël, L. Simultaneous determination of 31 elements in foodstuffs by ICP-MS after closed-vessel microwave digestion: Method validation based on the accuracy profile. *J. Food Compos. Anal.* **2015**, *41*, 35–41. [CrossRef]

28. Boumans, P.W.J.M. Basic concepts and characteristics of ICP-AES. In *Inductively Coupled Plasma Emission Spectroscopy. Part I. Methodology, Instrumentation, and Performance*; Boumans, P.W.J.M., Ed.; Wiley: New York, NY, USA, 1987; pp. 100–257.
29. Pavlović, J.L.J.; Mitić, S.M.; Mitić, M.N.; Pavlović, A.N.; Micić, R.J.; Stojković, M.B. Multielement Analysis of South Serbian Strawberry Cultivars by Inductively Coupled Plasma—Optical Emission Spectrometry. *Anal. Lett.* **2018**, *51*, 1417–1432. [CrossRef]
30. Commission Regulation (EU) No 488/2014 of 12 May 2014 Amending Regulation (EC) No 1881/2006 as Regards Maximum Levels of Cadmium in Foodstuffs. Available online: http://extwprlegs1.fao.org/docs/pdf/eur133613.pdf (accessed on 24 April 2018).
31. United States, Environmental Protection Agency, Integrated Risk Information System (US-EPA, IRIS). Cadmium. Available online: https://cfpub.epa.gov/ncea/iris2/chemicalLanding.cfm?substance_nmbr=141 (accessed on 24 April 2018).
32. United States, Environmental Protection Agency, Integrated Risk Information System (US-EPA, IRIS). Chromium(III), Insoluble Salts. Available online: https://cfpub.epa.gov/ncea/iris2/chemicalLanding.cfm?substance_nmbr=28 (accessed on 24 April 2018).
33. United States, Environmental Protection Agency, Integrated Risk Information System (US-EPA, IRIS). Chromium(VI). Available online: https://cfpub.epa.gov/ncea/iris2/chemicalLanding.cfm?substance_nmbr=144 (accessed on 24 April 2018).
34. United States, Environmental Protection Agency, Integrated Risk Information System (US-EPA, IRIS). Nickel, Soluble Salts. Available online: https://cfpub.epa.gov/ncea/iris2/chemicalLanding.cfm?substance_nmbr=271 (accessed on 24 April 2018).
35. Commission Regulation (EU) No 2015/1005 of 25 June 2015 Amending Regulation (EC) No. 1881/2006 as Regards Maximum Levels of Lead in Certain Foodstuffs. Available online: https://eur-lex.europa.eu/legal-content/EN/TXT/PDF/?uri=CELEX:32015R1005&from=EN (accessed on 24 April 2018).
36. Nayak, P. Aluminum: Impacts and Disease—Review. *Environ. Res.* **2002**, *89*, 101–115. [CrossRef] [PubMed]
37. Exley, C. What is the risk of aluminium as a neurotoxin? *Expert Rev. Neurother.* **2014**, *14*, 589–591. [CrossRef] [PubMed]
38. Exley, C. The toxicity of aluminium in humans. *Morphologie* **2016**, *100*, 51–55. [CrossRef] [PubMed]
39. Exley, C. Human exposure to aluminium. *Environ. Sci. Process. Impacts.* **2013**, *15*, 1807–1816. [CrossRef] [PubMed]
40. Başgel, S.; Erdemoglu, S.B. Determination of mineral and trace elements in some medicinal herbs and their infusions consumed in Turkey. *Sci. Total Environ.* **2006**, *359*, 82–89. [CrossRef] [PubMed]
41. Hua, Z.; Zhen-Yu, W.; Xin, Y.; Hai-Tian, Z.; Ying-Chun, Z.; Ai-Jun, D.; Jing, J.; Jing, W. Determination of free amino acids and 18 elements in freeze-dried strawberry and blueberry fruit using an Amino Acid Analyzer and ICP-MS with micro-wave digestion. *Food Chem.* **2014**, *147*, 189–194.
42. Patel, S. Rose hip as an underutilized functional food: Evidence-based review. *Trends Food Sci. Technol.* **2017**, *63*, 29–38. [CrossRef]
43. Alam, M.G.M.; Snow, E.T.; Tanaka, A. Arsenic and heavy metal contamination of vegetables grown in Samta village, Bangladesh. *Sci. Total Environ.* **2003**, *308*, 83–96. [CrossRef]
44. Duran, A.; Tuzen, M.; Soylak, M. Trace element levels in some dried fruit samples from Turkey. *Int. J. Food Sci. Nutr.* **2008**, *59*, 581–589. [CrossRef] [PubMed]
45. Mutuma, S.; Amuna, P.; Shukla, H.; Sumar, S. Chromium in food, nutrition and health—An introduction. *Nutr. Food Sci.* **1999**, *2*, 81–88. [CrossRef]
46. Lim, T.H.; Sargent, T.; Kusubov, N. Kinetics of trace element chromium(III) in the human body. *Am. J. Physiol. Regul. Integr. Comp. Physiol.* **1983**, *244*, 445–454. [CrossRef] [PubMed]
47. Anderson, R.A.; Bryden, N.A.; Polansky, M.M. Chromium content of selected breakfast cereals. *J. Food Compos. Anal.* **1988**, *1*, 303–308. [CrossRef]
48. Anderson, R.A.; Bryden, N.A.; Polansky, M.M. Dietary chromium intake: Freely chosen diets, institutional diets, and individual diets. *Biol. Trace Elem. Res.* **1992**, *32*, 117–121. [CrossRef] [PubMed]
49. Agency for Toxic Substances and Disease Registry (ATSDR). Toxicological Profile for Nickel. Available online: https://www.atsdr.cdc.gov/ToxProfiles/tp15.pdf (accessed on 24 April 2018).

50. Zeiner, M.; Juranović Cindrić, I.; Lovrenčić Mikelić, I.; Medunić, G.; Kampić, Š.; Tomašić, N.; Stingeder, G. The determination of the extractability of selected elements from agricultural soil. *Environ. Monit. Assess.* **2013**, *185*, 223–229. [CrossRef] [PubMed]
51. Pöykiö, R.; Mäenpää, A.; Perämäki, P.; Niemelä, M.; Välimäki, I. Heavy Metals (Cr, Zn, Ni, V, Pb, Cd) in Lingonberries (*Vaccinium vitis-idaea* L.) and Assessment of Human Exposure in Two Industrial Areas in the Kemi-Tornio Region, Northern Finland. *Arch. Environ. Contam. Toxicol.* **2005**, *48*, 338–343. [CrossRef] [PubMed]

13

A Cross-Flow Ultrasound-Assisted Extraction of Curcuminoids from *Curcuma longa* L.: Process Design to Avoid Degradation

Arianna Binello [1], Giorgio Grillo [1], Alessandro Barge [1], Pietro Allegrini [2], Daniele Ciceri [2] and Giancarlo Cravotto [1,*]

1 Dipartimento di Scienza e Tecnologia del Farmaco, University of Turin, Via P. Giuria 9, 10125 Turin, Italy; arianna.binello@unito.it (A.B.); giorgio.grillo@unito.it (G.G.); alessandro.barge@unito.it (A.B.)
2 INDENA S.pA., Viale Ortles, 12, 20139 Milan, Italy; pietro.allegrini@indena.com (P.A.); daniele.ciceri@indena.com (D.C.)
* Correspondence: giancarlo.cravotto@unito.it

Abstract: Rhizomes of *Curcuma longa* L. are well known for their content of curcuminoids, which are compounds with interesting biological activity against various inflammatory states and diseases. Curcuminoids can degrade during processing. This piece of work investigates fast, efficient and cost-effective metabolite recovery from turmeric under ultrasound-assisted extraction (UAE). An analytical evaluation of curcuminoid stability under sonication in different solvents is reported for the first time. HPLC and quantitative ^{1}H-NMR were used. Under the applied conditions, EtOAc was found to be the optimal extraction medium, rather than EtOH, due to its lower radical generation, which facilitates better curcuminoid stability. Kinetic characterization, by means of the Peleg equation, was applied for single-step UAE on two different rhizome granulometries. Over a time of 90 min, maximum extraction yields were 25.63% and 47.56% for 6 and 2 mm matrix powders, respectively. However, it was observed that the largest portion of curcuminoid recovery was achieved in the first 30 min. Model outcomes were used as the basis for the design of a suitable multi-step cross-flow approach that supports and emphasizes the disruptive role of cavitation. The maximum curcuminoid yield was achieved over three steps (92.10%) and four steps (80.04%), for lower and higher granulometries, respectively. Finally, the central role of the solvent was further confirmed by turmeric oleoresin purification. The EtOAc extract was purified via crystallization, and a 95% pure curcuminoid product was isolated without any chromatographic procedure. No suitable crystallization was observed for the EtOH extract.

Keywords: *Curcuma longa* L.; curcuminoid stability; multi-step extraction; ultrasound-assisted extraction; extraction kinetic

---

1. Introduction

In recent years, non-synthetic and biologically active compounds from vegetal sources have gained increasing interest because of their important role in health-care systems worldwide; pigments have found use as additives or supplements in food, pharmaceutical and cosmetic industries. One of the most widely studied sources of natural pigment plants is *Curcuma longa* L., a perennial rhizomatous shrub belonging to the *Zingiberaceae* family. Also known as turmeric, this plant is commonly used as a coloring and flavoring agent in the food industry and, in particular, it is known as the 'golden spice of life', and constitutes the main component of curry [1]. However, it is also appreciated in traditional medicine for its biological properties, which are mainly related to its curcuminoids; chemical components that include curcumin (CUR), demethoxycurcumin (DMC) and bisdemethoxycurcumin (BDMC) (Figure 1) [2].

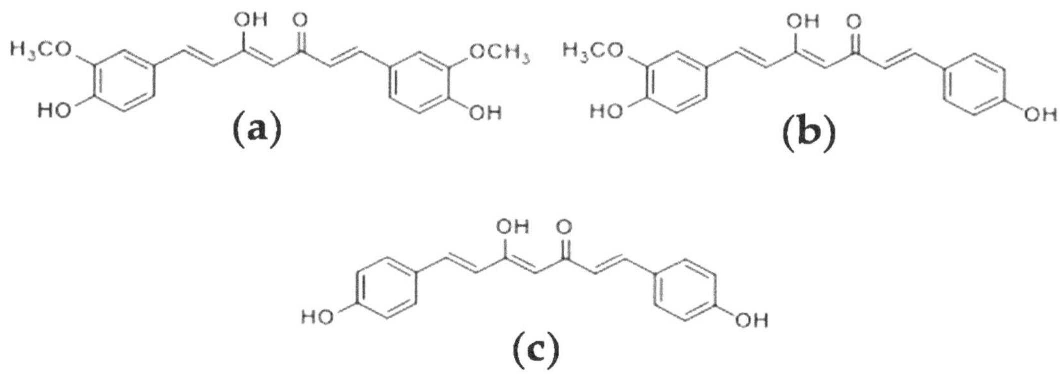

Figure 1. Chemical structures of curcuminoids: (**a**) curcumin (CUR), (**b**) demethoxycurcumin (DMC), (**c**) bisdemethoxycurcumin (BDMC).

Although these substances can be chemically synthesized, is worth noting that the Joint FAO/WHO Expert Committee on Food Additives (JECFA) specifications only allow curcuminoids extracted from natural source material to be used as food additives [3].

Turmeric has curcuminoid contents of 2–9% depending on its growing conditions and origin. The main biological activities exhibited by these compounds are antioxidant, anti-inflammatory, antibacterial, antiviral, antifungal, anticancer, immune-stimulatory and neuroprotective [4–7].

The most common techniques to obtain curcuminoids from turmeric involve solvent extraction followed by column chromatography. However, the choice of the solvent and the conditions applied must take into account the safety of the final use to which these compounds are intended, for example, their acceptability in the food industry. Soxhlet, ultrasonic-assisted extraction (UAE) and microwave-assisted extraction (MAE) are currently the most commonly used methods, although other techniques, such as pulsed-ultrasonic and supercritical-fluid extraction, are also reported to be efficient processes [8–10]. It is worth noting that UAE is now widely used in vegetal-matrix extraction thanks to its efficiency, which allows it to: (1) enhance extraction yields and rates; (2) make use of alternative solvents; (3) reduce costs and required extraction times; and (4) preserve heat-sensitive compounds. Several cavitation devices are available, including bath, probe and flow systems, and can provide huge process-type flexibility, such as scalability, and the use of counter-current or co-current systems [11]. In particular, the co-current cross-flow approach is a relatively simple sequential-step protocol that is used in solid/liquid extraction for different purposes, such as decontamination, leaching and food washing. Essentially, the solid to be extracted is mixed with fresh solvent, then recovered and drained. The extracted solid then meets fresh solvent and is recovered and drained again. The higher the number of steps, the more the matrix approaches depletion [12].

A great deal of metabolite-extraction research has investigated this mechanism, and it has been described using Fick's second law of diffusion [13,14]. However, a deeper modeling of the kinetics underlying the solid–liquid extraction processes is not common in literature; mathematical assets, such as useful engineering tools, are needed to cast light on method applicability and optimization. Desorption processes, on the other hand, are commonly interpreted by mathematical descriptors in literature, and Peleg's Model is one of the best known [15]. The common principles shared by dehydration/rehydration phenomena and metabolite extraction allow the ability of this model to render UAE kinetics to be investigated [16,17].

An equation that describes the time-dependent trend of a system can ensure the understanding necessary to define the fastest extraction rates and the best compromises between metabolite yield and time consumption. This approach is fundamental from an industrial point of view, where the main focus is the simultaneous maximization of productivity and the minimization of energy consumption and costs.

The international authorities (FAO/WHO and European Commission), which monitor the development and commercialization of food additives, have decreed that a limited number of solvents are permitted for use in the preparation of curcuminoid-based products [18]. The permitted polar and non-polar organic solvents include acetone, methanol, ethanol, iso-propanol, hexane, ethyl acetate and supercritical carbon dioxide, which have been accepted for use in the extraction of CUR and its analogues from turmeric.

S.R. Shirsath et al. [17] have thoroughly explored, using kinetic investigations, CUR extraction from *Curcuma amada*, and have screened parameters, such as time, temperature, solvent and granulometry. Ethanol was confirmed as the solvent of choice (vs. methanol, acetone and ethyl acetate), as it was able to give 72% of total metabolites. A direct comparison confirmed the efficiency of UAE, compared to silent conditions which provided an average yield decrease of 10%. A broader screening can be performed to consider other aspects that are involved in the industrial processing of vegetal matrixes, such as easier-to-handle granulometries (less prone to filter clogging), final product stability and purification steps.

In fact, although ethanol is the preferred solvent, the choice of extraction solvent must also be evaluated in relation to the curcuminoid purification steps. On the laboratory scale, chromatography is the technique most commonly employed to isolate curcuminoids, for example using silica as the stationary phase and various organic-solvent mixtures (e.g., chloroform:methanol) as eluents [19]. Crystallization can be included among the purification steps. Ukrainczyk et al. have reported the influence that process conditions have on the purification of crude curcumin via successive cooling crystallizations with isopropanol. [20]. Concentrated oleoresin was selected for the formation of crystals via the slow addition of petroleum ether, water and hexane. The best crystal quality was found when petroleum ether was used, whereas the crystals prepared using the other two solvents were sticky in nature [21]. A non-classical crystallization pathway for curcumin particles was found by Alpana et al. when this process is carried out in sonochemical conditions with or without stabilizers [22].

The quantification of curcuminoid content in turmeric extracts is commonly performed using chromatographic techniques (HPLC, UPLC and capillary electrophoresis). However, in recent years, the quali-quantitative control of herbal products have also been carried out with spectroscopic fingerprinting [23,24]. In particular, Gad and Bouzabata [25] have recently investigated the use of UV, FT-IR and ^{1}H NMR for the quality control of *Curcuma longa* by comparing data with those obtained from HPLC analyses. They observed that NMR shows good potential efficiency.

Although curcuminoids possess health-promoting factors, they have found limited application in the food and pharmaceuticals industries because of their low water solubility, poor bioavailability and poor stability in in-vivo and in-vitro environment [26]. These compounds undergo degradation by acidic or alkaline hydrolysis, oxidation, photo-degradation and, moreover, are sensitive to light. The autoxidative degradation of CUR at physiological pH gives bicyclopentandione as the major product, while vanillin, ferulic acid and are reported to be minor compounds [27,28]. The same compounds have been identified as curcuminoid photodegradation products [29].

A systematic stability study of curcuminoids has been carried out by Peram et al. using a RP-HPLC method that analyzes their behavior under different stress degradation conditions (i.e., acidic, alkaline, oxidative, photolytic, and thermal degradation) [30]. The authors reported that the order of stability of curcuminoids was: BDMC, followed by DMC and CUR. This proves that these compounds possess a precise synergistic stabilizing mechanism when present in a mixture, as compared to their pure forms.

The processing conditions that are applied to obtain purified bioactives from turmeric, can affect the labile stability and the biological activity of curcuminoids. Hence, it is of the utmost importance that degradation behavior be considered. Valuable information can be obtained from stability studies in order to understand how the health-promoting effects can be retained. Although mechanistic studies have been published on curcuminoid degradation under different stress conditions [30,31], the kinetics under US treatment, both when curcuminoids are present in standard solutions and when they are found in turmeric extracts, have yet to be reported to the best of our knowledge.

In our work, the rapid and exhaustive UAE of turmeric has been defined using multi step extractions and kinetic studies in ethyl acetate media. Solvent choice was subordinated to bioactive stability and the effectiveness of purification by crystallization.

For the sake of comparison, curcuminoid stability under US irradiation in ethanol and ethyl acetate systems has been tested both on optimized extracts and in standard mixtures.

NMR has been applied as both a qualitative and quantitative analytical method to monitor the stability of curcuminoids under sonication treatment. The degradation products were also detected using HPLC and UPLC-MS analyses.

2. Materials and Methods

2.1. Chemicals

Ethyl acetate (ACS grade, ≥99%) (Sigma-Aldrich, Milan, Italy) was used in the extraction procedures. Acetonitrile CHROMASOLV® (gradient grade, for HPLC, ≥99.9%) for HPLC analysis was purchased from Sigma-Aldrich, while Milli-Q H_2O was obtained in the laboratory using a Milli-Q Reference A+System (Merck Millipore). Standards of Curcumin (87.02% Curcumin, 12.98% other curcuminoids) were purchased from Sigma Aldrich.

2.2. *Curcuma longa* L. Matrix

Curcuma longa L. mother rhizomes (India), which were cured and sun dried, were kindly provided by Indena SpA (Milano, Italy), in two different average granulometries, 6 and 2 mm. Biomass was ground in a hammer mill. The biomass was stored at room temperature (RT) in a dry and dark environment to avoid metabolite degradation.

2.3. Curcuminoid Stability Tests under US

The lability of curcuminoids under US irradiation was evaluated in EtOAc, and compared with the most common GRAS solvent for *Curcuma longa* L. extraction, namely EtOH. In order to evaluate the stabilizing effects of co-extracted molecules, both the dry extract (see conventional extraction, Section 2.5) and the curcumin standard were subjected to sonication prior to solubilization. A total of 2 mL of each solvent were used to dissolve 5 mg of the chosen sample in an analytical tube. The solutions were sonicated in a cup-horn PEX 3 Sonifier (24 kHz, 200W, REUS, Contes, France) for either 30, 60, 90 or 120 min. In order to preserve cavitation efficiency, an average temperature of 40 °C was maintained during the tests thanks to a double layered mantle that was crossed by cooling tap water. After treatment, the samples were dried under vacuum for HPLC analyses.

2.4. Ultrasound-Assisted Extraction (UAE)

Curcuma longa L. rhizome powder (10 g) was transferred into a 100 mL glass vessel. Ethyl acetate was added, based on a previous screening, to maintain a solid/liquid (S/L) ratio of 1:5. Extraction was performed in two US devices: a probe system equipped with a titanium horn (20.5 kHz, 350-500W, HNG-20500-SP, Hainertec Suzhou, China), and a cup-horn PEX 3 Sonifier (see Section 2.3). The US horn requires an ice bath to control the temperature in the medium. In both systems, an average temperature of 40 °C was maintained to preserve cavitation efficiency while a time screening was performed.

Extractions were conducted in one or multiple sequential steps (cross-flow), with the aim of depleting the matrix.

A technique comparison was performed with 6 mm rhizome powder and the optimized parameters were then transposed to a 2 mm matrix in order to evaluate the incisiveness of mass transport and the physical effects of US treatment. The crude extract was filtered on sintered glass and dried under vacuum.

2.5. Conventional Extraction

Classical extractions were carried out as a benchmark for technique screening and for overall curcuminoid yield definition. For the sake of comparison, parameters that were used for optimized UAE, such as temperature, solvent (EtOAc), extraction steps and time, were transposed to the 2 mm matrix in a conventional magnetic stirred system (silent conditions). The maximum curcuminoid yield, which corresponded to matrix depletion, was defined in accordance to the rhizome-powder extraction conditions (S/L ratio 1:5), with homogenization pre-treatment (5 min, OV5 rotor-stator homogenizer Velp Scientifica, Usmate Velat, Italy) and 4 sequential extraction steps (4h each) for a total of 16 h, at RT and under protection from light. For the sake of comparison, the same conditions were used with EtOH as a substitute for EtOAc.

2.6. Curcuminoid Determination

Total curcuminoids (TC) were determined by HPLC using the external standard method. Analyses were performed on a Waters 1525 binary pump equipped with a 2998 PDA, and a Phenomenex Kinetex® Column (5 µm C18 100 Å, 250 × 4.6 mm). Data acquisition was accomplished using Empower PRO (Waters Associates, Milford, MA). A CH_3CN 5% acetic acid aqueous solution was used as the mobile phase. Chromatographic separation was performed in isocratic (50:50 *v/v*) at 25 °C, and a flow rate of 1 mL/min. The injection volume was 10 µL, while sample detection was carried out at 425 nm. Before injection, all samples were dissolved in MeOH, giving concentrations of between 1 and 2 mg/mL. All the samples were passed through 0.2-µm membrane filters before injection into the HPLC apparatus. The calibration curve was obtained using curcumin standard solutions (from 0.02 to 2 mg/mL); a linear regression with $R^2 > 0.999$ was obtained. The reported results express both TC (expressed as curcumin equivalents, percentage yields and mg/mL amounts) and curcumin concentration. Relative limit of detection (LOD) and limit of quantification (LOQ) were determined as 0.005 mg/mL and 0.02 mg/mL, respectively.

2.7. NMR Quali and Quantitative Analyses

The quali-quantitative determination of curcumin was also carried out by ^{1}H-NMR. NMR spectra were acquired on a Jeol ECZR 600 spectrometer, operating at a 14 T magnetic field strength and equipped with a Jeol Royal standard probe. Signal acquisition and FID processing were carried out using Jeol DELTA software. Qualitative evaluation, via spectrum analysis and a comparison with the spectra of standards, allowed the different curcuminoids that were present in the whole extract to be identified. The quantitation of curcumin was achieved by acquiring ^{1}H-NMR spectra in MeOD using a 90° pulse, ^{13}C decoupling and a repetition time that was longer than 7-times the longest T1 (typically 40–60 s). FIDs were processed with a zero-filling that was double that of the experimental point (64K+64K), an exponential apodization function with 0.1Hz width and final interactive baseline correction. Specific peak area at 7.20 ppm was compared with the area of potassium terephthalate (analytical standard grade) D_2O solution (with a precisely known concentration) signals, which were obtained in the same receiver gain conditions as the sample spectra. A set of three different concentrations of standard curcumin was used to verify the correspondence between real curcumin concentration and curcumin concentration as determined by NMR quantitation (using an external standard). Relative LOD and LOQ were calculated as 0.5 mg/mL and 1 mg/mL, respectively.

2.8. Statistic Treatment

To validate reproducibility and give soundness to the experimental section, every procedure (Sections 2.3–2.5) was performed in triplicate and percentual standard deviation was consequently calculated. The results are expressed as the mean ± %SD in Tables as well as in Appendix A. Degradation percentage SDs are graphically depicted in Figures 4–6. Moreover, upper and lower SD envelopes have been extrapolated by means of linear regression, describing deviation trends where possible.

2.9. Kinetic Model

The hyperbolic model of Peleg (see Equation (1)) was used to evaluate the extraction kinetics and to determine the point of maximum extraction rate by means of the related constants.

$$C(t) = C_0 + \frac{t}{k_1 + k_2 t} \tag{1}$$

C(t) is the concentration of the extract after extraction time t, whilst C_0 is equal to 0 at the beginning of the process. The *Peleg Initial Extraction Rate* (k_1) is correlated to the starting extraction rate (B_0, Equation (2)) and can be exploited to calculate the relative extraction rate in each moment of the extraction (B_t).

$$B_0 = \frac{1}{k_1} \tag{2}$$

This parameter can be used to calculate the instant of maximum extraction speed, the critical point for cross-flow and counter-current extractions, which is fundamental for an industrial transposition of the process.

The *Peleg Capacity Constant* (k_2) is correlated to the highest extraction yield at the steady state (Y_s, Equation (3)). For an ideal process, it can be used to evaluate the number of sequential extraction steps necessary to deplete the matrix. From a graphical point of view, this parameter represents an horizontal asymptote.

$$c_0 = c_{eq} = Y_s = \frac{1}{k_2} \tag{3}$$

Two extraction sets were conducted using a probe system, according to Section 2.4, on the 2 and 6 mm Curcuma powder. Sampling was performed at 2, 5, 7.5, 15, 30, 60, 90 and 120 min, by collecting 1 mL of solution. The crude extract was filtered on sintered glass and dried under vacuum for HPLC analyses.

Equation (1) can be conveniently linearized in Equation (4), thus providing a fast and easy way to extrapolate k_1 and k_2 as the intercept and slope, respectively. Hence, the kinetic constants can be calculated by linear interpolation of the experimental yields at different extraction times (see Table A1 and Figure A1 for 6 mm and Table A2 and Figure A2 for 2 mm), and then inputted into general Equation (1).

$$\frac{t}{C(t)} = k_1 + k_2 t \tag{4}$$

The obtained hyperbolic curve describes a time-dependent extraction trend. This model is a useful means to display the horizontal asymptote of Y_s and the extraction rates (slope of the curve). Furthermore, the knee-point can be exploited to determine the best trade-off between productivity and process extent.

2.10. Crystallization

The ethyl acetate extracts were concentrated under vacuum to approximately an S/L ratio of 1:1. The mixture was stirred at 20–25 °C for 24 h, then the suspension was filtered under vacuum. The wet solid was ground twice at 20–25 °C for 1 h with 3 volumes of isopropanol 90% (3 mL per g of wet solid). The wet solid was finally dried at 50 °C under vacuum.

3. Results

3.1. Curcuminoid Stability Test

In order to investigate the effect of the US irradiation of EtOH and EtOAc on curcumin, a known amount of analytical standard was subjected to prolonged sonication in both solvents. Due to the small volume of the samples, the cup-horn was thought to be the most suitable device for the degradation treatments. As shown in Table 1, the effective stability of the pure curcumin was determined by monitoring its concentration every 30 min using HPLC. Total treatment lasted two hours, and the results are reported in Table 1 as curcumin amounts and degradation percentages.

Table 1. Curcumin standard US degradation tests.

Irradiation (min)	EtOH			EtOAc		
	Curcumin conc. ± %SD (mg/mL)		Degradation (%)	Curcumin conc. ± %SD (mg/mL)		Degradation (%)
0	4.21	± 0.42	-	4.21	± 0.39	-
30	3.70	± 0.61	12.11	3.89	± 0.46	7.60
60	3.35	± 0.54	20.43	3.73	± 0.65	11.40
90	3.17	± 0.73	24.70	3.52	± 0.63	16.39
120	2.96	± 0.68	29.69	3.48	± 0.70	17.34

Curcumin concentration quantified by HPLC.

In order to evaluate the degradation that was only related to solvent sonolysis, the degradation test was repeated in EtOAc, in closed vials. The curcumin concentration was determined by quantitative NMR. Although its sensitivity is much lower than HPLC-UV or HPLC-MS, NMR quantitation offers the opportunity to determine the analyte concentration using an external standard whose spectrum is acquired once. The external standard can be a different molecule to the analyte as the only requirement is that its concentration is exactly known, and that its NMR spectrum is acquired following the quantitative protocol. Table 2 reports the comparison between the two behaviors.

Table 2. Sonolysis: the influence of air on curcumin degradation.

Irradiation (min)	Closed Vessel	
	Degradation (%)	± SD
30	4.00	± 0.58
60	8.54	± 0.72
90	10.72	± 1.14
120	11.27	± 1.48

Solvent: EtOAc; NMR quantifications.

Many studies have indicated that curcuminoid stability increases when all of the components are studied together, rather than in their pure form [30]. In order to verify whether this statement can also be applied to US-mediated degradation, the screening that was performed on the analytical standard was repeated on the extract obtained from the conventional procedure (see Section 2.5). The results for EtOH and EtOAc are reported in Table 3.

Table 3. Curcuminoid extract US degradation tests.

Irradiation (min)	EtOH			EtOAc		
	Curcumin conc. ± %SD (mg/mL)		Degradation (%)	Curcumin conc. ± %SD (mg/mL)		Degradation (%)
0	3.83	± 0.39	-	4.79	± 0.51	-
30	3.71	± 0.61	3.13	4.74	± 0.70	1.05
60	3.54	± 0.57	7.57	4.61	± 0.38	3.76
90	3.16	± 0.49	17.49	4.48	± 0.57	6.47
120	3.08	± 0.67	19.58	4.25	± 0.44	11.27

HPLC quantification of curcuminoids, expressed as curcumin equivalents.

3.2. *Total Curcuminoid Content Determination*

The quantification of total curcuminoids in plant rhizomes was obtained by performing sequential extractions with EtOAc under ultraturrax® treatment (as described in Section 2.5), with the aim of achieving full matrix depletion.

The triplicate average gives a total amount of 67.14 mg/g_{Matrix} curcuminoids (HPLC determination, expressed as curcumin equivalents), which corresponds to 6.71% of raw material and 40.49 mg/g_{Matrix} (4.05%) of curcumin. The same extraction protocol was performed with EtOH as the solvent and revealed a negligible difference of +0.84% compared to the previous test. This evidence indicates that no difference subsists between the two solvent systems in terms of total matrix depletion, and that 67.14 mg/g_{Matrix} is admissible as the maximum metabolite yield.

3.3. *US-Assisted Extraction*

This work investigated the UAE of *Curcuma longa* L. and did so by varying US equipment and process parameters. A multi-step approach, which depended on single-step kinetics, was finally defined and investigated for two different granulometries.

Firstly, the effects of acoustic cavitation were studied on the coarse particle size (6 mm), using two different US reactors: an immersion horn (two power intensity, 350 W and 500 W) and a cup-horn (200 W). In order to emphasize the difference between each run, sequential UAE was adopted, and this allowed a preventive evaluation of multi-step feasibility to be performed. Solvent volumes were minimized to an S/L ratio of 1:5, in order to move towards a more sustainable process. Hence, three sequential extractions of 30 min were chosen for each step, in order to avoid the overheating of the system due to solution consistency.

3.3.1. Kinetic Model—Single-Step UAE

Once the extraction device was selected, a better understanding of the kinetic was required. For this purpose, 6 mm rhizomes (treated in Table 4) were considered together with smaller particles (size of 2 mm) for extraction modeling. Samples were gathered after a suitable time-span of 120 min of UAE.

Table 4. US technology screening.

US Technology	Dry Extract (%)	Curcuminoid conc. ± %SD (%)	
Cup-horn [a]	19.19	72.07	± 0.22
Horn [b]	16.31	68.43	± 0.38
Horn [c]	18.17	72.68	± 0.29

Matrix 6 mm, 3 steps of 30 min, S/L ratio 1:5. [a] 200W; [b] 350W; [c] 500W. Curcuminoid yields, calculated as percentage recovery of total curcuminoid-content. HPLC quantification expressed as curcumin equivalents.

Detected yields were processed according to the Peleg Model, as described in Section 2.9. Using linearization (see Equation (4)), it was possible to extrapolate the kinetic constants by building a general equation for the system that expressed a curve with a physical meaning that describes curcuminoid recovery in a time-dependent function.

The plots (see Figure 2 for 6 mm and Figure 3 for 2 mm) and relative equations (see Equation (5) for 6 mm and Equation (6) for 2 mm) are reported below. Linearization gave impressive interpolation of the experimental values, with R^2 being over 0.99 (see Figures A1 and A2). It is possible to observe how the theoretical curves match the experimental points.

Extraction rates (B_t) can be expressed as the inverse of k_1, which is computable for every instant (t). Bt expresses the extraction efficiency at any precise moment and is mathematically depicted as the slope of a line tangent to the model curve at time t (see Tables A1 and A2).

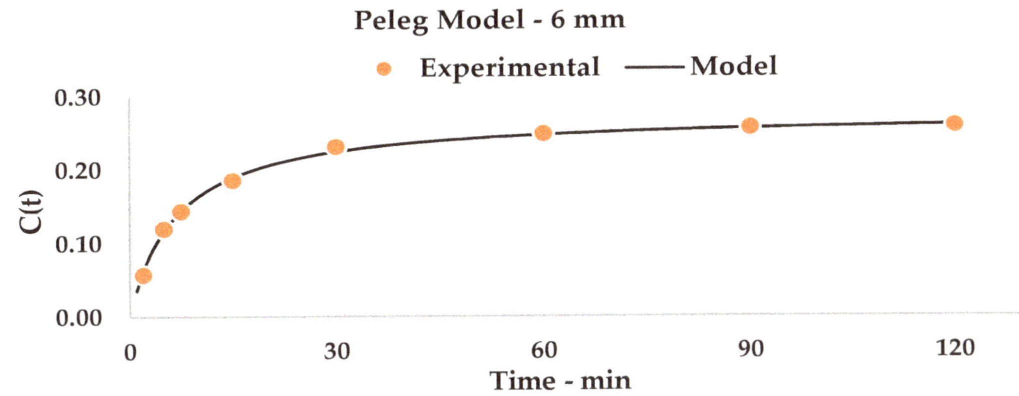

Figure 2. Kinetic Model, 6 mm matrix. Statistical values are reported in Table A1.

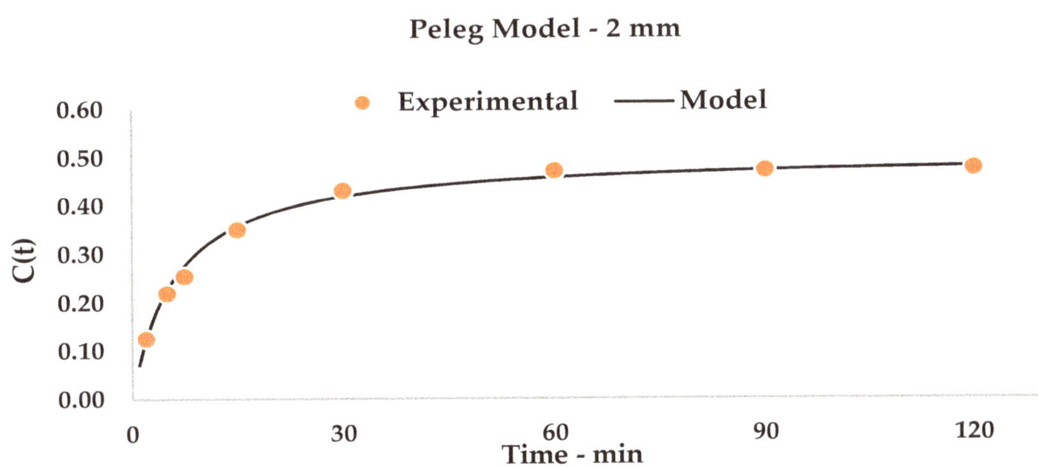

Figure 3. Kinetic Model, 2 mm matrix. Statistical values are reported in Table A2.

$$C(t) = \frac{t}{24.5350 + 3.6412\,t} \tag{5}$$

$$C(t) = \frac{t}{12.2240 + 1.9831\,t} \tag{6}$$

3.3.2. Cross-Flow UAE

The kinetic model gave 30 min as the best extraction time, as it allowed a good compromise between curcuminoid yield and process duration to be achieved. This discovery was developed and sequential UAE was tested. Biomass was recovered run-by-run and was then submitted to a next stage, feeding fresh EtOAc. Both particle sizes were used in order to define granulometry dependency. Multi-step yields and yield increases are reported in Table 5.

Table 5. Cross-flow UAE, step screening with 6 and 2 mm matrix.

Step	6 mm Matrix				2 mm Matrix			
	Dry Extract %	Curcuminoid Yield ± %SD %		Yield Increase %	Dry Extract %	Curcuminoid Yield ± %SD %		Yield Increase %
1	5.44	23.10	± 0.37	-	11.19	43.10	± 0.22	-
2	12.24	51.97	± 0.33	28.87	22.56	88.54	± 0.19	45.44
3	18.17	72.68	± 0.29	20.71	23.77	92.10	± 0.30	3.56
4	20.12	80.04	± 0.31	7.36	-	-	-	-

30 min per step, S/L ratio 1:5, Horn 500 W; Curcuminoid yields, calculated as percentage recovery of total curcuminoid content. HPLC quantification expressed as curcumin equivalents.

The results depict, as expected, that the extraction rate of the 2 mm rhizome powder was clearly higher than that of the 6 mm analogue. Furthermore, the maximum extraction yield was reached with one extraction step of difference.

3.4. Crystallization

The reported procedure (Section 3.1) was applied to both the EtOH and EtOAc extracts. It was found that the alcoholic turmeric oleoresin could not be purified via crystallization, whilst EtOAc successfully led to suitable precipitate formation. HPLC quantification confirmed an average curcuminoid purity of 95%, expressed as curcumin equivalents.

4. Discussion

4.1. Curcuminoid Stability Test

The fact that radicals are formed as a consequence of acoustic cavitation has been well described in the literature [32]. Although the production of $OH^{\bullet}$ and $H^{\bullet}$ in aqueous media is something that researchers have robustly proven using different approaches [33], very few demonstrations of radical production in organic solvents (as methanol) are available [34]. Radical concentration is frequency-dependent in linear mode, but it is also possible to produce traces of $R^{\bullet}$ or $RO^{\bullet}$ at low frequencies [35]. The extreme reactivity of curcumin-like compounds means that the presence of radicals in the extraction media can affect extraction yields, triggering the degradation and the accumulation of oxidized compounds, which will reduce general shelf-life and product activity.

These considerations can inform the choice of solvent for curcuminoid extractions. Ethanolic solutions are, according to the literature, the best performing media, as they achieve high yields in relatively short times. It is reasonable to think that ethanol is subjected to the formation of $CH_3CH_2^{\bullet}$ and $OH^{\bullet}$ radicals, due to his polarity, as these reactive species can also be generated from MeOH and H_2O [34]. On the other hand, EtOAc (a GRAS approved solvent) is less likely to undergo the same fate.

In the case of pure curcumin, the onset of degradation after 30 min supports the low stability of the compound in both solvent systems (Figure 4, related to Table 1). Tests were performed using a common US set-up, in order to be a closer fit for the usual extraction conditions. This approach cannot avoid oxidation due to atmospheric O_2, which overlaps with the other degradation pathways. In any case, it is possible to state that EtOH media display higher sensitivity than EtOAc media. Furthermore, the time increase leads to a stronger lowering of curcumin concentration in alcoholic systems, reaching a maximum degradation of 29.76% vs. 17.28%.

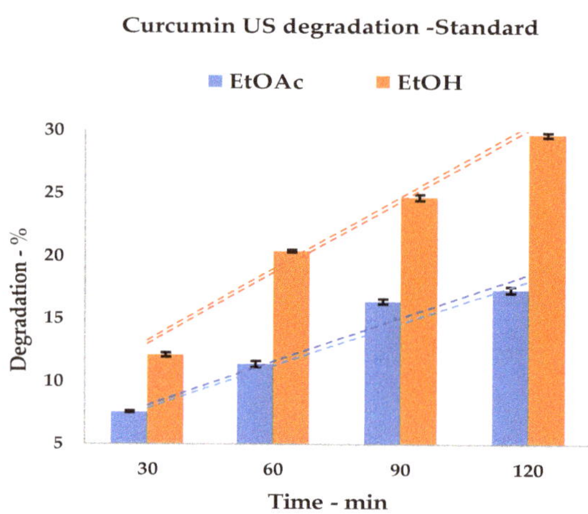

Figure 4. Curcumin standard degradation trend; dashed lines represent upper and lower SD envelopes.

Tests were repeated in sealed vials to investigate the correlation between atmospheric oxygen and overall degradation. With strongly reduced O_2 presence, stability should be principally affected by solvent sonolysis.

The concentration of curcumin decreases over sonication time, but its decrease is lower than when measured in open air. The graph depicted in Figure 5 shows behavior over time.

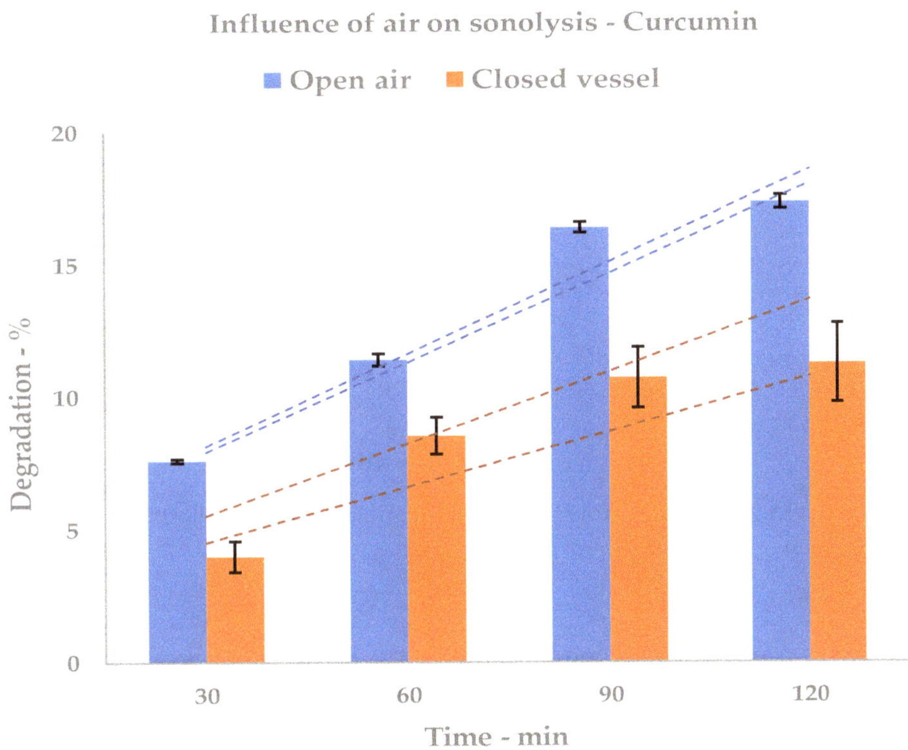

Figure 5. Sonolysis: influence of air. HPLC (open air) and NMR (closed vessel) quantification; dashed lines represent upper and lower SD envelopes.

According to the literature, the stability of curcuminoids increases if all three components (CUR, DMC, BDMC) are mixed together [29].

The stability screenings were therefore repeated with the analytical standard of curcumin being replaced with a conventional extract (see procedure in Section 2.5). US extracts were not used as the starting material to ensure that the matrix was not affected by previous irradiation.

The results fit into the abovementioned studies; general stability was increased by the presence of curcuminoids and co-extracts, shifting from 29.76% to 19.56% and from 17.28% to 11.14%, for EtOH and EtOAc media, respectively. It was possible to observe a sudden increase in the degradation of the alcoholic solvent between 60 and 90 min. This is a retardant effect that is possibly due to the presence of the co-extracts (see Figure 6), as this behavior is not visible in pure curcumin (see Figure 4).

These results demonstrate that different solvent media can give rise to different radical concentrations, which is critical for labile molecules, such as curcumin-like compounds, that are well known for their low shelf-life and stability. EtOAc appeared to provide milder radical generation, which paves the way for a suitable multistep extraction protocol that reduces the degradation that is caused by prolonged treatments. These tests are fundamental for a preliminary characterization of a multistep extraction protocol, and for the design of an efficient procedure for the complete depletion of turmeric rhizome powder.

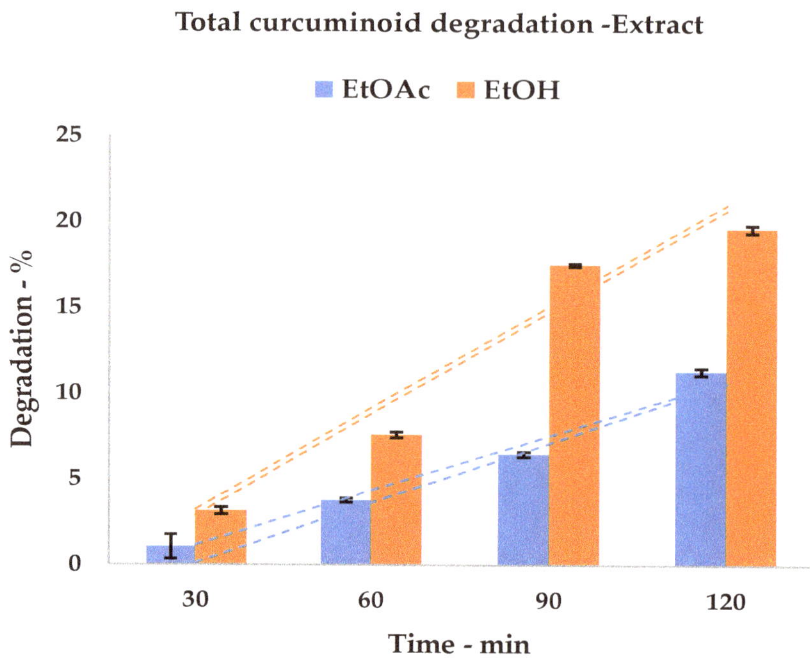

Figure 6. Curcumin extract degradation trend; A: ethanol; B: ethyl acetate; dashed lines represent upper and lower SD envelopes.

4.2. US-Assisted Extraction

In the early stages, the work was focused on 6 mm turmeric powder and finding the best US device for extraction. This approach was a useful means to properly evaluate the physical effects of sonication upon a roughly milled biomass. Bulky powders require better extraction performance for successful cellular/particle cleavage and convenient mass-transfer enhancement to occur. In summary, this type of matrix requires harsher conditions and is a challenge for UAE intensification.

The first approach to UAE was the screening of several US devices, and, in particular, two different set-ups: the immersion horn and cup-horn. The results indicate that a slightly better result was obtained for the horn at 500 W (see Table 4). Technology comparisons shed light on handling and process suitability, and thus influence the practical applications. The cup-horn results were promising even if the power was lower, although temperature control appeared to be very challenging. The sedimentation of heavier particles on the cavitating surface, causing overheated spots, is a possible explanation for this. The media was not easily cooled by the water jacket because of the consequent poorer mass-transfer. The immersion horn, on the other hand, appeared to efficiently transmit energy to the media, despite the radiating surface being smaller. The free tip maintained high mass transfer and easier heat transfer towards the cooling bath. Hence, this first screening facilitated the choice of the 500 W immersion horn as the elective device for the subsequent investigations.

4.2.1. Kinetic Model—Single Step UAE

After the selection of the US device, and before performing the cross-flow extraction design, it was essential to characterize a single-step process. Thus, UAE was approached from a kinetic point of view in order to shed light on the dependency of yield on treatment time. The influence of granulometry was considered for this screening, including the smaller particle size of 2 mm.

When comparing the two kinetic systems (Figures 2 and 3), some considerations can be pointed out. Firstly, the k_2 constants describe, as their inverse values, the curcuminoid total yield in the steady state, and are translated by a horizontal asymptote on the graph. This quantity is, as expected, higher for the 2 mm matrix than for the 6 mm one, as it describes, for the first, a matrix that is more prone to extraction (Y_S: 50.43% vs. 27.46%). Interestingly, the chosen step-length for the initial optimization

(30 min, Table 4) appears to give results that are near the maximum yield for single-step extractions: 23.10% vs. 27.46% for 6 mm and 43.10% vs. 50.43% for 2 mm, respectively. This information allowed us to focus our attention on a dedicated multistep cross-flow UAE design.

4.2.2. Cross-Flow UAE

Sequential UAE was based on a value of 30 min as it was the best productivity trade-off for single-step extractions.

The highest curcuminoid yield for the 6 mm matrix was obtained with four-step cross-flow, and a progressive reduction in extraction rate was observed (Table 5). As expected, the curcuminoid output steeply increased with the cross-flows and was more pronounced in the 2 mm than in the 6 mm matrix; the yield was almost double at the first extraction (43.10% vs. 23.10%, also see Figure 7).

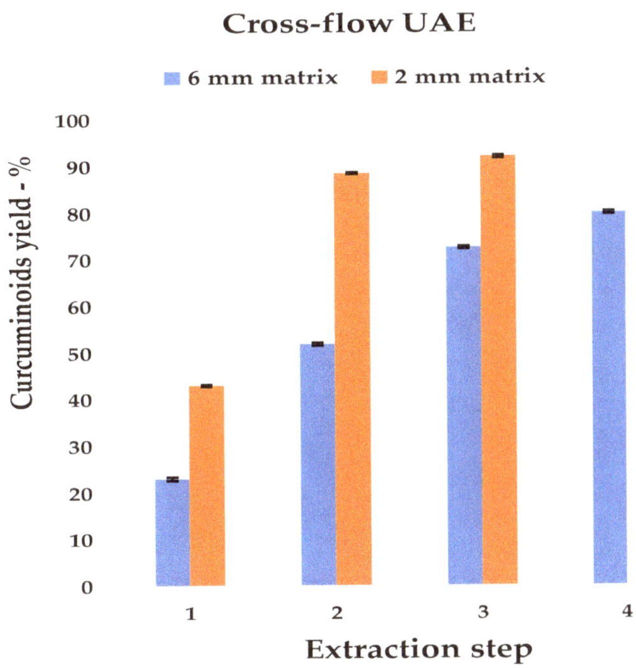

Figure 7. Cross-flow trend for 6 and 2 mm particle sizes.

It is interesting to observe that the maximum production achieved for the coarse material after 4 steps, is overcome by 2 mm turmeric with two UAE steps (80.04% vs. 88.54%, respectively). A fourth extraction was not suitable with the finer Curcuma powder because of the poor quantity of the curcuminoids left in the rhizomes (less than 8%); an additional step was considered not to be cost-effective.

Due to its nature, it is possible to describe a cross-flow protocol (fresh solvent on recycled biomass), as a progressive sum of single extraction stages. Starting from this approximation, experimental yields can be compared step-by-step to the Peleg Model as extrapolated by a reiteration of every stage. In Table 6, it is possible to observe the progress of the model experiments in relation to their step number.

The prediction of sequential extractions using model reiterations clearly shows dramatic discrepancies from the experimental data. This behavior can be explained by the constant biomass modification produced by acoustic cavitation. A likely explanation may be the capacity of US to physically disaggregate vegetal materials [36]. UAE reduces particle size during extraction, enhancing the mass-transfer and modifying the kinetic profile, and thus explains the underestimation of the theoretical approach.

Table 6. Model fitting to cross-flow screening.

Step	6 mm Curcuminoid Yield		2 mm Curcuminoid Yield	
	Model %	Experimental %	Model %	Experimental %
1	22.92	23.10	41.06	43.10
2	40.59	51.97	65.26	88.54
3	54.20	72.68	79.52	92.10
4	64.70	80.04	87.93	-

Thirty min per step, S/L ratio 1:5, Horn 500 W; curcuminoid yields, calculated as the percentage recovery of total curcuminoid content. HPLC quantification expressed as curcumin equivalents.

Moreover, it is possible to state that the observed trend for sequential extraction changes when approaching matrix depletion. This is possibly due to the increase in the magnitude of the degradation mechanisms in light of the accumulation of the treatment periods. Indeed, according to Section 3.1, the total degradation for curcuminoid extracts in EtOAc, is 6.45% after 90 and 11.14% after 120 min (equal to 3 and 4 steps). As a matter of fact, the peculiarities of UAE, such as particle comminution and mass-transfer enhancement, helped to provide a constant yield increase, thus balancing and overcoming curcuminoid degradation.

The granulometric variation can also be investigated by means of B_t values (Equation (2), Tables A1 and A2). As the most relevant evolution of this parameter occurs in the initial moments of extraction (see Figure 8), the first three points sampled for Peleg's Model are taken as being descriptive.

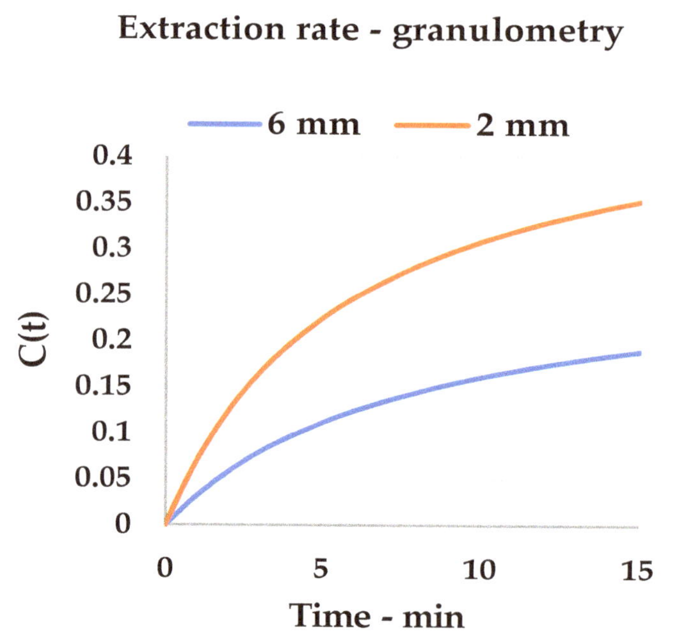

Figure 8. Granulometry comparison: slope variation during initial extraction phase, Peleg model.

Hence, it is possible to appreciate the slope divergence by plotting the *extraction rates* (Figure 9) for 2, 5 and 7.5 min for every particle size. The reverse proportionality of B_t to extraction time was confirmed. It is important to underline that the dashed bars represent the average *extraction rate*, which results from the accumulation of all the rates and is extrapolated from the kinetic constant k_1, of the systems (B_0). Therefore, although cavitation can progressively reduce particle size during extraction, the early stages, in which disaggregation is negligible, have a prevailing influence on the overall process.

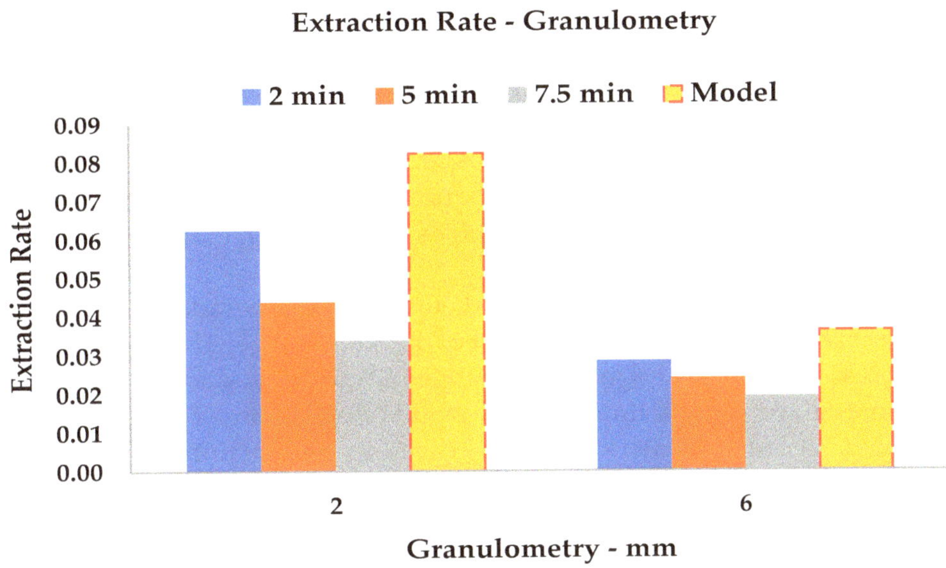

Figure 9. Extraction-rate variation with granulometry. B_t of 2, 5 and 7.5 min compared to overall B_0.

4.3. Crystallization

The features and the advantages of using EtOAc as the extraction solvent were also investigated in terms of metabolite purification via crystallization. In addition to the better stability against curcuminoid radical degradation, the acetate showed an interesting predisposition to crystal generation, which was not observed in the EtOH extracts, under the tested conditions. The efficiency of this purification step is crucial if complex and expensive protocols, such as chromatography, are to be avoided. The final detected purity of the crystallized curcuminoids is 95% on average, as determined by HPLC and expressed as curcumin equivalents.

5. Conclusions

A thorough investigation into curcuminoid stability under US interactions has been performed with the elective solvent and its derived ester (EtOH vs. EtOAc). The latter was expected to be less prone to sonolysis. For the sake of comparison, both pure curcumin and a curcuminoid mixture were tested, and HPLC-UV and quantitative ^{1}H-NMR were used to determine metabolite degradation. The main advantage of this method is the mere requirement of one analysis for the defined external reference, and no specific and expensive standard for each target analyte. Stability tests have shown that solvents can lead to different radical concentrations, which is crucial for labile molecules, such as curcumin-like compounds. US irradiation caused milder radical generation in the presence of EtOAc than in EtOH, even in relation to the presence of oxygen. In particular, the use of a sealed vessel decreased curcumin degradation by 6.01% after 2 h of US irradiation in the aprotic solvent. The synergistic stabilizing mechanism of curcuminoids when present in a mixed form was confirmed for both the systems, as reported in the literature. In detail, degradation behavior was strongly prevalent in EtOH, with 12.48% and 8.42% decreases being observed for curcumin and the curcuminoids mix, respectively, after 2 h.

These preliminary tests have designated EtOAc as the elective solvent for efficient US-assisted extraction procedures that aim for curcuminoid stability and maximum yields. The use of UAE was first evaluated in a single-step extraction of *Curcuma longa* L., and the impact of different granulometries, namely 6 and 2 mm, was screened. Particular attention was paid to the kinetic features. UAE equilibria were reached before rhizome depletion (25.63% and 47.56% for 6 and 2 mm, respectively), proving the need for a multi-step protocol. Stability tests supported the use of EtOAc, even for sequential extractions, due to its capacity to minimize degradation. Hence, the multi-step co-current cross-flow approach was exploited to maximize yield, and the influence of granulometry is to be highlighted.

Maximum curcuminoid recovery was achieved in three steps (92.10%) and four steps (80.04%), respectively, for the 2 and 6 mm rhizome powders.

Comparisons between single extraction kinetics and the cross-flow trend highlight the disaggregation power of US, which is able to accelerate extraction through sequential stages. EtOAc has shown interesting results and applicability in the UAE of *Curcuma longa* L. but also has a crucial role in the final purification steps. In this work, EtOAc produced an extract that is prone to the crystallization process, and an average of 95% pure product was achieved, whilst ethanol prevented crystal formation from the oleoresin extract.

In conclusion, the effect of US on curcumin and curcuminoid degradation has been screened with EtOH and EtOAc for the first time herein, to the best of our knowledge. The results were then exploited to define a UAE process for *Curcuma longa* L. A Peleg kinetic model was used to describe the single-step extraction, and the optimized process time was the groundwork upon which the multi-step cross-flow process was designed. A small particle size was crucial to the obtaining of improved final yields. Finally, the efficacy of the applied solvent was then confirmed by means of suitable curcuminoid purification via crystallization.

Author Contributions: Conceptualization, D.C., A.B. (Arianna Binello) and G.C.; methodology, G.G. and A.B. (Alessandro Barge); validation, A.B. (Alessandro Barge); formal analysis, A.B. (Arianna Binello); investigation, G.G.; data curation, D.C., P.A.; writing—original draft preparation, A.B. (Arianna Binello) and G.G.; writing—review and editing, G.C. and D.C.; supervision, P.A. and G.C.; All authors have read and agreed to the published version of the manuscript.

Acknowledgments: G.G. acknowledges Indena SpA for their research scholarship.

Appendix A

Table A1. Experimental UAE of 6 mm *Curcuma longa* L.

t min	Curcuminoid Yield ± %SD %		B_t [a]	Linearization t/C_t [b]
0	-		-	-
2	0.057	± 0.23	0.0285	35.0877
5	0.12	± 0.21	0.024	41.6667
7.5	0.144	± 0.32	0.0192	52.0833
15	0.1857	± 0.29	0.0124	80.7754
30	0.231	± 0.37	0.0077	129.8701
60	0.2482	± 0.33	0.0041	241.7405
90	0.2563	± 0.24	0.0028	351.1510
120	0.2588	± 0.24	0.0022	463.6785

[a] Extraction rates; [b] linearization values.

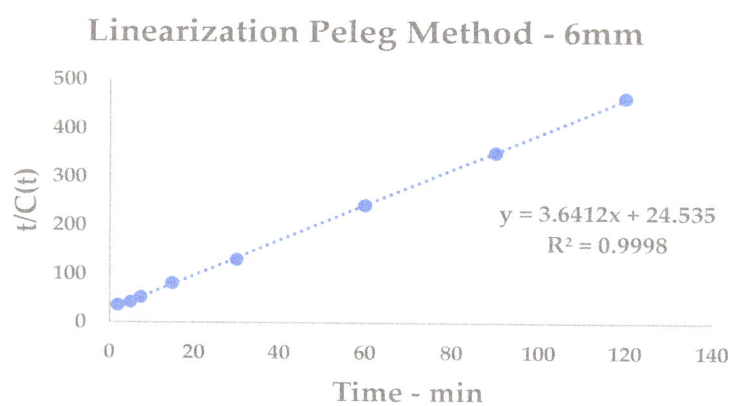

Figure A1. Peleg Model linearization, 6 mm.

Table A2. Experimental UAE of 2 mm *Curcuma longa* L.

t min	Curcuminoid Yield ± %SD %		B_t [a]	Linearization t/C_t [b]
0	-		-	-
2	0.1252	± 0.33	0.0625	15.9744
5	0.2188	± 0.25	0.0438	22.8519
7.5	0.2543	± 0.26	0.0339	29.4927
15	0.3501	± 0.30	0.0233	42.8449
30	0.4310	± 0.22	0.0144	69.6056
60	0.4710	± 0.31	0.0079	127.3885
90	0.4718	± 0.28	0.0052	190.7588
120	0.4756	± 0.29	0.004	252.3129

[a] Extraction rates; [b] linearization values.

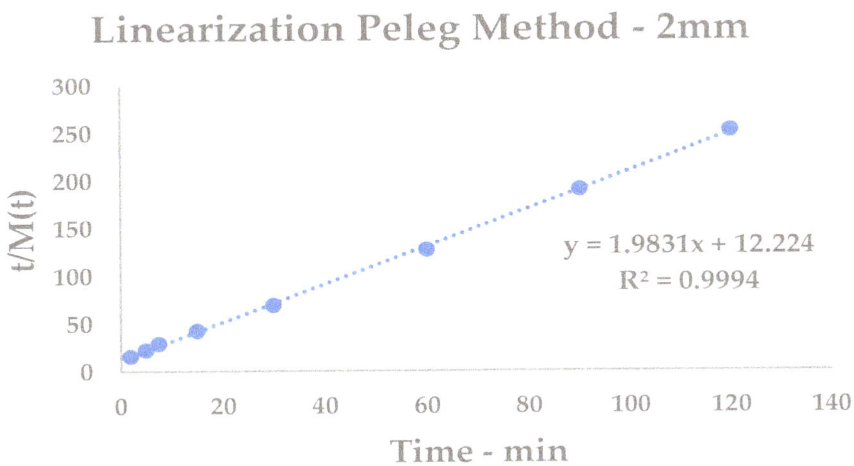

Figure A2. Peleg Model linearization, 2 mm.

References

1. Ravindran, P.N. *Turmeric—The golden spice of life, In Turmeric: The Genus Curcuma*; Ravindran, P.N., Babu, K.N., Sivaraman, K., Eds.; CRC Press: Boca Raton, FL, USA, 2007; Volume 1, pp. 1–14.
2. Dao, T.T.; Nguyen, P.H.; Won, H.K.; Kim, E.H.; Park, J.; Won, B.Y.; Oh, W.K. Curcuminoids from *Curcuma longa* and their inhibitory activities on influenza a neuraminidases. *Food Chem.* **2012**, *134*, 21–28. [CrossRef]
3. Nguyen, T.T.H.; Si, J.; Kang, C.; Chung, B.; Chung, D.; Kim, D. Facile preparation of water soluble curcuminoids extracted from turmeric (*Curcuma longa* L.) powder by using steviol glucosides. *Food Chem.* **2017**, *214*, 366–373. [CrossRef] [PubMed]
4. Sueth-Santiago, V.; Mendes-Silva, G.P.; Decoté-Ricardo, D.; de Lima, M.E.F. Curcumin, the golden powder from turmeric: Insights into chemical and biological activities. *Quim. Nova.* **2015**, *38*, 538–552. [CrossRef]
5. Esatbeyoglu, T.; Huebbe, P.; Ernst, I.M.; Chin, D.; Wagner, A.E.; Rimbach, G. Curcumin—From molecule to biological function. *Angew. Chem. Int. Ed.* **2012**, *51*, 5308–5332. [CrossRef] [PubMed]
6. Niranjan, A.; Prakash, D. Chemical constituents and biological activities of turmeric (*Curcuma longa* L.)—A review. *J. Food Sci. Technol.* **2008**, *45*, 109–116.
7. Bessone, F.; Argenziano, M.; Grillo, G.; Ferrara, B.; Pizzimenti, S.; Barrera, G.; Cravotto, G.; Guiot, C.; Stura, I.; Cavalli, R.; et al. Low-dose curcuminoid-loaded in dextran nanobubbles can prevent metastatic spreading in prostate cancer cells. *Nanotechnology* **2019**, *30*, 214004–214016. [CrossRef] [PubMed]
8. Priyadarsini, K.I. The chemistry of curcumin: From extraction to therapeutic agent. *Molecules* **2014**, *19*, 20091–20112. [CrossRef] [PubMed]
9. Patra, J.K.; Das, G.; Lee, S.; Kang, S.-S.; Shin, H.-S. Selected commercial plants: A review of extraction and isolation of bioactive compounds and their pharmacological market value. *Trends Food Sci. Technol.* **2018**, *82*, 89–109. [CrossRef]

10. Chemat, F.; Vian, M.A.; Fabiano-Tixier, A.-S.; Nutrizio, M.; Jambrak, A.R.; Munekata, P.E.S.; Lorenzo, J.M.; Barba, F.J.; Binello, A.; Cravotto, G. A review of sustainable and intensified techniques for extraction of food and natural products. *Green Chem.* **2020**, *22*, 2325–2353. [CrossRef]
11. Grillo, G.; Boffa, L.; Binello, A.; Mantegna, S.; Cravotto, G.; Chemat, F.; Dizhbite, T.; Lauberte, L.; Telysheva, G. Cocoa bean shell waste valorization; extraction from lab to pilot-scale cavitational reactors. *Food Res. Int.* **2019**, *115*, 200–208. [CrossRef]
12. Berk, Z. *Extraction, In Food Process Engineering and Technology*, 3rd ed.; Taylor, S.L., Ed.; Elsevier Inc.: Amsterdam, The Netherlands, 2018; Volume 1, pp. 289–310.
13. Cacace, J.E.; Mazza, G. Optimization of extraction of anthocyanins from black currants with aqueous ethanol. *J. Food Sci.* **2003**, *68*, 240–248. [CrossRef]
14. Herodez, S.S.; Hadolin, M.; Skerget, M.; Knez, Z. Solvent extraction study of antioxidants from Balm (Melissa officinalis L.) leaves. *Food Chem.* **2003**, *80*, 275–282. [CrossRef]
15. Peleg, M. An empirical model for the description of moisture sorption curves. *J. Food Sci.* **1988**, *54*, 1216–1217. [CrossRef]
16. Bucić-Kojić, A.; Planinić, M.; Tomas, S.; Bilić, M.; Velić, D. Study of solid—Liquid extraction kinetics of total polyphenols from grape seeds. *J. Food Eng.* **2007**, *81*, 236–242. [CrossRef]
17. Shirsath, S.R.; Sable, S.S.; Gaikwad, S.G.; Sonawane, S.H.; Saini, D.R.; Gogate, P.R. Intensification of extraction of curcumin from *Curcuma amada* using ultrasound assisted approach: Effect of different operating parameters. *Ultrason. Sonochem.* **2017**, *38*, 437–445. [CrossRef]
18. Aguilar, F.; Dusemund, B.; Galtier, P.; Gilbert, J.; Gott, D.M.; Grilli, S.; Gürtler, R.; König, J.; Lambré, C.; Larsen, J.-C.; et al. Scientific opinion on the re-evaluation of curcumin (E 100) as a food additive. *EFSA J.* **2010**, *8*, 1–46. [CrossRef]
19. Revanthy, S.; Elumalai, S.; Benny, M.; Antony, B.J. Isolation, purification and identification of curcuminoids from turmeric (*Curcuma longa* L.) by column chromatography. *Exp. Sci.* **2011**, *2*, 21–25.
20. Ukrainczyk, M.; Hodnett, B.K.; Rasmuson, A.C. Process parameters in the purification of curcumin by cooling crystallization. *Org. Process Res. Dev.* **2016**, *20*, 1593–1602. [CrossRef]
21. Joshi, P.; Jain, S.; Sharma, V. Turmeric (*Curcuma longa*) a natural source of edible yellow colour. *Int. J. Food Sci. Technol.* **2009**, *44*, 2402–2406. [CrossRef]
22. Thorat, A.A.; Dalvi, S.V. Particle formation pathways and polymorphism of curcumin induced by ultrasound and additives during liquid antisolvent precipitation. *CrystEngComm* **2014**, *16*, 11102–11114. [CrossRef]
23. Kuhnen, S.; Bernardi Ogliari, J.; Dias, P.F.; da Santos, M.S.; Ferreira, A.N.G.; Bonham, C.C.; Maraschin, M. Metabolic fingerprint of brazilian maize landraces silk (stigma/styles) using NMR spectroscopy and chemometric methods. *J. Agric. Food Chem.* **2010**, *58*, 2194–2200. [CrossRef] [PubMed]
24. Yang, S.-O.; Shin, Y.-S.; Hyun, S.-H.; Cho, S.; Bang, K.-H.; Lee, D.; Choi, H.-K. NMR-based metabolic profiling and differentiation of ginseng roots according to cultivation ages. *J. Pharm. Biomed. Anal.* **2012**, *58*, 19–26. [CrossRef] [PubMed]
25. Gad, H.A.; Bouzabata, A. Application of chemometrics in quality control of Turmeric (*Curcuma longa*) based on Ultra-violet, Fourier transform-infrared and ^{1}H NMR spectroscopy. *Food Chem.* **2017**, *237*, 857–864. [CrossRef] [PubMed]
26. Li, B.; Konecke, S.; Wegiel, L.A.; Taylor, L.S.; Edga, K.J. Both solubility and chemical stability of curcumin are enhanced by solid dispersion in cellulose derivative matrices. *Carbohydr. Polym.* **2013**, *98*, 1108–1116. [CrossRef]
27. Gordon, O.N.; Luis, P.B.; Sintim, H.O.; Schneider, C. Unraveling curcumin degradation: Autoxidation proceeds through spiroepoxide and vinylether intermediates en route to the main bicyclopentadione. *J. Biol. Chem.* **2015**, *290*, 4817–4828. [CrossRef]
28. Schneider, C.; Gordon, O.N.; Edwards, R.L.; Luis, P.B. Degradation of curcumin: From mechanism to biological implications. *J. Agric. Food Chem.* **2015**, *63*, 7606–7614. [CrossRef] [PubMed]
29. Khurana, A.; Ho, C.T. High performance liquid chromatographic analysis of curcuminoids and their photo-oxidative decomposition compounds in *Curcuma longa* L. *J. Liq. Chromatogr.* **1988**, *11*, 2295–2304. [CrossRef]
30. Peram, M.R.; Jalalpure, S.S.; Palkar, M.B.; Diwan, P.V. Stability studies of pure and mixture form of curcuminoids by reverse phase-HPLC method under various experimental stress conditions. *Food Sci. Biotechnol.* **2017**, *26*, 591–602. [CrossRef]

31. Price, L.C.; Buescher, R.W. Kinetics of alkaline degradation of the food pigments curcumin and curcuminoids. *J. Food Sci.* **1997**, *62*, 267–269. [CrossRef]
32. Didenko, Y.T.; Suslick, K.S. The energy efficiency of formation of photons, radicals and ions during single-bubble cavitation. *Nature* **2002**, *418*, 394–407. [CrossRef]
33. Riesz, P.; Kondo, T. Free radical formation induced by ultrasound and its biological implications. *Free Radic. Biol. Med.* **1992**, *13*, 247–270. [CrossRef]
34. Riesz, P.; BerdahI, D.; Christman, C.L. Free radical generation by ultrasound in aqueous and nonaqueous solutions. *Environ. Heal. Perspect.* **1985**, *64*, 233–252. [CrossRef] [PubMed]
35. Riesz, P.; BerdahI, D.; Christman, C.L. Dissolved gas and ultrasonic cavitation—A review. *Ultrason. Sonochem.* **2013**, *20*, 1–11. [CrossRef]
36. Toma, M.; Vinatoru, M.; Paniwnyk, L.; Mason, T.J. Investigation of the effects of ultrasound on vegetal tissues during solvent extraction. *Ultrason. Sonochem.* **2001**, *8*, 137–142. [CrossRef]

14

Optimization of Sample Preparation for Detection of 10 Phthalates in Non-Alcoholic Beverages in Northern Vietnam

Thanh-Thien Tran-Lam [1], Yen Hai Dao [1], Duong Thanh Nguyen [1], Hoi Kim Ma [2], Trung Quoc Pham [1] and Giang Truong Le [1,*]

1 Institute of Chemistry, Vietnam Academy of Science and Technology (VAST), 18 Hoang Quoc Viet, Cau Giay, Hanoi 100000, Vietnam; thanhthien307@gmail.com (T.-T.T.-L.); dhy182@gmail.com (Y.H.D.); ntduong182@gmail.com (D.T.N.); phamquoctrung0811@gmail.com (T.Q.P.)

2 University of Science, Vietnam National University HCMC, Linh Trung Ward, Thu Duc District, Ho Chi Minh City 720400, Vietnam; makimhoi@gmail.com

* Correspondence: hoasinhmoitruong.vast@gmail.com

Abstract: A novel method was developed for the sensitive, cheap and fast quantitation of 10 phthalates in non-alcoholic beverages by liquid–liquid extraction (LLE) combined with gas chromatography tandem mass spectrometry (GC-MS/MS). The best results were obtained when *n*-hexane was used as extraction solvent. A central composite design (CCD) was applied to select the most appreciated operating condition. The method performance was evaluated according to the SANTE/11945/2015 guidelines and was linear in the 0.1 to 200 μg/L range for 10 phthalate compounds, with $r^2 > 0.996$ and individual residuals <15%. Repeatability (RSD_r), within-laboratory reproducibility (RSD_{wr}), and the trueness range were from 2.7 to 9.1%, from 3.4 to 14.3% and from 91.5 to 118.1%, respectively. The limit of detection (LOD) was between 0.5 to 1.0 ng/L and the limit of quantitation (LOQ) was between 1.5 to 3.0 ng/L for all 10 compounds. The developed method was successfully applied to the analysis of non-alcoholic beverages.

Keywords: phthalate; non-alcoholic beverages; liquid–liquid extraction; response surface methodology; GC-MS/MS

---

1. Introduction

Vietnam is a developing country with a high demand for soft drinks. According to the Vietnam beverage association (VBA) the average consumption of non-alcoholic beverages is over 43 L/person/year and the market is expected to develop from 4 billion liters in 2017 to an estimated 5 billion liters in 2020. With the positive growth of beverages market, consumers increasingly pay more attention to the quality of bottles along with the impact on human health due to long storage time and high temperature conditions.

Polyvinyl chloride (PVC) and polyethylene (PE) are main raw materials of plastic bottles that are widely used in a broad variety of beverages in many countries around the world [1]. In the manufacture process, phthalates are used to produce flexible and durable plastics. It is estimated that in 2017 the global consumption of phthalate compounds was nearly 8 million tons [2]. Because phthalates do not form chemical bonds in the plastic's network, they are easily released and migrate into food and beverages at different stages such as packaging, bottling and production [3–5]. As a result, the consumers can inevitably be exposed to phthalates via eating and drinking. The presence of phthalates was detected in more than 95% of human urine samples in numerous countries around the world [6,7]. Bioaccumulative potential, toxicity and adverse effects of phthalates on experimented animals have been reported in previous studies [8,9]. Phthalates are known as endocrine disrupters, severely affecting respiratory, hepatic and reproductive organs [10–13].

In 1999, the United States Environmental Protection Agency (U.S. EPA) added 8 phthalates to the list of hazardous chemicals, of which di (2-ethylhexyl) phthalate (DEHP) was in the category of carcinogenic substances of level II [14]. The Chemicals Act (REACH) of the European Council have introduced three phthalate compounds including DEHP, di-*n*-butyl phthalate (DBP) and benzyl butyl phthalate (BzBP) into Section 52, Appendix XVII (Annex XVII of the REACH Regulation) since 2007 in order to restrain the production, trade and use of these compounds [15]. Since then, the list of prohibited phthalate substances has increased steadily. RoHS 2 EU/65/2011 amended in 2016 remains the restrictions of DEHP, BzBP, and DBP concentration <0.1% (enacted since 2011) and will be forbidden in all electrical and electronic devices as of 22/7/2019.

Currently, there has been no research and statistics in Vietnam about phthalates concentrations and the risk of exposure to consumers due to these compounds being contained in plastic bottles. To support consumers becoming more aware of the hazards of these toxins and selecting healthy food, the assessment of phthalates in types of non-alcoholic beverage drinks is vitally important.

Therefore, in this study, the presence of 10 phthalates in 148 samples of non-alcoholic beverages divided into 6 groups in Vietnam was investigated by liquid–liquid extraction and gas chromatography-tandem mass spectrometry (GC-MS/MS) [16]. We also applied response surface methodology (RSM) to plan the sample preparations.

2. Materials and Methods

2.1. Chemicals and Materials

Individual neat crystal phthalates standards, including Dimethyl phthalate (DMP), diethyl phthalate (DEP), dipropyl phthalate (DPP), diisobutyl phthalate (DiBP), benzyl butyl phthalate (BzBP), di-*n*-hexyl phthalate (DnHP), di (2-ethylhexyl) phthalate (DEHP), di-*n*-octyl phthalate (DnOP), dicyclohexyl phthalate (DCHP) and di-*n*-butyl phthalate (DBP) and three isotope titrants (dimethyl phthalate-3,4,5,6-d4 (DMP-d4), diisobutyl phthalate-3,4,5,6-d4 (DiBP-d4) and di (2-ethylhexyl) phthalate-3,4,5,6-d4 (DEHP-d4)) were obtained from Sigma (St. Louis, MO, USA). The purities of phthalate standards and isotope internal standards were guaranteed above 98%. Only glassware was used in all analytical procedure. All the containers such as volume flasks, centrifuge tubes, pipettes and extraction funnels are rinsed carefully by methanol, ethyl acetate and *n*-hexane. Standard stock solution of 10 phthalate compounds and the isotope titrants were prepared by dissolving each compound in *n*-hexane to obtain solutions with concentration of 1000 mg/L and then diluted to 10 mg/L. The internal isotope solutions were prepared at the same procedure to the concentration of 100 mg/L. Standard solutions containing 0.1, 1.0, 5.0, 10.0, 20.0, 50.0, 100.0 and 200.0 µg/L of the 10 phthalates in *n*-hexane were prepared daily and used for the preparation of calibration curves.

Methanol and acetonitrile were purchased from Thermo Fisher (Waltham, MA, USA). Dichloromethane, *n*-hexane, sodium sulfate, sodium chloride and sodium hydroxide were obtained from Merck (Waltham, MA, USA) with purity of above 95%. Ultrapure water was prepared by Milli-Q® Gradient A10 (Merck Millipore, Burlington, MA, USA).

Fourteen commercial mineral water, 17 carbonated drinks, 29 functional drinks, 32 juice drinks, 33 tea drinks and 23 fermented milks were purchased from the retail market in Hanoi, Vietnam. All non-alcoholic beverages were assigned with unique marks and protected from the light until needed.

2.2. Instrumentation and Chromatographic Conditions

All the phthalates determination was performed using a GC-MS/MS system (Thermo Fisher Scientific, Waltham, MA, USA), a Trace GC 1310 gas chromatograph, a TriPlus RSH Autosampler and TSQ 8000 mass spectrometer (Thermo, Waltham, MA, USA) and controlled by a computer running

TraceFinder software. A DB5-MS (30 m × 0.25 mm × 0.25 μm) gas chromatography column from Agilent (Santa Clara, CA, USA) was used to separate phthalates. Oven temperature was set initially at 100 °C (hold for 1 min), then increased to 280 °C at 10 °C/min and to 310 °C at 5 °C/min. At 310 °C, temperature was maintained for 5 min. Helium was used as a carrier gas in a constant flow of 1 mL/min and the injection volume was 1 μL with an autosampler in splitless mode. The total of analysis time was 20 min. Solvent delay was 1 min. The GC was interfaced by a heated transfer liner (310 °C) to the mass spectrometer in electron ionization mode with an electron energy of 70 eV. Inlet temperature was 290 °C and inject volume was 1 μL. The criteria for the identification of phthalates were based on both the same retention times as the standard within ±2% and correctly relative abundance of two characteristic ions within ±15%. Data processing was done by TraceFinder software from Thermo Fisher Scientific. Identifying and quantifying ions, retention time, and collision energy are listed in Table S1.

2.3. Sample Preparation Procedure

We injected 5.00 mL of samples into a 15 mL centrifuge tube, and then added 10 μL of internal isotope and 1.5 mL of methanol. The mixtures were mixed well by vortex and transferred to the extraction funnel. Next, 15 mL *n*-hexane was added to the funnel, and the mixtures were shaken vigorously for 7 min. After standing for 5 min to separate phases, 0.5 mL of 10% NaCl solution was added to remove the emulsion. The *n*-hexane solvent phases were transferred to 50 mL centrifuge tube. The procedure was repeated one more time, then the solutions after 2 extractions were transferred into erlenmeyer flasks and mixed vigorously. Next, 15 g Na_2SO_4 was added and shaken seriously to remove water completely. The remaining solutions were evaporated to about 5 mL by a rotary evaporator, and then dried by nitrogen until dry. Finally, the dried samples were dissolved in 1 mL *n*-hexane, filtered through a 0.22 μm Polytetrafluoroethylene (PTFE) filter and analyzed by means of the GC-MS/MS.

2.4. Experimental Design

Several trials were conducted to optimize a liquid–liquid extraction process for the quantitative analysis of phthalates in soft drinks. The D-optimal was selected to confirm the significant variables ($V_{solvent}/V_{sample}$ ratio, number of extractions, NaCl concentration and extraction time). The response was the sum of all phthalate peak areas. Two different full factorial designs were created at three levels: low (−1), medium (0) and high (+1). MODDE 12.1 software was used to design experimental matrices, calculate regression values and analytical variance. The D-optimal planning method was used with three continuous variables and one intermittent variable. A total of 29 experiments were done and the experimental results are shown in Table 1.

Table 1. Independent variable and experiment level.

Coded	Independent Variable	Level Experiment		
		−1	**0**	**1**
X1	$V_{solvent}/V_{sample}$ ratio	5.0	6.0	7.0
X2	Number of extractions	1	2	3
X3	NaCl (M)	0.1	0.2	0.3
X4	Time of extraction	5	10	15

The relationship between the response function Y and the coded variables (X1, X2, X3, and X4) is indicated in the following equation:

$$Y = \beta_0 + \beta_i \sum x_i + \beta_{ii} \sum x^2{}_i + \beta_{ij} \sum x_i x_j \quad (1)$$

where Y is a response function; x_i and x_j are independent variables; β_0 is a constant; and β_i, β_{ii}, and β_{ij} are linear, quadratic, and interactive coefficients, respectively.

The appropriate fitting model for the response was selected based on the comparison of various statistical parameters such as R^2, Q^2, lack of fit and adequate precision.

2.5. *Figure of Merit*

Validating the analysis method in this research followed the instruction of European SANTE 11945/2015. The parameters evaluated in the validating process for carbonated beverages and fat drink samples are linearity, linear range, recovery, precision, limit of detection (LOD) and limit of quantitation (LOQ). Quantification was carried out by the internal calibration method. To assess the specificity, blank samples were tested based on the extraction process in order to evaluate false positive phenomenon and contamination of the chemicals. To appraise the linear range, 7 values of mix standard solution of 10 phthalate compounds with the concentration of 1.0, 5.0, 10.0, 20.0, 50.0, 100.0 and 200.0 μg/L were prepared. LOD is defined as the three times the standard deviation of eleven consecutive blank injections divided by the slope of the calibration curve (LOD = (3 × (SD_{blank}) (slope of the calibration cure))) and LOQ is calculated based on the lowest spike level for which the criteria for trueness (i.e., 70–120%) and precision (<20%) met. Precision is calculated using 15 determinations (i.e., three concentration levels in quintuplicate). The repeatability (RSD_r) is calculated from the results of four replicate experiments in a single day of standard 1, 10 and 100 μg/L and the within-laboratory reproducibility (RSD_{wr}) is calculated from results obtained over four consecutive days. The trueness is calculated depending on the method of standard addition with the help of three different concentrations (1, 10 and 100 μg/L).

3. Results and Discussion

3.1. *Selecting Extraction Solvent and Optimizing the Method of Solvent Evaporation*

The analytical method of phthalates in beverage samples is based on the liquid–liquid extraction technique, and thus the recovery of the compounds relies on two fundamental factors: (i) removing solvent to extract compounds out of the matrix, and then eliminating the matrix and (ii) evaporating the solvent to concentrate the samples after extraction.

3.1.1. Solvent Evaporation Method

One of the most common disadvantage of the liquid–liquid extraction technique is a high volume of extraction solvent. It is, therefore, required to have a method that evaporates only the solvent but not the analytical substance. Herein, we implemented the evaluation of three solvent evaporation methods: (1) using nitrogen to remove the solvent, (2) using the rotary evaporator system, and (3) combining these two methods. The standard solution was mixed in 30 mL *n*-hexane, which was carried out in experimental conditions as mentioned above. The result was reflected through the recovery of 10 phthalate compounds (Figure 1).

The recovered efficiency of 10 phthalates when using the vacuum rotary evaporator to evaporate solvent ranged from 12% to 62%; while using the nitrogen gas flow, the figure was from 32% to 72%. However, when combining these methods, the recoveries of all 10 phthalates were higher than single methods, ranging from 91% to 105%. This combination saved analysis time, and reduced the evaporation of substances as well as contact time between the substances and surrounding atmosphere. Therefore, we incorporated vacuum rotary evaporator into nitrogen gas to evaporate the solvent in the sample preparation process.

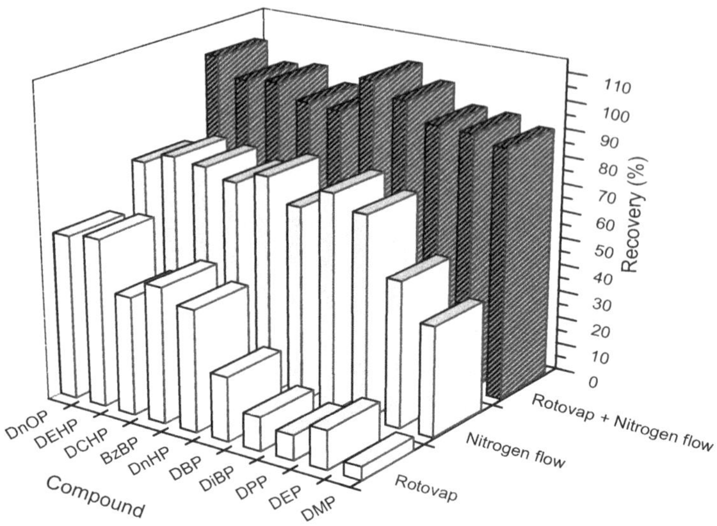

Figure 1. Comparison of recoveries between variable solvent evaporation methods. Note: Dimethyl phthalate, DMP; Diethyl phthalate, DEP; Dipropyl phthalate, DPP; Diisobutyl phthalate, DiDP; Benzyl butyl phthalate, BzBP; di-*n*-hexyl phthalate, DnHP; di(2-ethylhexyl) phthalate, DEHP; di-*n*-octyl phthalate, DnOP; Dicyclohexyl phthalate, DCHP and di-*n*-butyl phthalate, DPBP.

3.1.2. Selecting the Extraction Solvent

The requirements of extraction solvents using in liquid-liquid extraction technique are dissolving well the analytical compounds, having strong affinity to the compounds and preventing matrix effect. In this study, we assessed the extraction ability of numerous solvents such as *n*-hexane, chloroform ($CHCl_3$), dichloromethane (CH_2Cl_2) and ethyl acetate ($CH_3COOC_2H_5$). The isotope standard solution was prepared and added to the drinks which contain fats. These mixtures and each of the mentioned solvents were evaporated by the combination of the rotary evaporator and nitrogen gas. The results of solvent selection were based on the sum of the chromatographic peak area of the three isotopes, and are shown in Figure 2.

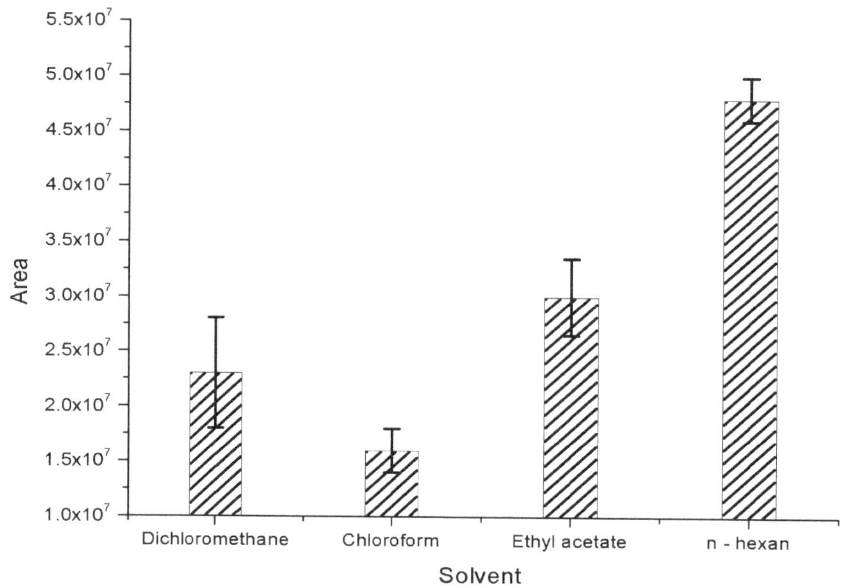

Figure 2. Total peak area of 3 isotope internal standards when using different solvents.

As shown in Figure 2, the total peak area of three internal standard substances of using *n*-hexane as an extraction solvent was much higher than those of dichloromethane, chloroform and ethyl acetate.

Although chloroform is a perfect candidate for extraction of many substances, in this case, when using this solvent in the extraction of beverage samples, other chemicals are also extracted into the organic phase leading to a decrease of the internal standard peak. Therefore, it cannot detect the signal of the internal standard peak. This problem is similar to that of dichloromethane and ethyl acetate solvents [17–20]. Nonetheless, the obtained area when using *n*-hexane was the highest one because this solvent has better capability to extract the analytical compounds to the organic phase and more relatively eliminates the matrix effect than utilizing dichloromethane. To sum up, we decided to use *n*-hexane as the extraction solvent in this study.

3.2. D-Optimal

Analysis of variance is widely used to predict the suitability of a model with experiment results. The obtained results (Table 2) indicated that the predicted values of the model were not conflict with the experiments. The coefficient of determination of R^2 was 0.932 and the coefficient of determination adjustment R^2_{adj} was 0.910. The suitability of the model was also shown in *P* values and Fisher test. $P_{regression}$ value was 0.000 (<0.05), and $P_{Lack\ of\ fit}$ was 0.221 (>0.005), which showed that the obtained model was consistent with the experiment.

Table 2. Analysis of variance (ANOVA).

Recovery	DF	SS	MS (Variance)	F	*P*	SD
Total corrected	28	1199.89	42.8532			6.54624
Regression	7	1118.59	159.798	41.2734	**0.000**	12.6411
Residual	21	81.3055	3.87169			1.96766
Lack of Fit (Model error)	19	79.1989	4.16836	3.9573	**0.221**	2.04166
Pure error (Replicate error)	2	2.10667	1.05333			1.02632
	N = 29	Q^2 =	0.857	Cond. no. =	1.752	
	DF = 21	R^2 =	0.932	RSD =	1.968	
	Comp. = 2	$R^2_{adj.}$ =	0.910			

Note: degrees of freedom, DF; sum of squares, SS; mean square, MS; Fisher, F; probability value, P; and standard deviation, SD.

The three-dimensional response surface shows the effect and interaction of the two factors on the target function. Figure 3a shows the combined effect of the $V_{solvent}/V_{sample}$ ratio and NaCl concentration. Figure 3b shows the image effect of NaCl concentration and time of extraction. Interaction between the $V_{solvent}/V_{sample}$ ratio and time of extraction is shown in Figure 3c. In general, when the value of the variables increases, the efficiency of the phthalate extraction rises and eventually reaches equilibrium.

The contribution of these factors on the extraction efficiency is shown in Figure 3d. The solvent/sample ratio was the biggest influence (54.8%), followed by extraction time (35.8%), and final NaCl concentration (9.4%).

The optimal tool of MODDE 12.1 software was used for the optimization. The results are shown in Table 3. Experimental result was obtained at optimum conditions, yield was 92.51 (95% confidence). This proves that the model was highly meaningful, allowing good experimental results.

Table 3. Optimization of phthalate extraction process.

$V_{solvent}/V_{sample}$ Ratio	Number of Extractions	NaCl Concentration (M)	Solvent	Drying Method	Time of Extraction (Min)	% Recovery Efficiency	
						Predicted	Experiment
6.5	2	0.42	*n*-hexan	Rotovap + nitrogen flow	14	90.7	91.1

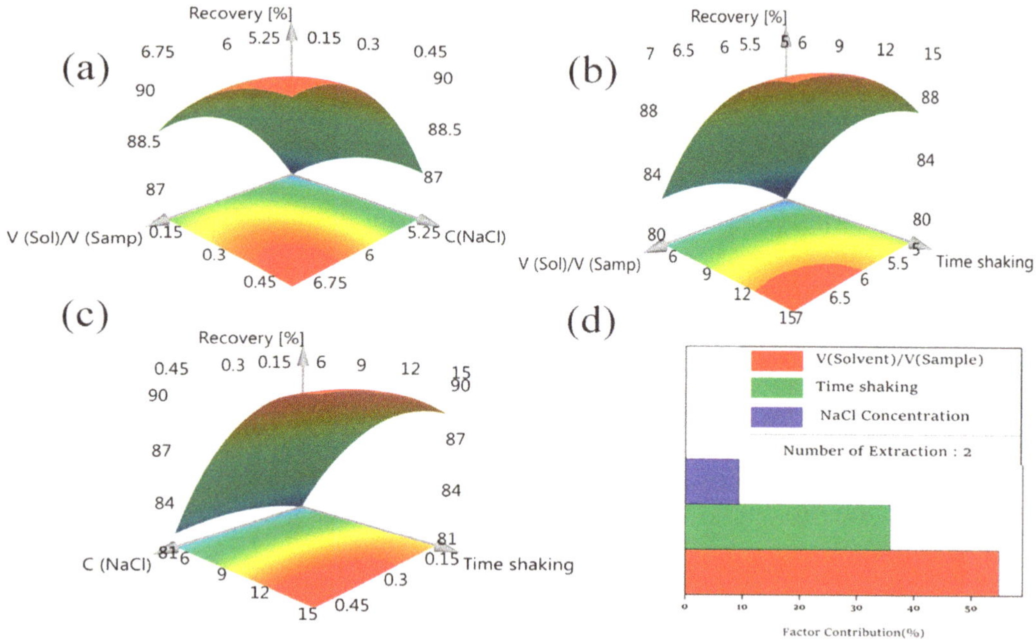

Figure 3. Response surface plots for the central composite design (CCD). Note: (**a**) $V_{solvent}/V_{sample}$ ratio vs. NaCl concentration; (**b**) NaCl concentration vs. time of extraction; (**c**) $V_{solvent}/V_{sample}$ ratio vs. time of extraction and (**d**) The phthalate extraction rises vs. eventually reaches equilibrium.

3.3. Method Performance

To evaluate the efficiency of the proposed method, a number of parameters of the method were investigated and manifested in Tables 4–6. The triple quadrupole detector provided a high degree of selectivity. The linear ranges of these phthalate compounds were built up from 1 to 200 μg/L. Additionally, the weights $1/x^2$ shown through the correlation coefficient (r^2) were greater than 0.996, a non-significant lack of fit and individual residuals deviation of <13% proved the quality of the method. The lowest LOD of these phthalate substances was 0.5 ng/L and the highest was 3.0 ng/L. The maximum of retention time was ±0.06 min, which was below the maximum tolerance deviation stated in SANTE guidelines (±0.1 min). The repeatability (RSD_r) and within-laboratory reproducibility (RSD_{wr}), which expressed percent relative standard deviation (%RSD), ranged from 1.0 to 9.1% and from 2.7 to 12.3%, respectively. All detected RSD values were smaller than 15% that meet the SANTE guideline of RSD $\leq$ 20%. The trueness of this method was appraised through the recovery values by adding standard solution to carbonated beverage and fat beverage samples at three different concentrations. The average recoveries of 10 phthalate compounds are demonstrated in Table 3 and are within the range required by the SANTE guidelines (between 70% and 120%).

Table 4. Linear dynamic range (μg/L), determination coefficients (r^2), residuals, retention times, limit of detection (LOD) and limit quantitation (LOQ).

Compound	Linear Dynamic Range (μg/L)	r^2	Maximum Individual Residual (%)	Retention Times (min)	LOD (ng/L)	LOQ (ng/L)
DMP	0.1–200	0.997	10.3	6.58	1.0	3.0
DEP	0.1–200	0.996	11.2	8.16	1.0	3.0
DPP	0.1–200	0.998	9.8	10.12	1.0	3.0
DiBP	0.1–200	0.999	11.5	12.00	0.5	1.5
BzDP	0.1–200	0.997	7.3	15.52	1.0	3.0
DnHP	0.1–200	0.997	12.9	15.42	1.5	4.5
DEHP	0.1–200	0.999	9.1	17.02	0.5	1.5
DnOP	0.1–200	0.999	8.2	18.42	1.0	3.0
DCHP	0.1–200	0.996	10.2	16.90	1.0	3.0
DBP	0.1–200	0.999	11.1	12.33	1.0	3.0

Table 5. Repeatability (RSD_r) and within-laboratory reproducibility (RSD_{wr}) for peak areas evaluated at three concentration levels.

Compound	RSDr (n = 5)			RSDwr (n = 5 × 4 Days)		
	1 μg/L	10 μg/L	100 μg/L	1 μg/L	10 μg/L	100 μg/L
DMP	3.6	1.0	3.2	9.1	4.1	5.1
DEP	6.3	4.3	3.8	6.7	4.7	3.4
DPP	7.5	5.4	4.3	8.5	3.3	3.8
DiBP	2.6	1.7	5.7	14.3	3.3	3.3
BzDP	1.6	2.8	3.6	8.6	2.7	3.3
DnHP	5.6	6.0	3.0	12.2	2.8	2.7
DEHP	2.4	3.7	4.4	10.2	7.0	2.8
DnOP	3.8	5.7	8.2	11.2	6.7	4.1
DCHP	5.2	7.4	9.1	12.3	8.6	4.7
DBP	1.9	2.3	8.8	9.9	6.7	3.3

data is presented as % RSD.

Table 6. Trueness results for 10 phthalate compounds in non-alcoholic beverages matrices.

Spiking Level Compound	1		10		100	
	Gas (M/R)	Fat (M/R)	Gas (M/R)	Fat (M/R)	Gas (M/R)	Fat (M/R)
DMP	109.2/7.3	103.2/3.4	94.9/4.3	100.6/9.1	99.2/3.8	103.3/3.8
DEP	97.2/3.8	92.6/6.3	98.3/1.0	98.9/8.8	108.3/1.0	109.3/5.7
DPP	102.2/4.2	101.2/7.5	91.5/1.9	101.4/2.3	94.0/1.7	103.1/5.6
DiBP	107.1/9.1	102.8/2.6	92.3/5.2	105.2/7.4	102.2/4.3	114.7/10.1
BzDP	99.6/6.9	99.3/1.0	113.3/3.8	101.3/5.7	108.5/1.9	97.5/1.6
DnHP	105.6/3.6	100.5/1.9	111.1/2.4	96.5/3.7	100.2/2.8	108.9/8.8
DEHP	114.7/10.1	105.1/2.4	101.1/5.6	99.7/6.0	101.1/5.7	103.2/7.5
DnOP	110.5/8.6	102.5/3.7	99.5/1.6	1002/2.8	94.3/5.2	118.1/12.3
DCHP	117.6/6.9	103.0/1.7	102.8/2.6	103.0/1.7	107.7/6.0	96.6/6.3
DBP	118.1/12.3	109.2/2.8	103.2/7.5	96.1/5.4	100.6/3.6	105.4/8.2

Note: M: Mean (% recovery); R: Relative standard deviation (%).

3.4. Levels of Phthalates in Samples

Non-alcoholic drink samples were analyzed based on the above sample preparation method. The results are shown in Table 7 and Figure 4. As described in Table 4, DBP and DEHP were also detected in all of the 148 collected beverage samples which were analyzed, while DnOP were found in 33% of samples. The appearance of phthalate compounds ranged from 1% to 100% so that almost all of the samples were contaminated by phthalates. As can be seen in Figure 4, DEHP was the phthalate which primarily presents in the samples (>35%), followed by DBP and DEP in mineral water, fruit juice, tea, fermented milk and functional drink. Conversely, in carbonated drink samples, DnOP was the most abundant phthalate substance (>50%). In relation to fermented milk, DMP and DEHP were comparatively in the same proportion (45.5 and 47.6%). It was recognizable that there was the extensive appearance of DMP, DnOP and DBP. DnHP virtually did not appear in these kinds of beverage drinks. DnOP was chiefly found in carbonated drink samples (54.3%). Additionally, DMP was mainly detected in fermented milk samples (45.5%).

The concentrations of phthalate compounds presenting in non-alcoholic beverages are also illustrated in Table 8. In 6 groups of experimental samples, DEHP was the phthalate substance containing the highest mean as well as medium value among all of the samples. The mean and medium were 91.6 and 64.5 μg/L. The variation of concentrations of 148 samples ranged from 0.092 to 466.6 μg/L, which were much higher than those of DBP (22.1 and 18.8 μg/L, the variation varies from 0.093 to 73.5 μg/L). The DMP, BzBP, DPP. DiBP, DnOP and DCHP contents were nd–131.9 μg/L, 0.30–21.5 μg/L, nd –0.52 μg/L, nd–1.9 μg/L, nd–200.4 μg/L and nd–0.60 μg/L, respectively.

Table 7. Detection of frequency [n (%)] of phthalates in non-alcoholic beverages in Hanoi.

Non-Alcoholic Beverages	N	DMP	DEP	DBP	BzBP	DEHP	DnHP	DPP	DiBP	DnOP	DCHP
Mineral water	14	14 (100)	14 (100)	14 (100)	14 (100)	14 (100)	0	14 (100)	14 (100)	0	14 (100)
Carbonated drinks	17	14 (82)	17 (100)	17 (100)	16 (94)	17 (100)	0	17 (100)	17 (100)	17 (100)	17 (100)
Functional drinks	29	19 (66)	26 (90)	29 (100)	28 (97)	29 (100)	0	25 (86)	29 (100)	0	29 (100)
Juice drinks	32	31 (97)	32 (100)	32 (100)	32 (100)	32 (100)	0	32 (100)	0	0	0
Tea drinks	33	0	33 (100)	33 (100)	33 (100)	33 (100)	0	0	0	0	0
Fermented milks	23	23 (100)	23 (100)	23 (100)	23 (100)	23 (100)	0	0	0	0	0
Total	148	101 (68)	145 (98)	148 (100)	146 (99)	148 (100)	0	88 (59)	60 (41)	49 (33)	60 (41)

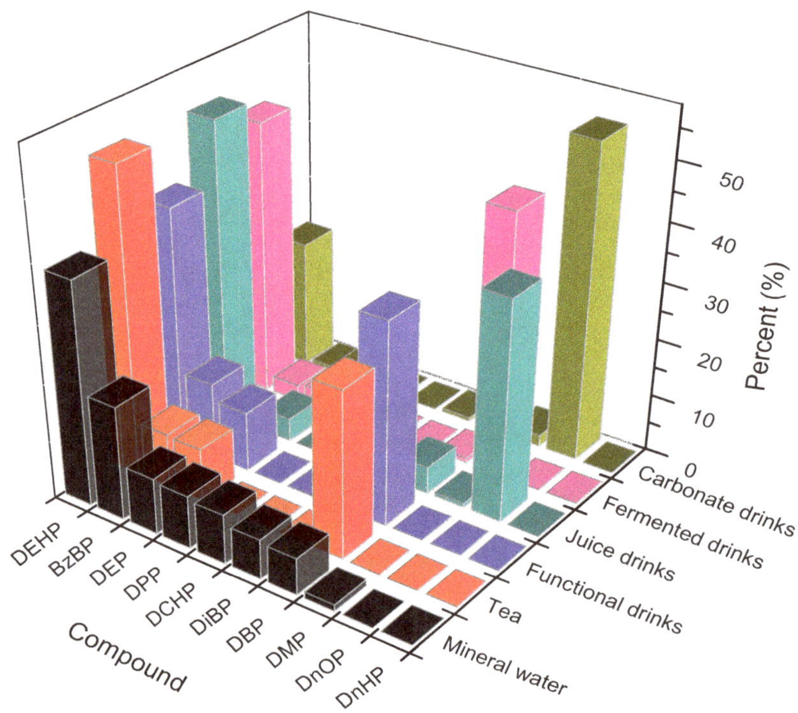

Figure 4. The distribution of 10 phthalate compounds in different kinds of samples. Note: Dimethyl phthalate, DMP; Diethyl phthalate, DEP; Dipropyl phthalate, DPP; Diisobutyl phthalate, DiDP; Benzyl butyl phthalate, BzBP; di-*n*-hexyl phthalate, DnHP; di (2-ethylhexyl) phthalate, DEHP; di-*n*-octyl phthalate, DnOP; Dicyclohexyl phthalate, DCHP and di-*n*-butyl phthalate, DPBP.

Table 8 also shows that the concentrations of phthalate compounds studied on different targets were considerably different. The DEHP in fruit juice samples had the highest mean (230.8 μg/L) and median (222.7 μg/L) among other types of beverage drinks in this experiment. Moreover, the mean and median of DEP (17.9 and 17.3 μg/L) in fruit juice drink were also far higher than other beverages. In terms of fermented milk, DMP content was detected in a range of 12.3 to 131.9 μg/L, and the average and median were 68.0 and 65.7 μg/L, respectively.

Table 8. Phthalate concentrations in different types of non-alcoholic beverages (μg/L).

Non-Alcoholic Beverages		Mineral Water	Carbonated Drinks	Functional Drinks	Juice Drinks	Tea Drinks	Fermented Milk
DMP	Range	0.02–0.05	1.1–5.5	0.04–0.08	0.1–5.6		12.3–131.9
	Mean	0.03	3.3	0.06	3.3	Nd	68
	Median	0.04	3.4	0.05	3.2		65.7
DEP	Range	0.05–0.57	0.38–13.0	1.6–17.9	3.0–33.4	0.9–21.1	0.6–10.2
	Mean	0.3	4.8	10	17.9	11.2	5
	Median	0.29	3.4	11	17.3	10.3	3.8
DBP	Range	0.09–0.95	0.18–43.5	0.16–64.7	1.6–73.4	6.0–4.3	0.81–1.35
	Mean	0.61	19.8	30.3	20.6	34.9	1.1
	Median	0.59	19	26.3	17.8	34.7	1.1
BzBP	Range	0.30–0.95	0.31–3.4	0.41–21.5	0.4–6.4	0.8–21.9	1.1–8.4
	Mean	0.61	1.4	11.8	3.1	11.5	4.2
	Median	0.59	1.2	11.7	3	11.3	3.7
DEHP	Range	0.46–1.8	0.09–71.0	0.72–96.1	27.7–466.6	15.4–87.8	8.3–151.2
	Mean	1.2	28.6	35.9	230.8	63.4	71.2
	Median	1.2	20.5	21	222.7	72.9	67.6
DnHP	Range						
	Mean	Nd	Nd	Nd	Nd	Nd	Nd
	Median						
DPP	Range	0.13–0.44	0.05–0.52	Nd–0.06	0.03–0.21		
	Mean	0.28	0.3	0.01	0.11	Nd	Nd
	Median	0.28	0.33	0.01	0.1		
DiBP	Range	0.14–0.41	0.04–1.9	Nd–0.14			
	Mean	0.22	0.76	0.07	Nd	Nd	Nd
	Median	0.15	0.83	0.06			
DnOP	Range		0.98–200.4		136–197		
	Mean	Nd	69.2	Nd	169.1	Nd	Nd
	Median		56.9		169		
DCHP	Range	0.13–0.60	0.02–0.09	Nd–0.15			
	Mean	0.26	0.07	0.02	Nd	Nd	Nd
	Median	0.16	0.07	0.01			

The distribution of the total phthalate concentration in non-alcoholic drinks was different among the sample matrices. Juice drinks had the highest phthalate concentration, followed by fermented milk and tea. As shown in Figure 5, DEHP was a major contributor leading to the phthalate contamination in non-alcoholic beverage, similar to the previous study [21]. The contamination of phthalates depended on the characteristics of the samples. The sample containing preservatives (potassium benzoate) had higher phthalate concentration than that which did not use preservatives [22]. Furthermore, the sample carrying high fat content was easier to contaminate by phthalate [23]. When comparing the data above, the identification of the sources of phthalate contamination was ambiguous because of other factors such as temperature, pH, light, turbidity and storage time [5,24,25].

Among all phthalates, DEHP is the most popular substance appearing in non-alcoholic beverages in similar studies. Figure 6 illustrates the degree of DEHP contamination in recent studies and the container of these products is not necessarily made from plastic. According to the research of Ustun et al., DEHP concentrations of soda, lemonade, mineral water and high-taste water in Turkey ranged from 73 to 2312 ng/g and the highest DEHP concentration was found in Cola soft drink [21]. In contrast, based on the study of Sireli et al., DEHP concentration in fruit juice drink varies from 1.1 to 44.3 ng/g, which is much lower than Ustun' research [26]. Wu et al. reported that the DEHP content in energy drink and tea ranged from 15 to 83 ng/g [27]. DEHP concentration was remarkably high in the study of Truong et al. of chocolate and high-fat drinks (111–1753 ng/g) [23]. In our research,

DEHP concentration varied from 0.1 to 466.6 ng/g, remained within the range of the above studies and predominantly concentrated in milk-containing fruit juice sample.

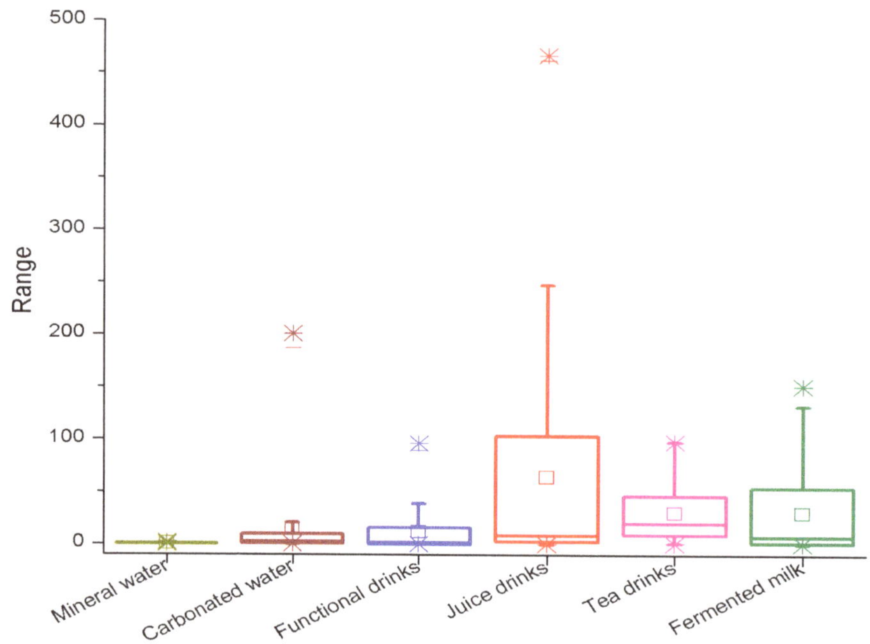

Figure 5. The distribution of the total phthalate concentration in non-alcoholic drinks.

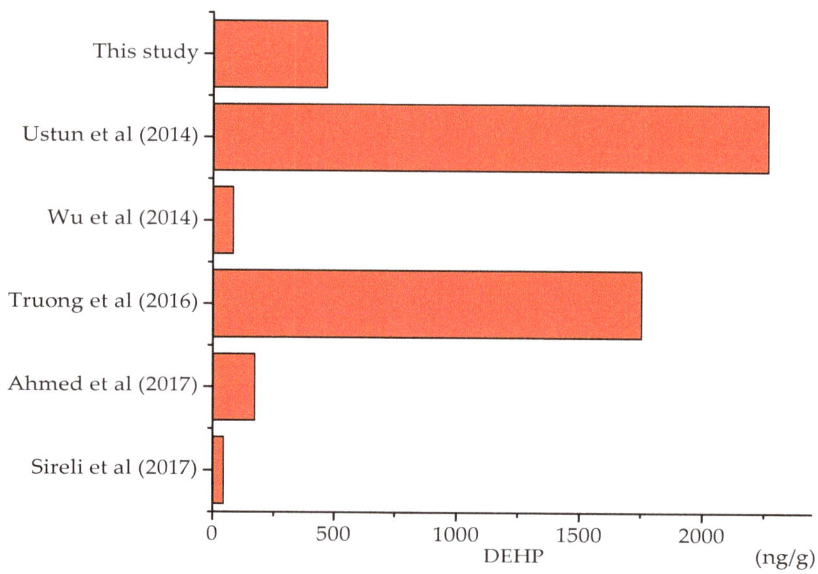

Figure 6. Concentrations of DEHP in similar studies.

3.5. Exposure to Phthalates

Assessing phthalate concentration in non-alcoholic beverages has been investigated by many researchers around the world. However, in Vietnam, there are no specific statistics on phthalate content in daily beverage drinks. Identification of the existence as well as frequency of the occurrence of phthalate compounds in the matrices totally depends on instrument detection limit (IDL) and method detection limit (MDL) of the study, but comparison of phthalate contamination in non-alcoholic beverages still has scientific meaning.

Relying on the studies of Guo [28] and Sireli [26], we calculated the daily intake of DEP, DBP, BzBP and DEHP in Vietnam following the formula below:

$$\mathrm{EDI} = \frac{\mathrm{CQ}}{\mathrm{bw}} r_{\mathrm{uptake}} \tag{2}$$

where EDI (μg/kg × day) is the estimated daily intake from drinking beverages, C (ng/g) is the phthalate concentration in beverages, r is the gastrointestinal uptake factor and bw (kg) is the body weight. In this study, average beverages intake was 150 g/day, r_{uptake} was 1 and an average bw of 50 kg was used for Vietnam population. The result is shown in the Table 9.

Table 9. Characteristics of the investigated phthalates.

Compound	EDI (μg/kg × Day)	TDI (μg/kg × Day)
DEP	1.50×10^{-4}–0.1002	0.800 (U.S. EPA)
DBP	2.70×10^{-4}–0.2202	0.100 (U.S. EPA)
BzBP	9.00×10^{-4}–0.0645	0.800 (U.S. EPA)
DEHP	2.70×10^{-4}–1.3998	0.020 (U.S. EPA)

The daily intake of DEHP when investigating phthalates in beverages in Vietnam was higher than TDI (U.S. EPA), but the contamination of DEP, DBP and BzBP was significantly lower than the threshold of regulation. The phthalate concentration in non-alcoholic beverages did not give rise to serious consequences for adult health. However, the beverages such as fruit juice and fermented milk, which were analyzed, are consumed daily by pregnant women. Because of this, there is likely to be a mother-to-child exposure through the placenta [29] leading to the phenomenon of hormonal disturbance in children [30].

4. Conclusions

In this research, we focused on the assessment of phthalate compounds in beverages, products whose consumption has grown dramatically in Vietnam, and thus the phthalate contamination factor for non-alcoholic drink was not exactly reflected the exposure level. Phthalate exposure in daily life possibly originates from different sources such as air [31–33], food [28,34], beverages [35,36] as well as cosmetics [37]. As a consequence, in this study, we solely concentrated on evaluating phthalate in beverage drinks, products which are consumed in huge quantities in Vietnam.

Liquid–liquid extraction and the GC-MS/MS analysis technique were optimized and conducted successfully in determining 10 phthalate compounds in different kinds of non-alcoholic drinks. The good recoveries (70–120%), RSDs of all the analysis samples and matrices were lower than 15% and low LOQ (0.5 ng/L) was confirmed. This method was utilized to analyze 10 phthalate substances in 148 non-alcoholic drink samples. The result showed that 100% of the samples were contaminated by DEHP and DEP, and almost all samples were polluted by phthalates. The result of phthalate contamination in this study did not reflect accurately the exposure of phthalates in beverage drinks because of other influencing factors. Therefore, it is necessary to implement more in-depth research to assess properly phthalate contamination during the production process, storage conditions, and when the human body is exposed to these products.

Author Contributions: All authors have equivalent contributions in this study.

Acknowledgments: Thanks to the Vietnam Academy of Science and Technology for financial support.

References

1. Green, R.; Hauser, R.; Calafat, A.M.; Weuve, J.; Schettler, T.; Ringer, S.; Huttner, K.; Hu, H. Use of di (2-ethylhexyl) phthalate-containing medical products and urinary levels of mono (2-ethylhexyl) phthalate in neonatal intensive care unit infants. *EHP* **2005**, *113*, 1222–1225. [CrossRef] [PubMed]
2. Deep, T. *Global Trends in PVC Resin Applications and Additives Usage*; Townsend Solutions: Houston, TX, USA, 2018.
3. Heudorf, U.; Mersch-Sundermann, V.; Angerer, J. Phthalates: Toxicology and exposure. *Int. J. Hyg. Environ. Health* **2007**, *210*, 623–634. [CrossRef] [PubMed]
4. Koch, H.M.; Calafat, A.M. Human body burdens of chemicals used in plastic manufacture. *Philo. Trans. R Soc. Lond. B Biol. Sci.* **2009**, *364*, 2063–2078. [CrossRef] [PubMed]
5. Amiridou, D.; Voutsa, D. Alkylphenols and phthalates in bottled waters. *J. Hazard. Mater.* **2011**, *185*, 281–286. [CrossRef] [PubMed]
6. Sakhi, A.K.; Lillegaard, I.T.; Voorspoels, S.; Carlsen, M.H.; Loken, E.B.; Brantsaeter, A.L.; Haugen, M.; Meltzer, H.M.; Thomsen, C. Concentrations of phthalates and bisphenol A in Norwegian foods and beverages and estimated dietary exposure in adults. *Environ. Int.* **2014**, *73*, 259–269. [CrossRef] [PubMed]
7. Guo, Y.; Alomirah, H.; Cho, H.S.; Minh, T.B.; Mohd, M.A.; Nakata, H.; Kannan, K. Occurrence of phthalate metabolites in human urine from several Asian countries. *Environ. Sci. Technol.* **2011**, *45*, 3138–3144. [CrossRef] [PubMed]
8. Meeker, J.D. Exposure to environmental endocrine disruptors and child development. *Arch. Pediatr. Adolesc. Med.* **2012**, *166*, E1–E7. [CrossRef] [PubMed]
9. Yang, J.; Hauser, R.; Goldman, R.H. Taiwan food scandal: The illegal use of phthalates as a clouding agent and their contribution to maternal exposure. *Food Chem. Toxicol.* **2013**, *58*, 362–368. [CrossRef] [PubMed]
10. Foster, P.M. Disruption of reproductive development in male rat offspring following in utero exposure to phthalate esters. *Int. J. Androl.* **2006**, *29*, 140–147; discussion 181–185. [CrossRef] [PubMed]
11. Gray, L.E., Jr.; Barlow, N.J.; Howdeshell, K.L.; Ostby, J.S.; Furr, J.R.; Gray, C.L. Transgenerational effects of Di (2-ethylhexyl) phthalate in the male CRL:CD(SD) rat: Added value of assessing multiple offspring per litter. *Toxicol. Sci.* **2009**, *110*, 411–425. [CrossRef] [PubMed]
12. van Wezel, A.P.; van Vlaardingen, P.; Posthumus, R.; Crommentuijn, G.H.; Sijm, D.T.H.M. Environmental Risk Limits for Two Phthalates, with Special Emphasis on Endocrine Disruptive Properties. *Ecotoxicol. Environ. Safe* **2000**, *46*, 305–321. [CrossRef] [PubMed]
13. Wormuth, M.; Scheringer, M.; Vollenweider, M.; Hungerbuhler, K. What are the sources of exposure to eight frequently used phthalic acid esters in Europeans? *Risk Anal.* **2006**, *26*, 803–824. [CrossRef] [PubMed]
14. Doull, J.; Cattley, R.; Elcombe, C.; Lake, B.G.; Swenberg, J.; Wilkinson, C.; Williams, G.; van Gemert, M. A cancer risk assessment of di (2-ethylhexyl)phthalate: Application of the new US EPA Risk Assessment Guidelines. *Regul. Toxicol. Pharmacol.* **1999**, *29*, 327–357. [CrossRef] [PubMed]
15. ECHA. ANNEX XVII TO REACH 2007. Available online: https://echa.europa.eu/documents/10162/13645/guideline_interpretation_concept_mouth_en.pdf/6f0869f4-1ff1-45ad-ac13-f623d9717718 (accessed on 10 October 2018).
16. Russo, M.V.; Avino, P.; Perugini, L.; Notardonato, I. Extraction and GC-MS analysis of phthalate esters in food matrices: A review. *RSC Adv.* **2015**, *5*, 37023–37043. [CrossRef]
17. Otero, P.; Saha, S.K.; Moane, S.; Barron, J.; Clancy, G.; Murray, P. Improved method for rapid detection of phthalates in bottled water by gas chromatography-mass spectrometry. *J. Chromatogr. B* **2015**, *997*, 229–235. [CrossRef] [PubMed]
18. Jara, S.; Lysebo, C.; Greibrokk, T.; Lundanes, E. Determination of phthalates in water samples using polystyrene solid-phase extraction and liquid chromatography quantification. *Anal. Chim. Acta* **2000**, *407*, 165–171. [CrossRef]
19. Prapatpong, P.; Kanchanamayoon, W. Determination of Phthalate Esters in Drinking Water using Solid-phase Extraction and Gas Chromatography. *J. Appl. Sci.* **2010**, *10*, 1987–1990. [CrossRef]
20. Russo, M.V.; Notardonato, I.; Avino, P.; Cinelli, G. Fast determination of phthalate ester residues in soft drinks and light alcoholic beverages by ultrasound/vortex assisted dispersive liquid–liquid microextraction followed by gas chromatography-ion trap mass spectrometry. *RSC Adv.* **2014**, *4*, 59655–59663. [CrossRef]

21. Ustun, I.; Sungur, S.; Okur, R.; Sumbul, A.T.; Oktar, S.; Yilmaz, N.; Gokce, C. Determination of Phthalates Migrating from Plastic Containers into Beverages. *Food Anal. Methods* **2015**, *8*, 222–228. [CrossRef]
22. Bošnir, J.; Puntarić, D.; Galić, A.; Škes, I.; Dijanić, T.; Klarić, M.; Grgić, M.; Čurković, M.; Šmit, Z. Migration of phthalates from plastic containers into soft drinks and mineral water. *Food Technol. Biotechnol.* **2007**, *45*, 91–95.
23. Truong, N.; Karakilic, V.; Soulsbury, K. Detection of DEHP in Beverages Sold in Canada Using Gas Chromatography-Mass Spectrometry. 2016. Available online: https://pdfs.semanticscholar.org/ad43/3e188a091783fa311fd0d8e778d3aaa36b0d.pdf (accessed on 10 October 2018).
24. Biscardi, D.; Monarca, S.; De Fusco, R.; Senatore, F.; Poli, P.; Buschini, A.; Rossi, C.; Zani, C. Evaluation of the migration of mutagens/carcinogens from PET bottles into mineral water by Tradescantia/micronuclei test, Comet assay on leukocytes and GC/MS. *Sci. Total Environ.* **2003**, *302*, 101–108. [CrossRef]
25. Zaki, G.; Shoeib, T. Concentrations of several phthalates contaminants in Egyptian bottled water: Effects of storage conditions and estimate of human exposure. *Sci. Total. Environ.* **2018**, *618*, 142–150. [CrossRef] [PubMed]
26. Sireli, U.T.; Filazi, A.; Yurdakok-Dikmen, B.; Iplikcioglu-Cil, G.; Kuzukiran, O.; Orhan, C.E. Determination of Phthalate Residues in Different Types of Yogurt by Gas Chromatography-Mass Spectrometry and Estimation of Yogurt-Related Intake of Phthalates. *Food Anal. Methods* **2017**, *10*, 3052–3062. [CrossRef]
27. Wu, P.-G.; Pan, X.-D.; Ma, B.-J.; Wang, L.-Y.; Zhang, J. Determination of phthalate esters in non-alcoholic beverages by GC–MS and optimization of the extraction conditions. *Eur. Food Res. Technol.* **2014**, *238*, 607–612. [CrossRef]
28. Guo, Y.; Zhang, Z.; Liu, L.; Li, Y.; Ren, N.; Kannan, K. Occurrence and profiles of phthalates in foodstuffs from China and their implications for human exposure. *J. Agric. Food Chem.* **2012**, *60*, 6913–6919. [CrossRef] [PubMed]
29. Tomita, I.; Nakamura, Y.; Yagi, Y. Phthalic acid esters in various foodstuffs and biological materials. *Ecotoxicol. Environ. Safe* **1977**, *1*, 275–287. [CrossRef]
30. Bigsby, R.; Chapin, R.E.; Daston, G.P.; Davis, B.J.; Gorski, J.; Gray, L.E.; Howdeshell, K.L.; Zoeller, R.T.; vom Saal, F.S. Evaluating the effects of endocrine disruptors on endocrine function during development. *Environ. Health Perspect.* **1999**, *107* (Suppl. 4), 613–618. [CrossRef]
31. Pei, X.Q.; Song, M.; Guo, M.; Mo, F.F.; Shen, X.Y. Concentration and risk assessment of phthalates present in indoor air from newly decorated apartments. *Atmos. Environ.* **2013**, *68*, 17–23. [CrossRef]
32. Song, M.; Chi, C.; Guo, M.; Wang, X.; Cheng, L.; Shen, X. Pollution levels and characteristics of phthalate esters in indoor air of offices. *J. Environ. Sci.* **2015**, *28*, 157–162. [CrossRef] [PubMed]
33. Bu, Z.; Zhang, Y.; Mmereki, D.; Yu, W.; Li, B. Indoor phthalate concentration in residential apartments in Chongqing, China: Implications for preschool children's exposure and risk assessment. *Atmos. Environ.* **2016**, *127*, 34–45. [CrossRef]
34. Fierens, T.; Vanermen, G.; Van Holderbeke, M.; De Henauw, S.; Sioen, I. Effect of cooking at home on the levels of eight phthalates in foods. *Food Chem. Toxicol.* **2012**, *50*, 4428–4435. [CrossRef] [PubMed]
35. Kanchanamayoon, W.; Prapatpong, P.; Chumwangwapee, S.; Chaithongrat, S. Analysis of phthalate esters contamination in drinking water samples. *Afr. J. Biotechnol.* **2012**, *11*, 16263–16269. [CrossRef]
36. Ahmed, M.B.M.; Abdel-Rahman, G.N.E.; Zaghloul, A.H.; Naguib, M.M.; Saad, M.M.E.D. Phthalates' releasing pattern in low pH beverages of fermented milk, fruit juice, and soft drink packaged in plastic bottles. *Biosci. Res.* **2017**, *14*, 513–524.
37. Koniecki, D.; Wang, R.; Moody, R.P.; Zhu, J. Phthalates in cosmetic and personal care products: Concentrations and possible dermal exposure. *Environ. Res.* **2011**, *111*, 329–336. [CrossRef] [PubMed]

15

Validation of a HILIC UHPLC-MS/MS Method for Amino Acid Profiling in Triticum Species Wheat Flours

Emmanouil Tsochatzis [1,*], Maria Papageorgiou [2] and Stavros Kalogiannis [3]

1 Department of Chemical Engineering, Aristotle University of Thessaloniki, GR-54124 Thessaloniki, Greece
2 Department of Food Science and Technology, International Hellenic University, Sindos Campus, P.O. Box 141, GR-57400 Thessaloniki, Greece; mariapapage@food.teithe.gr
3 Department of Nutritional Sciences and Dietetics, International Hellenic University, Sindos Campus, P.O. Box 141, GR-57400 Thessaloniki, Greece; kalogian@nutr.teithe.gr
* Correspondence: tsochatzism@gmail.com

Abstract: Amino acids are essential nutritional components as they occur in foods either in free form or as protein constituents. An ultra-high-performance (UHPLC) hydrophilic liquid chromatography (HILIC)-tandem Mass Spectrometry (MS) method has been developed and validated for the quantification of 17 amino acids (AA) in wheat flour samples after acid hydrolysis with 6 M HCl in the presence of 4% (*v*/*v*) thioglycolic acid as a reducing agent. The developed method proved to be a fast and reliable tool for acquiring information on the AA profile of cereal flours. The method has been applied and tested in 10 flour samples of spelt, emmer, and common wheat flours of organic or conventional cultivation and with different extraction rates (70%, 90%, and 100%). All the aforementioned allowed us to study and evaluate the variation of the AA profile among the studied flours, in relation to other quality characteristics, such as protein content, wet gluten, and gluten index. Significant differences were observed in the AA profiles of the studied flours. Moreover, AA profiles exhibited significant interactions with quality characteristics that proved to be affected based mainly on the type of grain. A statistical and multivariate analysis of the AA profiles and quality characteristics has been performed, as to identify potential interactions between protein content, amino acids, and quality characteristics.

Keywords: amino acid profiling; hydrophilic interaction chromatography (HILIC); tandem mass spectrometry; Triticum species flours; flour quality characteristics

1. Introduction

Amino acids (AA) are essential nutritional components present in foods either in their free form (FAA) or as protein constituents. They directly contribute to the flavor of foods as they are precursors of aroma compounds and colors formed by thermal or enzymatic reactions during production, processing, and storage of food. Hence, information on the profile and amount of free AA is highly needed in food science and nutrition studies [1–5].

Analysis of AA, in either free form or in protein profile, was highly challenging due to their structural and polarity differences. Ion exchange liquid chromatography has found wide application in AA analysis, especially through the commercial amino acid analyzers, which utilize cation-exchange chromatography followed by post-column derivatization with a chromophore or fluorophore derivatizing agent [6–8]. On the other hand, conventional reversed-phase (RP) High Pressure Liquid Chromatography (HPLC) proved to be time-consuming since most AAs are highly polar and cannot be determined without pre- or post-column derivatization [2,9,10]. Recently the advancement of hydrophilic interaction chromatography (HILIC) and new analytical columns provided alternative paths for the profiling of AA by liquid chromatography (LC). HILIC is generally known to enhance the sensitivity of electrospray ionization-mass spectrometry (ESI–MS) detection, and it is increasingly employed in the analysis of polar analysis in various matrices, [3,5,11–15].

So far, the application of HILIC-MS has found limited use in the analysis of AA in food samples. To our knowledge, three methods have been reported on the determination of FAA in liquid food matrices such as juice, beer, honey, or tea [3], ginkgo seeds [5], and fruits of Ziziphus jujuba [14]. Only recently, it was developed a HILIC-MS/MS method for the determination of either free AA or amino acids profile in high protein content food matrix (mussels) [15]. Moreover, the analytical methods that determine amino acid profile (TAA) were developed for the application in pure protein samples, such as collagen [16] or bovine serum albumin (BSA) and angiotensin I [13]. In all these methods, single ion monitoring-MS (SIM-MS) detection [5] or tandem MS [4,13] were applied. By using tandem MS detection, the separation of isobaric AA was feasible in most cases, whereas the use of single MS detection led to increased total analysis time since longer chromatographic runs were needed for the separation of all compounds. Prior to the analysis of amino acids, proteins need to be hydrolyzed in order to release their constituting AA. The most commonly applied method is hydrolysis by digestion with a strong inorganic acid [13,15–17].

Cereals are considered as one of the basic foods consumed by humans and animals. The carbohydrates they contain provide approximately 50% of the total daily calories, whereas the proteins one-third of the total protein need [17]. The composition of amino acids varies among the proteins of the different cereal grains or flours. Wheat (*Triticum* species) is the 3rd most-produced cereal worldwide. Wheat proteins are known to be low in some amino acids that are considered essential for the human diet, especially lysine (the most deficient amino acid) and threonine (the second limiting amino acid) [18–21]. On the other hand, they are rich in glutamine and proline (s), the functional amino acids in dough formation. Tetraploid Emmer wheat (*Triticum turgidum* species, dicoccum, genomes AABB), also known as emmer, faro or zea in different countries, is hulled wheat that differs from the domestic species on the fact that the ripened seed head of the wild species shatters and spreads the seed onto the ground while in the domesticated counterpart the seed head remains intact, making it easier for humans to harvest the grain. It is considered as the ancestor of bread wheat and durum wheat growing in the margin of the Mediterranean area [22]. Hexaploid spelt, *Triticum aestivum* variety spelta (genomes AABBDD) is also a hulled cereal grain with high resistance to environmental factors (diseases, stress) showing good yields under disadvantageous conditions [23]. It is suitable for organic farming and contributes to agro-diversity [24]. Spelt is becoming widely used in the growing natural food market. It has been reported that spelt protein content showed a great variation depending on the genotype [21,25,26], but it is higher than in common wheat [25,27]. The amino acid composition of the proteins from spelt differs slightly from one of the modern bread and pasta wheats [21]. It has been suggested that spelt-based products could be potentially more digestible than those from common wheat. Certain ancient wheats, einkorn, spelt, emmer, and Khorasan are currently of particular interest for use in selected bakery products [21]. The development of new cultivars has been attempted with the aim to improve the content of all essential amino acids [17,25].

In the present work a single HILIC-MS/MS method was developed, validated and applied for the determination of 17 amino acids, in a comparative study for the determination of the TAA of 'ancient' wheat species such as emmer, spelt, and bread wheat (*Triticum aestivum*) with different extraction rates and cultivated under different practices (organic or conventional) in a fast and reliable way without the need of derivatization, preceded only by a simple hydrolysis procedure. Finally, analysis of variances and multivariate analysis were performed in order to explore potential interactions of the TAA with flour quality characteristics of samples under investigation.

2. Materials and Methods

2.1. Chemicals

Standards of amino acids (AA) namely phenylalanine (Phe), tryptophan (Trp), isoleucine (Ile), leucine (Leu), asparagine (Asn), methionine (Met), valine (Val), proline (Pro), tyrosine (Tyr), alanine (Ala), threonine (Thr), glycine (Gly), serine (Ser), glutamic acid (Glu), aspartic acid (Asp), arginine (Arg), glutamine (Gln), lysine (Lys), histidine (His), cysteine (Cys) and cystine (Cys2),were purchased from Sigma-Aldrich (Steinheim, Germany). Acetonitrile (ACN) and water (LC-MS grade) was purchased from Carlo Erba (Milan, Italy). Formic acid (LC-MS additive), TGA (thioglycolic acid; ≥98%), and ammonium formate (NH_4HCO_2) were purchased from Sigma-Aldrich (Steinheim, Germany). Analytical grade hydrochloric acid (HCl) 37% w/w was also supplied from Carlo Erba (Milan, Italy). The chromatographic column HILIC amide BEH, Acquity UPLC 1.7 m, 2 × 150 mm (Waters) were used.

2.2. Standard Solutions

Stock solutions of the compounds were prepared in 0.1 M HCl at a concentration of 10 mg mL^{-1} and stored in amber vials at −20 °C. Working standards were prepared from the stock solutions by appropriate dilution with ACN/water 95:5 (*v*/*v*) and stored at −20 °C.

2.3. Sample Preparation for Amino Acid Profile Analysis

Ten (10) Triticum flours were selected from the Greek market, to be tested for their AA concentration and quality characteristics. The commercial flour samples differed in their extraction rate, type of cultivation (organic or conventional), and type of wheat, as shown in Table 1. The first letter of codification corresponds to wheat type ('S' for spelt, 'B' for bread wheat and 'E' for Emmer) followed by a 2-digit number revealing the flour extraction rate and the character 'O' in the case of organically cultivated wheat.

Table 1. Studied Triticum flours with their extraction rates, type of cultivation, and type of wheat.

No.	Extraction Rate	Cultivation	Wheat Type
1	70%	Conventional	Spelt (S-70)
2	70%	Organic	Spelt (S-70-O)
3	90%	Organic	Bread wheat (B-90-O)
4	100%	Organic	Emmer (E-100-O)
5	100%	Organic	Spelt (S-100-O)
6	70%	Conventional	Emmer (E-70)
7	100%	Conventional	Spelt (S-100a)
8	100%	Conventional	Spelt (S-100b)
9	70%	Conventional	Bread wheat (B-70)
10	90%	Conventional	Bread wheat (B-90)

An amount of 10 g of flour was selected from 4 different positions of the package and mixed properly to create a homogeneous material that was then kept in a chemical dryer until analysis. The remaining original sample was kept at 2–4 °C. A modified method reported previously from Tsochatzis et al. [15] has been applied for the determination of amino acid mass fraction (g/100 g protein) in the aforementioned homogenized flour samples. In brief, for the AA profile 10.0 ± 0.5 mg of dried flour was placed in a hydrolysis tube along with 100 μL of 6 M HCl containing 4% (*v*/*v*) TGA as the reducing agent. The tube was flushed with N_2 gas to establish oxygen-free conditions, sealed, and heated at 110 °C for 18 h. Then, the mixture was transferred to a centrifuge tube with 0.5 mL HCl 0.1 M and 0.5 mL water and centrifuged at 4200× *g* for 5 min. The supernatant was collected and filtered through a 0.22 μm polytetrafluoroethylene (PTFE) syringe filter. An amount of 0.5 mL of the clear extract was diluted with 4.5 mL ACN/water 95:5 *v/v* and injected to the UHPLC-HILIC-ESI MS/MS system.

2.4. UHPLC-MS/MS Analysis

UHPLC–tandem mass spectrometry was based on the method described by Tsochatzis et al. [15]. In brief, the analysis was performed on an Accela TSQ Quantum TM Access MAX Triple Quadrupole Mass Spectrometer system (Thermo Scientific, San Jose, CA, USA) operating under XCalibur (Thermo Scientific, San Jose, CA, USA) Software. The mobile phase consisted of solvent A: ACN/5 mM $HCOONH_4$, pH = 3.0 adjusted with HCOOH 95:5 (*v/v*) and solvent B: ACN/5 mM $HCOONH_4$, pH = 3.0 adjusted with HCOOH, 40:60 (*v/v*). Elution was based on a linear gradient program of 13 min from 80% A:20% B to 62% A:38% B, followed by a 2 min equilibration step to the initial conditions prior to the next injection. The flow rate was 400 μL min^{-1}, and the total analysis time was 15 min.

Chromatographic separation was performed on a 2.1 mm × 150 mm ACQUITY UPLC 1.7 μm BEH HILIC amide column (Waters), equipped with an ACQUITY UPLC BEH Amide 1.7 μm Van-Guard Pre-column, maintained at 40 °C. Selected Reaction Monitoring (SRM) with Electrospray positive ionization mode (ESI +) was applied with spray voltage at 3000 V, capillary temperature: 300 °C, vaporizer temperature: 300 °C, sheath gas pressure at 40 arbitrary units (Arb), aux gas pressure at 10 Arb, ion sweep gas pressure at 2.0 Arb, ion source discharge current at 4.0 μA and collision gas pressure at 1.5 mTorr. Auto-samplers' temperature was set at 4 °C, and the injection volume set at 5 μL. Amino acid individual data regarding molecular formulas, monoisotopic masses, precursor-product ion for the aforementioned SRM, along with their respective retention times in standard solutions and after acidic hydrolysis, are given in the Supplementary material (Table S1).

2.5. Method Validation

Method linearity, precision, trueness, the limit of detection (LOD), and limit of quantification (LOQ) were calculated. The linearity of the method was firstly assessed by analyzing standard solutions mixtures at six concentration levels for all AA (0.5, 1, 10, 50, 100, 200 μg mL^{-1}), representing the working concentration range. Calibration curves were constructed by plotting the peak areas of the respective AAs followed by linear regression analysis (R^2), based on the standard addition method. LODs and LOQs were calculated according to the signal-to-noise ratio (S/N) and the slope (S), using the equations LOD = 3 SD/S and LOQ = 10 SD/S [28].

Precision and trueness were assessed in the B-90 flour sample, which was used as a reference sample. The sample was fortified at two concentration levels (20.0 and 40.0 mg/100 g) of all AAs tested. All calculations were performed using the concentration values expressed in g/100 g for each AA individually. For short-term repeatability, the fortified samples were analyzed in triplicates during the day while for intermediate precision, the aforementioned samples were analyzed in triplicates for three consecutive days. Relative standard deviation (RSD; %) and recoveries were calculated as ((amount found in the spiked sample—amount found in the sample)/amount added) × 100 [2]. All the results regarding precision and trueness are presented in Supplementary material's Table S2.

2.6. Quantitation and Matrix Effect

The calibration curves of the studied AAs, based on the linear regression coefficients (R^2) have been performed with the standard addition method. Flour samples were fortified at two concentration (5 and 10 μmol/100 g) levels of all AAs tested, followed by an analysis in triplicates. Calibration curves were constructed by linear regression analysis of the peak area (Y) versus the injected concentration (X), and they have been assessed based on the linear regression coefficients (R^2). Linear equations were established to determine the initial concentration of amino acids in the dried cereal flour samples. Evaluation of the matrix effect (ME, %) was performed by the slope comparison method as it was previously reported [2,15].

2.7. Protein Content and Flour Quality Parameters

The moisture, ash, total protein content, wet gluten, and gluten index (GI) were determined following ICC standard methods 109/1, 104/1, 105/2, and 155, respectively [29].

2.8. Data Processing and Statistical Analysis

Data were processed using the XCalibur application manager for the quantification of compounds. Regression analysis and statistics were performed using Microsoft Excel, and further statistical analysis, such as Analysis of Variance (ANOVA), followed by Tukey comparison test in all cases, has been performed with Minitab 18.0 statistical software (Minitab Inc., State College, PA, USA). The multivariate statistical analysis, combined with cluster analysis, has been performed with Simca 15 (Umetrics, Umea, Sweden).

3. Results

3.1. Acidic Hydrolysis and Antioxidant Agent

A properly performed hydrolysis is a prerequisite of a successful analysis regarding amino acid profiling in food matrices. The conditions selected were based according to the previously reported conditions by Fountoulakis et al. and the applied conditions in case of mussels from Tsochatzis et al. [15] or the ones reported by EZ Faast [30]. We selected the conditions of 110 °C for 18 h, in order to minimize (in combination with the antioxidant agent) degradation of specific amino acids, while we made a compromise in the recoveries of the more hydrophobic AA, such as valine and isoleucine and leucine [6,15,30]. In addition, the selection of this temperature was selected to minimize the potential cross-reactions of amino-acids with starch. The selection of the hydrolysis conditions was also based on the study of Tsochatzis et al. [15].

3.2. Analytical Method Development, Validation, and Optimization

The total analysis time was less than 12 min. The method exhibited good linearity in the concentration range of 0.5–200 μg mL^{-1} with a linear regression coefficient (R^2) of above 0.99 for each AA. The effects from the matrix were minimal in both cases of either standard solutions or amino acid determination after hydrolysis. A typical chromatogram of the AA analysis in flours is presented in Figure 1.

The present analytical method exhibited satisfactory sensitivity for all AA. LODs varied from 0.002 (valine, serine, leucine, isoleucine) to 0.009 g/100 g (threonine) and the LOQs from 0.007 (serine, leucine) to 0.024 g/100 g (threonine, lysine) and a minimal matrix effect was observed. The analytical figures of merit of the method are given in Table 2.

Regarding the trueness, it was assessed by the recoveries from spiked cereal bread wheat flour, after acidic hydrolysis, at two concentration levels. The resulting recoveries ranged from 85.7% (lysine) to 121.8% (leucine) in the intra-day assay and from 86.8% (lysine) to 123.3% (leucine) for the intermediate precision (Table S2). The respective precision expressed in relative standard deviation (RSD %) values ranged from 0.6% (Glutamic acid) to 13.9% (proline) and from 1.4% (serine) to 13.7% (lysine) for short-term repeatability and intermediate precision respectively. The results of the analytical method presented adequate precision and accuracy results.

3.3. Amino Acid Profile of Flour Samples

Statistical differences have been identified between the AA mass fractions (Table 3) among the studied flour samples. From Table 3, it could be concluded that phenylalanine, threonine, glycine, and histidine did not have statistical differences among the various flours. Instead, significant differences have been observed for isoleucine, serine, lysine, valine, methionine, proline, glutamic acid, and glutamine. The results were in accordance with previously reported work, regarding the study of AA content in ancient cereal grain wheat cultivars [21].

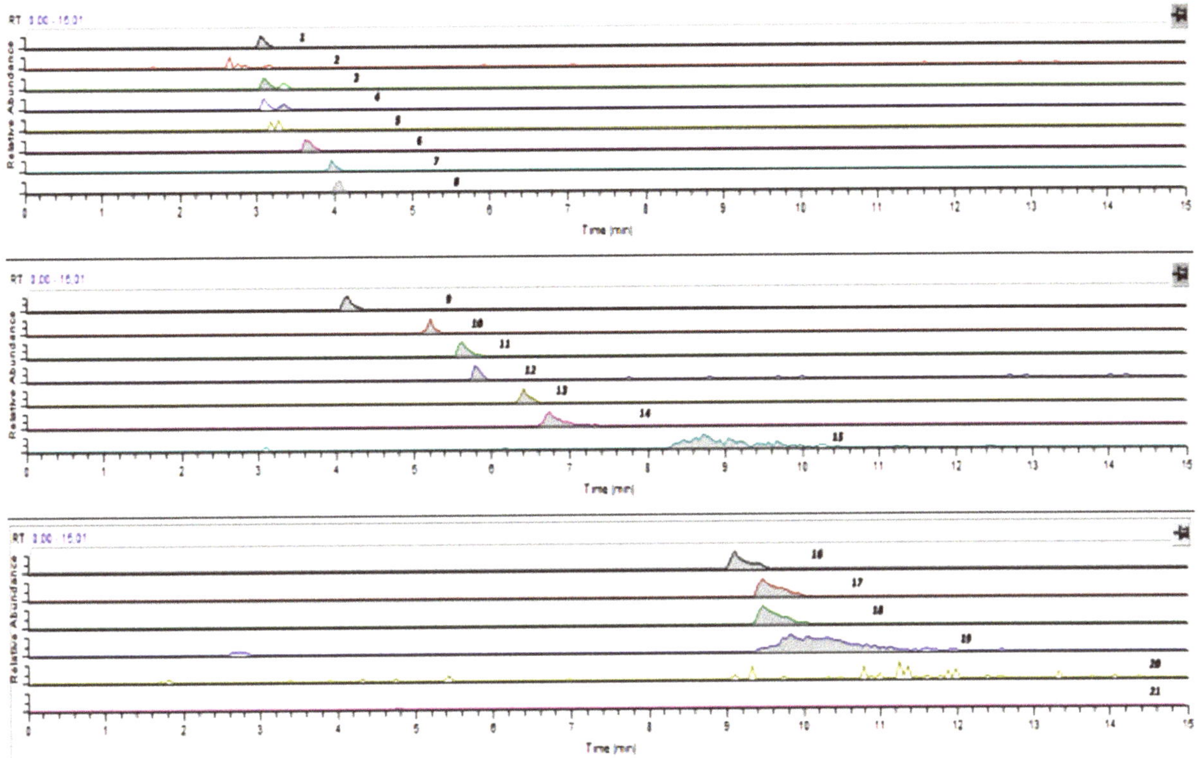

Figure 1. HILIC UHPLC–MS/MS chromatographic traces of the 17 amino acids quantified in the hydrolyzed flour sample B-90 (Elution order is 1: phenylalanine; 2: tryptophan; 3: Isoleucine; 4: leucine; 5: asparagine; 6: methionine; 7: valine; 8: proline; 9: tyrosine; 10: alanine; 11: threonine; 12: glycine; 13: serine; 14: glutamic acid; 15: aspartic acid; 16: arginine; 17: glutamine; 18: lysine; 19: histidine; 20: cystine; 21: cysteine).

Table 2. Limits of detection (LODs), limits of quantification (LOQs), linear ranges, and linear regression coefficients for the amino acids matrix match calibration in hydrolyzed cereal flour.

Amino Acid	Linear Regression (R^2)	LOD (g/100 g)	LOQ (g/100 g)	Matrix Effect (%) *
Glycine	0.993	0.005	0.010	94.6
Alanine	0.994	0.007	0.020	106.4
Serine	0.996	0.002	0.007	88.1
Proline	0.999	0.004	0.013	107.8
Valine	0.995	0.002	0.007	111.7
Threonine	0.993	0.009	0.024	108.8
Aspartic acid	0.993	0.007	0.020	102.8
Glutamic acid	0.996	0.007	0.022	92.2
Isoleucine	0.999	0.002	0.009	82.6
Leucine	0.999	0.002	0.007	88.5
Asparagine	-			-
Lysine	0.993	0.009	0.024	82.2
Methionine	0.994	0.007	0.020	86.3
Histidine	0.994	0.007	0.022	83.6
Phenylalanine	0.996	0.007	0.020	85.3
Arginine	0.997	0.004	0.016	110.1
Tyrosine	0.997	0.004	0.016	109.2
Cystine	-	-	-	-
Cysteine	-	-	-	-

* Calculated based on the slope ratio method [2].

Table 3. Average amino acids composition (g/100 g protein) of wheat flours under investigation.

Target AA	Flour Sample (g/100 g Protein; Dry Basis) *									
	1	2	3	4	5	6	7	8	9	10
	Spelta (S-70)	Spelta (S-70-O)	Bread Wheat (B-90-O)	Emmer (E-100-O)	Spelta (S-100-O)	Emmer (E-70)	Spelta (S-100a)	Spelta (S-100b)	Bread Wheat (B-70)	Bread Wheat (B-90)
Phe	3.73 ^{e}	3.94 ^{d}	4.23 ^{ab}	3.93 ^{d}	3.99 ^{cd}	3.52 ^{g}	4.13 ^{bc}	4.38 ^{a}	3.71 ^{ef}	3.54 ^{fg}
Ile	1.59 ^{cd}	1.75 ^{a}	1.74 ^{a}	1.55 ^{de}	1.57 ^{de}	1.50 ^{ef}	1.70 ^{ab}	1.65 ^{bc}	1.32 ^{g}	1.46 ^{f}
Leu	3.82 ^{d}	4.08 ^{c}	4.18 ^{bc}	4.03 ^{c}	4.09 ^{c}	3.76 ^{d}	4.28 ^{b}	4.49 ^{a}	3.76 ^{d}	3.46 ^{e}
Met	0.96 ^{a}	0.84 ^{c}	0.91 ^{b}	0.81 ^{cd}	0.89 ^{b}	0.81 ^{cd}	0.88 ^{b}	0.97 ^{a}	0.79 ^{d}	0.73 ^{e}
Tyr	1.72 ^{c}	1.88 ^{ab}	1.95 ^{a}	1.82 ^{b}	1.84 ^{b}	1.63 ^{d}	1.83 ^{b}	1.95 ^{a}	1.65 ^{cd}	1.64 ^{d}
Pro	6.18 ^{b}	6.87 ^{a}	5.44 ^{c}	4.44 ^{ef}	4.64 ^{de}	4.38 ^{f}	4.69 ^{d}	3.82 ^{g}	2.97 ^{i}	3.55 ^{h}
Val	0.96 ^{c}	1.10 ^{a}	1.05 ^{b}	0.87 ^{d}	0.89 ^{d}	0.81 ^{e}	0.95 ^{c}	0.82 ^{e}	0.69 ^{f}	0.73 ^{f}
Ala	1.91 ^{c}	1.94 ^{c}	2.09 ^{b}	1.68 ^{d}	2.05 ^{b}	1.88 ^{c}	2.38 ^{a}	1.12 ^{f}	0.96 ^{g}	1.52 ^{e}
Thr	2.29 ^{d}	2.59 ^{a}	2.51 ^{abc}	2.42 ^{c}	2.46 ^{bc}	2.13 ^{e}	2.45 ^{bc}	2.54 ^{ab}	2.11 ^{e}	1.86 ^{f}
Gly	2.67 ^{d}	2.07 ^{e}	3.48 ^{a}	2.02 ^{e}	3.00 ^{c}	2.13 ^{e}	2.72 ^{d}	3.44 ^{a}	3.47 ^{a}	3.15 ^{b}
Lys	1.57 ^{c}	1.43 ^{de}	1.46 ^{d}	1.28 ^{f}	1.36 ^{e}	1.25 ^{f}	1.43 ^{de}	1.42 ^{de}	2.38 ^{a}	1.89 ^{b}
Ser	1.27 ^{b}	1.43 ^{a}	1.39 ^{a}	1.28 ^{b}	1.16 ^{c}	1.25 ^{b}	1.29 ^{b}	1.12 ^{cd}	0.86 ^{e}	1.09 ^{d}
Arg	4.68 ^{e}	5.22 ^{bc}	5.37 ^{ab}	4.94 ^{d}	5.02 ^{cd}	4.60 ^{ef}	4.99 ^{d}	5.50 ^{a}	4.39 ^{f}	4.16 ^{g}
His	2.42 ^{c}	2.59 ^{a}	2.65 ^{a}	2.56 ^{a}	2.59 ^{a}	2.38 ^{cd}	2.58 ^{a}	2.54 ^{ab}	2.44 ^{bc}	2.30 ^{d}
Gln	1.08 ^{e}	1.30 ^{ab}	1.32 ^{a}	1.14 ^{d}	1.23 ^{c}	1.13 ^{d}	1.22 ^{c}	1.27 ^{bc}	1.09 ^{de}	1.09 ^{de}
Asp	4.36 ^{cd}	4.66 ^{b}	5.02 ^{a}	3.87 ^{f}	4.42 ^{c}	4.51 ^{bc}	4.16 ^{e}	4.04 ^{ef}	4.18 ^{de}	3.93 ^{f}
Glu	41.33 ^{e}	39.67 ^{e}	41.47 ^{e}	49.22 ^{c}	52.41 ^{ab}	50.10 ^{c}	54.28 ^{a}	44.83 ^{d}	40.90 ^{e}	41.25 ^{bc}

* Values are means of triplicate measurements. Values followed by different letters within a row are significantly different ($p < 0.05$).

3.4. Flour Quality Parameters

The results from all the studied quality characteristics of the flours are presented in Table 4. Gluten index and wet gluten have been evaluated, and ANOVA showed that the tested flours presented a significant difference with the" bread" wheat flour, also showed that flours do differ in their protein (%) content (data obtained from the Kjeldahl method), gluten index, and wet gluten. By comparing the set of ANOVA results, it could be identified that "bread" flours presented a large variety of total protein content, ranging from 12.3% (B-70) to 13.9 (B-90-O), while differences were also observed for their gluten index and wet gluten. In the case of spelta, the S-70 types, either organic or conventional, presented close results in all the studied quality characteristics, something which was also valid in case of "Emmer"-type.

Table 4. Protein content of the tested flours.

No.	Wheat Type	Protein (% dw) *	Gluten Index	Wet Gluten
1	Spelt (S-70)	13.5 ^{ab}	24 ^{a}	37.4 ^{d}
2	Spelt (S-70-O)	13.5 ^{ab}	42 ^{b}	39.8 ^{d}
3	Bread wheat (B-90-O)	12.5 ^{a}	94 ^{c}	27.5 ^{b}
4	Emmer (E-100-O)	13.1 ^{ab}	19 ^{a}	25.3 ^{b}
5	Spelt (S-100-O)	12.9 ^{ab}	20 ^{a}	26.7 ^{b}
6	Emmer (E-70)	14.1 ^{c}	18 ^{a}	15.1 ^{a}
7	Spelta (S-100a)	12.0 ^{a}	51 ^{b}	33.3 ^{c}
8	Spelta (S-100b)	12.0 ^{a}	76 ^{bc}	23.8 ^{b}
9	Bread wheat (B-70)	12.3 ^{a}	98 ^{c}	33.8 ^{c}
10	Bread wheat (B-90)	13.9 ^{c}	27 ^{a}	29.5 ^{bc}

* Values followed by different letters within a column are significantly different ($p < 0.05$).

4. Discussion

4.1. Acidic Hydrolysis

Regarding sample preparation and acidic hydrolysis, we studied the effect of the antioxidant agent as specific AA are susceptible to decomposition during acidic hydrolyses, such as tryptophan, as well as others like asparagine and glutamine to be converted to aspartic acid and glutamic acid, respectively. In this case, an antioxidant agent is needed to prevent the aforementioned decomposition or converting reactions. We selected to proceed with TGA, as it has been reported to be an effective antioxidant agent for amino acid analysis in food matrices [6,15,30]. In this case, the two levels of TGA have been tested, of 2% *v*/*v* [30] and 4% [15]. The results indicated that the presence of TGA at a level of 4% *v*/*v*, was more effective, and it was selected for the final protocol.

Even though it is a practice in acid hydrolysis to add an antioxidant agent to prevent the degradation of AAs, to our knowledge, such a protective step has not been previously applied prior to HILIC-MS analysis in case of cereal flour analysis. Moreover, no single set of conditions has been suggested so far for the effective prevention of all AA degradation in wheat flour samples.

4.2. Sample Preparation and Analytical Methods

By applying the present method, all AAs eluted prior to 11 min of run time. The selected chromatographic conditions, especially for the mobile phase, have been selected with the aim to provide as good peak shape for basic, neutral, and acidic AAs and certainly taking into consideration the effect of pH to HILIC peak shapes [15,31]. As reported previously by Tsochatzis et al., it was noticed that the retention times and the respective peak shapes of the analytes in the extracts are slightly shifted from the respective AA standard. This behavior could be due to the highly acidic conditions during hydrolysis, which affects the pH of the final injection sample and the separation of the analytes. Due to the contribution of ionic interactions in the retention mechanism, in HILIC, the charge state of the analyte affects the retention time of the compound [15,31]. To eliminate this effect, after trying different approaches, it was selected the dilution (10 times with ACN/water 95:5 *v/v* to obtain a final pH of 2.5–3) of the final solution before injection. The dilution did not present a limitation for the present method since the concentration of amino acids obtained after hydrolyzing the proteins is significantly higher than that of free amino acids present in most foods [15,32]. The Multiple Reaction Monitoring (MRM) transitions and the parameters in the tandem MS detection were selected after tuning for the optimum signal for each of the analytes.

Trueness assessed by the reported recoveries from spiked cereal flour after acidic hydrolysis. An important source of slight variations could be considered the acidic hydrolysis and especially the precision of the temperature that needs to be controlled and precisely assessed [32]. The reported results for both trueness and precision regarding the quantification of AAs are similar to the ones previously reported in other food matrices [3,5,14,15].

With the current method, only 17 amino acids could be determined following acid hydrolysis with TGA as an antioxidant agent. The amino acids tryptophan, cysteine, cystine, and asparagine were affected from the acid hydrolysis; tryptophan was unstable, asparagine and glutamine tended to convert to aspartic acid and glutamic acid, respectively, while cysteine and cystine were oxidized to cysteic acid [9,15,33].

4.3. Comparative Study of Amino Acid Profile of Flour Samples

Regarding stability and yields of amino-acids during acidic hydrolysis, serine and tyrosine are generated in low yields, methionine is sensitive to oxidation, due to acidic conditions it could be oxidized to its sulfone product, and finally, valine hydrolyzes in poor yields (longer time and temperatures are needed) [1,34,35].

In case of proline, Kapusniak et al. reported reactions of starch with α-amino acids, where proline, alanine, isoleucine, and valine most readily reacted with starch that could affect the hydrolysis [36] while, Ito et al. (2006) highlighted the significant effect of hydrolysis time on the yields of all amino acids, with changes in isoleucine, lysine and serine, while there was observed an existing binding of amino acids (glycine, alanine, and partial lysine) to starch chains [37].

In addition, partial losses of the amino acids tyrosine and serine while the other amino acids (valine, leucine, and isoleucine) are requiring longer hydrolysis time in order to obtain higher yields. Our results are in accordance with Rowan et al. [38].

It has been reported that spelt has a high concentration in methionine compared to wheat [25,39,40]. This fact was confirmed in the present study, where all tested spelt flours, presented significantly higher methionine content than emmer as well as bread wheat flours. The values of the methionine varied from 0.73–0.97 g/100 g protein (dry basis); spelt flours showed 10% higher values than the reference bread wheat flour. Aspartic acid content was significant higher in all studied spelt flours compared to whole grain wheat flour is in accordance with the results of [21,25,38].

In the case of alanine, the results indicated that all flours have significantly higher content compared to the B-90 except for E-100-O that showed similar value. The results suggest that hydrolysis has a great impact and effect on the release and analysis of the amino acids, while on the other hand, the side cultivation technique and location have a potential role in the final concentration of specific amino acids.

It is also reported in the literature that wheat proteins have low mass fractions of certain AA, especially lysine and threonine [18–21]. Contrary, in the current study, it was observed that the refined bread wheat flours and the whole grain wheat flour (reference) presented significantly higher contents of lysine (2.38 and 1.89 g/100 g protein) compared to other types of studied flours (emmer, spelt) that had significantly lower mass fractions [21]. It is also reported that wheat proteins have low mass fractions of certain AA, especially lysine and threonine [18–21].

Escarnot et al., reported that the spelt amino acid profile differs from that of bread wheat, supported by limited evidence of higher content for isoleucine, leucine, and glycine [25,40–43]. Our results are in accordance with the reported literature on higher protein content in spelt than in wheat grains under low nitrogen fertilization [25,40], although this has not been proved to be statistically significant, as it is also observed in our study. Pruska-Kedzior et al. found significantly higher protein content in spelt flour, but there should be considered that genotype and the cultivation conditions highly affect the protein content [27].

Statistical analysis (ANOVA) performed for each AA, indicated that there are significant statistical differences between all the studied flours (Table 3). Statistical analysis revealed some very interesting results about the interaction between the type of flour and the concentration of amino acids. In general, there was a significant interaction between the type of the flour studied and the amount of AA.

Glutamine and proline are the functional amino acids in dough formation [21]. For glutamine, the organic B-90 presented the higher mass fraction among the flours studied, followed by the S-70-O and conventional spelta S-100a. Especially in the case of glutamic acid (Glu) and proline, the whole grain wheat flour showed higher concentration than the B-90 (Table 3). White wheat flour and white whole grain flours showed lower proline and glutamic acid concentrations that the rest of flours studied, which is in accordance with the results of Abdel-Aal and Hucl and Escarnot et al. [21,25]. The aforementioned authors reported that the spelt is rich in proline, which is the major functional amino acid in dough formation. The data obtained showed that the organic bread wheat flour (B-90-O) showed much higher proline content than the two bread wheat flours B-70 and B-90).

4.4. *Interaction with Quality Parameters*

A multivariate analysis was performed to study potential interactions of the TAA with the quality characteristics of cereal flours under investigation. Thus, by developing this analytical tool that reveals the amino acid profile of flours, one could discover the amino acid profile of proteins of different Triticum species and use it for choosing the variety with a more balanced profile for the use in cereal product development in correlation with other quality characteristics. The performed multivariate analysis between the AA content and the flour quality characteristics showed that the studied cereal flours could be distinguished in groups, based on their origin. The score plot, along with the respective loadings is presented in Figure 2.

Results showed that the studied types of flours presented a specific AA content pattern and specific quality characteristics. From the principal component analysis (PCA) biplot and the respective score plots, it could be identified that there is a clear differentiation of the three different types of flours. Three different groups were identified as expected; the emmer grains (blue dots), the spelta (brown dots) and the bread wheat (purple dots). The right part represents the bread wheat flours, with high protein content (%), high gluten index, and lower falling number (FN). The Emmer type flours presented lower protein content and lower gluten index, while the spelt type flours are in between these two categories, presenting intermediate quality characteristics of the other two types of flours.

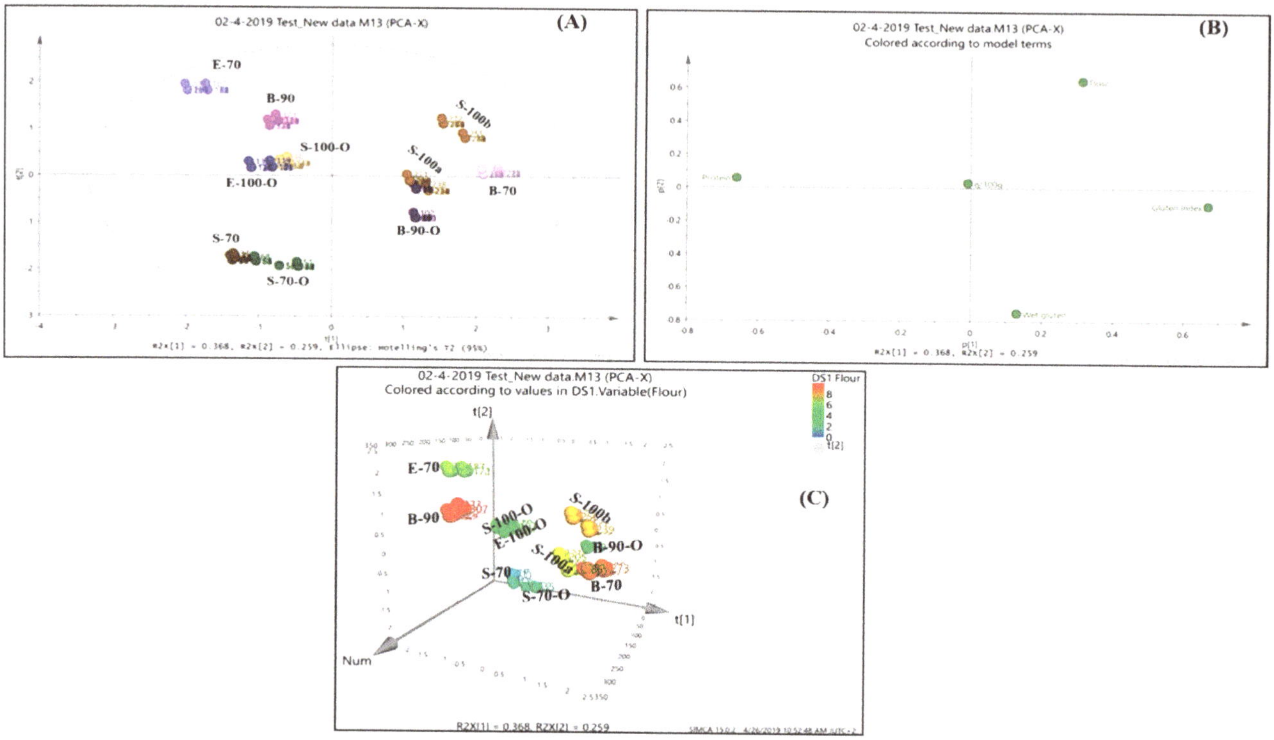

Figure 2. Principal components analysis of the studied amino acid profile and quality characteristics; (**A**) score plot and (**B**) loading plot and (**C**) 3D scatterplot (for the type studied cereal flour numbering, please see Table 1).

All the interactions and effects could be observed in the loading plot (Figure 2B). Briefly, the main effects resulted from the type of flour, and the protein content tends to play a significant factor in the clustering of the assessed grains which is resulting also to various quality characteristics and specific AA profiles, for which we have already identified statistical differences among the three types of flours. In addition, the organic cultivated grains tended to reach a higher protein content that is also reflecting to a higher content of certain AAs. On the other hand, a lower protein content reflected to a lower content of certain AA, but significant In concluding, it seems that the type of cultivation of the cereal grains affects the AA profile, as well as the quality characteristics of the flours and there is an indication of the potential effect of the cultivation (organic, conventional). Flours from organic cultivated grains seemed to be close in protein content (%) and eventually in the amino acids, but there is a clear differentiation in gluten index and wet gluten. In this case, the "spelt" flours are closer to "emmer" for the aforementioned characteristics, and all of them differentiated from the bread flour either type-70 or type-90. Spelt flours, either type-70 or 100, presented similar characteristics for either organic or conventional cultivated grains, while also "Emmer" type-70 or 100, presented also close characteristics. The biggest differentiation was noticed in the "bread" flours, where the cultivation and the extraction rate presented a significant effect for all the studied factors and the amino acid content (Table 3 and Figure 2).

5. Conclusions

A UHPLC-HILIC-tandem MS method has been developed and validated for the quantification of 17 amino acids in cereal flour samples after acid hydrolysis with HCl in the presence of a reducing agent. Tryptophan, cysteine, cystine, and asparagine were not possible to be quantified as they were degraded during hydrolysis due to harsh acidic conditions applied. The method proved to be a fast and reliable tool for acquiring information on amino acids profile from cereal flours. The developed analytical method has been applied in different flours such as spelt, dicoccum, whole grain wheat, and white wheat. Moreover, multivariate analysis showed that protein content and type of flour have the main contribution and effect, interacting with either the AA profile and with the studied quality characteristics. It has also been presented that not only with quality characteristics of the flours, but the type oz flour has significant interactions. A clear effect of an indication of the potential effect has been identified among the different flours studied.

Author Contributions: Conceptualization—E.T., M.P. and S.K.; methodology—E.T. and S.K.; formal analysis—E.T., M.P. and S.K.; investigation—E.T.; resources—E.T., M.P. and S.K.; data curation—E.T., M.P and S.K.; writing—original draft preparation—E.T.; writing—review and editing—E.T., M.P. and S.K.; visualization—E.T.; supervision—E.T. and S.K.; project administration—E.T., M.P. and S.K.

References

1. Belitz, H.-D.; Grosch, W.; Schieberle, P. *Food Chemistry*, 4th ed.; Springer: Berlin, Germany, 2009.
2. Jia, S.; Kang, Y.P.; Park, J.H.; Lee, J.; Kwon, S.W. Simultaneous determination of 23 amino acids and 7 biogenic amines in fermented food samples by liquid chromatography/quadrupole time-of-flight mass spectrometry. *J. Chromatogr. A* **2011**, *1218*, 9174–9182. [CrossRef] [PubMed]
3. Gökmen, V.; Serpen, A.; Mogol, B.A. Rapid determination of amino acids in foods by hydrophilic interaction liquid chromatography coupled to high-resolution mass spectrometry. *Anal. Bioanal. Chem.* **2012**, *403*, 2915–2922. [CrossRef] [PubMed]
4. Zhao, M.; Ma, Y.; Dai, L.-L.; Zhang, D.-L.; Li, J.-H.; Yuan, W.-X.; Li, Y.-L.; Zhou, H.-J. A high-performance liquid chromatographic method for simultaneous determination of 21 free amino acids in tea. *Food Anal. Methods* **2012**, *6*, 69–75. [CrossRef]
5. Zhou, G.; Pang, H.; Tang, Y.; Yao, X.; Mo, X.; Zhu, S.; Guo, S.; Qian, Y.; Su, S.; Zhang, L.; et al. Hydrophilic interaction ultra-performance liquid chromatography coupled with triple-quadrupole tandem mass spectrometry for highly rapid and sensitive analysis of underivatized amino acids in functional foods. *Amino Acids* **2013**, *44*, 1293–1305. [CrossRef] [PubMed]
6. Fountoulakis, M.; Lahm, H.-W. Hydrolysis and amino acid composition analysis of proteins. *J. Chromatogr. A* **1998**, *826*, 109–134. [CrossRef]
7. Kim, S.-L.; Lee, J.-E.; Kwon, Y.-U.; Kim, W.-H.; Jung, G.-H.; Kim, D.-W.; Lee, C.-K.; Lee, Y.-Y.; Kim, M.-J.; Hwang, T.-Y.; et al. Introduction and nutritional evaluation of germinated soy germ. *Food Chem.* **2013**, *136*, 491–500. [CrossRef] [PubMed]
8. Rombouts, I.; Lamberts, L.; Celus, I.; Lagrain, B.; Brijs, K.; Delcour, J.A. Wheat gluten amino acid composition analysis by high-performance anion-exchange chromatography with integrated pulsed amperometric detection. *J. Chromatogr. A* **2009**, *1216*, 5557–5562. [CrossRef]
9. Kelly, M.T.; Blaise, A.; Larroque, M. Rapid automated high performance liquid chromatography method for simultaneous determination of amino acids and biogenic amines in wine, fruit and honey. *J. Chromatogr. A* **2010**, *1217*, 7385–7392. [CrossRef]
10. Mayer, H.K.; Fiechter, G. Application of UHPLC for the determination of free amino acids in different cheese varieties. *Anal. Bioanal. Chem.* **2013**, *405*, 8053–8061. [CrossRef]
11. Bernal, J.; Ares, A.M.; Pól, J.; Wiedmer, S.K. Hydrophilic interaction liquid chromatography in food analysis. *J. Chromatogr. A* **2011**, *1218*, 7438–7452. [CrossRef]
12. Nováková, L.; Havlíková, L.; Vlacková, H. Hydrophilic interaction chromatography of polar and ionizable compounds by UHPLC. *Trends Anal. Chem.* **2014**, *63*, 55–64. [CrossRef]
13. Kato, M.; Kato, H.; Eyama, S.; Takatsu, A. Application of amino acid analysis using hydrophilic interaction liquid chromatography coupled with isotope dilution mass spectrometry for peptide and protein quantification. *J. Chromatogr. B* **2009**, *877*, 3059–3064. [CrossRef] [PubMed]
14. Guo, S.; Duan, J.; Qian, D.; Tang, Y.; Qian, Y.; Wu, D.; Su, D.; Shang, E. Rapid determination of amino acids in fruits of Ziziphus jujuba by hydrophilic interaction ultra-high-performance liquid chromatography coupled with triple-quadrupole mass spectrometry. *J. Agric. Food Chem.* **2013**, *61*, 2709–2719. [CrossRef] [PubMed]
15. Tsochatzis, E.D.; Begou, O.; Gika, H.G.; Karayannakidis, P.D.; Kalogiannis, S. A hydrophilic interaction chromatography-tandem mass spectrometry method for amino acid profiling in mussels. *J. Chromatogr. B* **2017**, *1047*, 197–206. [CrossRef] [PubMed]

16. Langrock, T.; Czihal, P.; Hoffmann, R. Amino acid analysis by hydrophilic interaction chromatography coupled on-line to electrospray ionization mass spectrometry. *Amino Acids* **2006**, *30*, 291–297. [CrossRef] [PubMed]
17. Papageorgiou, M.; Skendi, A. *Sustainable Recovery and Reutilisation of Cereal Processing By-Products*, 1st ed.; Elsevier: Amsterdam, The Netherlands, 2018; Chapter 1.
18. Kies, C.; Fox, H.M. Determination of the first-limiting amino acid of wheat and triticale grain for humans. *Cereal Chem.* **1970**, *47*, 615–622.
19. Fleming, S.E.; Sosulski, F.W. Nutritive value of bread fortified with concentrated plant proteins and lysine. *Cereal Chem.* **1977**, *54*, 1239–1248.
20. Abdel-Aal, E.-S.M.; Sosulski, F.W. Bleaching and fractionation of dietary fiber and protein from wheat-based stillage. *Lebensm. Wiss. Technol.* **2001**, *34*, 159–167. [CrossRef]
21. Abdel-Aal, E.-S.M.; Hucl, P. Amino Acid Composition and In Vitro Protein Digestibility of Selected Ancient Wheats and their End Products. *J. Food Comp. Anal.* **2002**, *15*, 737–747. [CrossRef]
22. Pflüger, L.A.; Martin, L.M.; Alvarez, J.B. Variation in the HMW and LMW glutenin subunits from Spanish accessions of emmer wheat (*Triticum turgidum* ssp. Dicoccum Schrank). *Theor. Appl. Genet.* **2001**, *102*, 767–772. [CrossRef]
23. Ruegger, A.; Winzerler, H.; Nosberger, J. Untersuchugen uber das keimungsverhalten von dinkel (*Triticum spelta* L.) und weizen (*Triticum aestivum* L.) unter erschwerten bedingungen. *Seed Sci. Technol.* **1990**, *18*, 311–320.
24. Kema, G.H.J.; Lange, W. Resistance in spelt wheat to yellow rust: II. Monosomic analysis of the Iranian accession 415. *Euphytica* **1992**, *63*, 219–224.
25. Escarnot, E.; Jacquemin, J.-M.; Agneessens, R.; Paquot, M. Comparative study of the content and profiles of macronutrients in spelt and wheat, a review. *Biotechnol. Agron. Soc. Environ.* **2012**, *16*, 243–256.
26. Wilson, J.D.; Bechtel, D.B.; Wilson, G.W.T.; Seib, P.A. Bread quality of spelt wheat and its starch. *Cereal Chem.* **2008**, *85*, 629–638. [CrossRef]
27. Pruska-Kedzior, A.; Kedzior, Z.; Klockiewicz-Kaminska, E. Comparison of viscoelastic properties of gluten from spelt and common wheat. *Eur. Food Res. Technol.* **2008**, *227*, 199–207. [CrossRef]
28. ICH. Guidance for Industry Q2B Validation of Analytical Procedures: Methodology. 1996. Available online: http://www.fda.gov/downloads/drugs/guidancecomplianceregulatoryinformation/guidances/ucm073384.pdf (accessed on 21 June 2018).
29. ICC-International Association for Cereal Science and Technology. 2019. Available online: https://www.icc-services.at/store/standard_methods (accessed on 31 January 2019).
30. Faast, E.Z. Phenomenex, Faast Amino Acid Analysis of Protein Hydrolysates, User's Manual. 2016. Available online: http://www.fortunesci.com/image/download2/USER%20GUIDE/Amino%20Acid%20Analysis%20of%20Protein%20Hydrolysates.pdf (accessed on 21 February 2019).
31. Gika, H.G.; Theodoridis, G.A.; Vrhovsek, U.; Mattivi, F. Quantitative profiling of polar primary metabolites using hydrophilic interaction ultra-high performance liquid chromatography-tandem mass spectrometry. *J. Chromatogr. A* **2012**, *1259*, 121–127. [CrossRef]
32. Alberto Lopes, J.; Tsochatzis, E.D.; Robouch, P.; Hoekstra, E. Influence of pre-heating of food contact polypropylene cups on its physical structure and on the migration of additives. *Food Packag. Shelf Life* **2019**, *20*, 100305. [CrossRef]
33. Zumwalt, R.W.; Gehrke, C.W. Amino Acid Analysis: A Survey of Current Techniques. In *Methods for Protein Analysis*; Cherry, J.P., Barford, R.A., Eds.; American Oil Chemists' Society (AOCS): Champaign, IL, USA, 1988; pp. 13–35.
34. Davidson, I. Hydrolysis of Samples for Amino Acid Analysis. In *Protein Sequencing Protocols (Methods in Molecular Biology)*; Smith, B.J., Ed.; Humana Press: Totowa, NJ, USA, 2003; Volume 211.
35. Otter, D. Standardised methods for amino acid analysis of food. *Br. J. Nutr.* **2012**, *108*, S230–S237. [CrossRef]
36. Kapusniak, J.; Ciesielski, W.; Koziol, J.J.; Tomasik, P. Reaction of starch with a-amino acids. *Eur. Food Res. Technol.* **1999**, *209*, 325–329. [CrossRef]
37. Ito, A.; Hattori, M.; Yoshida, T.; Watanabe, A.; Sato, R.; Takahashi, K. Regulatory effect of amino acids on the pasting behavior of porato starch is attrbutable to its binding to the starch chain. *J. Agric. Food Chem.* **2006**, *54*, 10191–10196. [CrossRef]

38. Rowan, A.M.; Moughan, P.J.; Wilson, M.N. Effect of hydrolysis time on the determination of amino acid composition of diet, ileal digesta and feces samples and on the determination of dietary amino acid digestibility coefficients. *J. Agric. Food Chem.* **1992**, *40*, 981–985. [CrossRef]
39. Ranhotra, G.S.; Gelroth, J.A.; Glaser, B.K.; Lorenz, K.J. Baking and nutritional qualities of a spelt wheat sample. *Lebensm. Wiss. Technol.* **1995**, *28*, 118–122. [CrossRef]
40. Bonafaccia, G.; Galli, V.; Francisci, R.; Mair, V.; Skrabanja, V.; Kreft, I. Characteristics of spelt wheat products and nutritional value of spelt wheat-based bread. *Food Chem.* **2000**, *68*, 437–441. [CrossRef]
41. Dvoracek, V.; Curn, V.; Moudry, J. Evaluation of amino acid content and composition in spelt wheat varieties. *Cereal Res. Commun.* **2002**, *30*, 187–193.
42. Matz, S.A. *Chemistry and Technology of Cereals as Food and Feed*; Springer Science & Business Media: New York, NY, USA, 1991; Chapter 4.
43. Matuz, J.; Bartok, T.; Morocz-Salamon, K.; Bona, L. Structure and potential allergenic character of cereal proteins. I. Protein content and amino acid composition. *Cereal Res. Commun.* **2000**, *28*, 263–270.

Permissions

All chapters in this book were first published by MDPI; hereby published with permission under the Creative Commons Attribution License or equivalent. Every chapter published in this book has been scrutinized by our experts. Their significance has been extensively debated. The topics covered herein carry significant findings which will fuel the growth of the discipline. They may even be implemented as practical applications or may be referred to as a beginning point for another development.

The contributors of this book come from diverse backgrounds, making this book a truly international effort. This book will bring forth new frontiers with its revolutionizing research information and detailed analysis of the nascent developments around the world.

We would like to thank all the contributing authors for lending their expertise to make the book truly unique. They have played a crucial role in the development of this book. Without their invaluable contributions this book wouldn't have been possible. They have made vital efforts to compile up to date information on the varied aspects of this subject to make this book a valuable addition to the collection of many professionals and students.

This book was conceptualized with the vision of imparting up-to-date information and advanced data in this field. To ensure the same, a matchless editorial board was set up. Every individual on the board went through rigorous rounds of assessment to prove their worth. After which they invested a large part of their time researching and compiling the most relevant data for our readers.

The editorial board has been involved in producing this book since its inception. They have spent rigorous hours researching and exploring the diverse topics which have resulted in the successful publishing of this book. They have passed on their knowledge of decades through this book. To expedite this challenging task, the publisher supported the team at every step. A small team of assistant editors was also appointed to further simplify the editing procedure and attain best results for the readers.

Apart from the editorial board, the designing team has also invested a significant amount of their time in understanding the subject and creating the most relevant covers. They scrutinized every image to scout for the most suitable representation of the subject and create an appropriate cover for the book.

The publishing team has been an ardent support to the editorial, designing and production team. Their endless efforts to recruit the best for this project, has resulted in the accomplishment of this book. They are a veteran in the field of academics and their pool of knowledge is as vast as their experience in printing. Their expertise and guidance has proved useful at every step. Their uncompromising quality standards have made this book an exceptional effort. Their encouragement from time to time has been an inspiration for everyone.

The publisher and the editorial board hope that this book will prove to be a valuable piece of knowledge for researchers, students, practitioners and scholars across the globe.

List of Contributors

Rafieh Fakhlaei
Food Safety and Food Integrity (FOSFI), Institute of Tropical Agriculture and Food Security, Universiti Putra Malaysia, Serdang 43400, Selangor, Malaysia

Jinap Selamat
Food Safety and Food Integrity (FOSFI), Institute of Tropical Agriculture and Food Security, Universiti Putra Malaysia, Serdang 43400, Selangor, Malaysia
Department of Food Science, Faculty of Food Science and Technology, Universiti Putra Malaysia, Serdang 43400, Selangor, Malaysia

Rashidah Sukor
Department of Food Science, Faculty of Food Science and Technology, Universiti Putra Malaysia, Serdang 43400, Selangor, Malaysia

Alfi Khatib
Pharmacognosy Research Group, Department of Pharmaceutical Chemistry, Kulliyyah of Pharmacy, International Islamic University Malaysia, Kuantan 25200, Pahang Darul Makmur, Malaysia
Faculty of Pharmacy, Airlangga University, Surabaya 60155, Indonesia

Ahmad Faizal Abdull Razis
Department of Food Science, Faculty of Food Science and Technology, Universiti Putra Malaysia, Serdang 43400, Selangor, Malaysia
Natural Medicines and Products Research Laboratory, Universiti Putra Malaysia, Serdang 43400, Selangor, Malaysia

Syahida Ahmad
Department of Biochemistry, Faculty of Biotechnology & Biomolecular Sciences, Universiti Putra Malaysia, Serdang 43400, Selangor, Malaysia

Arman Amani Babadi
School of Energy and Power Engineering, Jiangsu University, Zhenjiang 212013, China

Biancamaria Senizza, Gabriele Rocchetti and Luigi Lucini
Department for Sustainable Food Process, Università Cattolica del Sacro Cuore, Via Emilia Parmense 84, 29122 Piacenza, Italy

Murat Ali Okur, Gokhan Zengin and Gunes Ak
Department of Biology, Science Faculty, Selcuk University, Campus, 4213 Konya, Turkey

Evren Yildiztugay
Department of Biotechnology Science Faculty, Selcuk University, Campus, 4213 Konya, Turkey

Domenico Montesano
Department of Pharmaceutical Sciences, Food Science and Nutrition Section, University of Perugia, Via S. Costanzo 1, 06126 Perugia, Italy

Jitlada Vichapong
Creative Chemistry and Innovation Research Unit, Department of Chemistry and Center of Excellence for Innovation in Chemistry, Faculty of Science, Mahasarakham University, Maha Sarakham 44150, Thailand

Yanawath Santaladchaiyakit
Department of Chemistry, Faculty of Engineering, Rajamangala University of Technology Isan, Khon Kaen Campus, Khon Kaen 40000, Thailand

Rodjana Burakham and Supalax Srijaranai
Materials Chemistry Research Center, Department of Chemistry and Center of Excellence for Innovation in Chemistry, Faculty of Science, Khon Kaen University, Khon Kaen 40002, Thailand

Kristina Mastanjević, Jasmina Lukinac, Marko Jukić, Vinko Krstanović and Krešimir Mastanjević
Faculty of Food Technology Osijek, Josip Juraj Strossmayer University of Osijek, F. Kuhača 20, 31000 Osijek, Croatia

Bojan Šarkanj
Department of Food Technology, University North, University Center Koprivnica, Trg dr. Žarka Dolinara 1, 48000 Koprivnica, Croatia

Haocheng Liu, Kejing An, Siqi Su, Yuanshan Yu, Jijun Wu, Gengsheng Xiao and Yujuan Xu
Sericulture & Agri-Food Research Institute Guangdong Academy of Agricultural Sciences/Key Laboratory of Functional Foods, Ministry of Agriculture and Rural Affairs/Guangdong Key Laboratory of Agricultural Products Processing, Guangzhou 510610, China

Nadia Smadi and Nada El Darra
Faculty of Heath Sciences, Beirut Arab University, Tarik El Jedidah-Beirut, Riad EL Solh, Beirut 1107 2809, Lebanon

Adla Jammoul
Food Department, Lebanese Agricultural Research Institute, Fanar Beirut 1107 2809, Lebanon

Agha Waqar Yunus, Nida Imtiaz and Haider Khan
Animal Sciences Institute, PARC National Agricultural Research Center, Park Road, Islamabad 45500, Pakistan

Mohammed Nawaz Mohammed Ibrahim
International Livestock Research Institute, ILRI-Pakistan, Park Road, Islamabad 45500, Pakistan

Yusuf Zafar
Pakistan Agricultural Research Council, G-5/1, Islamabad 44000, Pakistan

Bo Wang
College of Veterinary Medicine, Yangzhou University, Yangzhou 225009, China
Joint International Research Laboratory of Agriculture & Agri-Product Safety, Yangzhou University, Yangzhou 225009, China

Yajuan Wang, Zhixiang Diao, Kaizhou Xie, Genxi Zhang, Tao Zhang and Guojun Dai
Joint International Research Laboratory of Agriculture & Agri-Product Safety, Yangzhou University, Yangzhou 225009, China
College of Animal Science and Technology, Yangzhou University, Yangzhou 225009, China

Xing Xie
Institute of Veterinary Medicine, Jiangsu Academy of Agricultural Sciences, Key Laboratory of Veterinary Biological Engineering and Technology, Ministry of Agriculture, Nanjing 210014, China

Alberto Nuñez, Yelena Sapozhnikova and Steven J. Lehotay
US Department of Agriculture, Agricultural Research Service, Eastern Regional Research Center, 600 East Mermaid Lane, Wyndmoor, PA 19038, USA

Mélanie Roué
Institut de Recherche pour le Développement (IRD)—UMR 241 EIO, 98716 Pirae, Tahiti, French Polynesia

Hélène Taiana Darius, André Ung, Jérôme Viallon and Mireille Chinain
Institut Louis Malardé (ILM), Laboratory of Toxic Microalgae—UMR 241-EIO, 98713 Papeete, Tahiti, French Polynesia

Manoella Sibat, Philipp Hess and Zouher Amzil
IFREMER, Phycotoxins Laboratory, F-44311 Nantes CEDEX, France

Konstantina Ntrallou
Department of Chemical Engineering, Aristotle University of Thessaloniki, 54124 Thessaloniki, Greece

Helen Gika
Laboratory of Forensic Medicine & Toxicology, Department of Medicine, Aristotle University of Thessaloniki, 54124 Thessaloniki, Greece
BIOMIC AUTH Center for Interdisciplinary Research of the Aristotle University of Thessaloniki, Innovation Area of Thessaloniki, 57001 Thermi, Greece

Emmanouil Tsochatzis
Department of Chemical Engineering, Aristotle University of Thessaloniki, 54124 Thessaloniki, Greece
BIOMIC AUTH Center for Interdisciplinary Research of the Aristotle University of Thessaloniki, Innovation Area of Thessaloniki, 57001 Thermi, Greece

Michaela Zeiner
Man-Environment-Technology Research Centre, School of Science and Technology, Örebro University, Gymnastikgatan 1, 70182 Örebro, Sweden
Division of Analytical Chemistry, Department of Chemistry, BOKU—University of Natural Resources and Life Sciences, Muthgasse 18, 1190 Vienna, Austria

Iva Juranović Cindrić
Department of Chemistry, Faculty of Science, University of Zagreb, Horvatovac 102a, 10000 Zagreb, Croatia

Arianna Binello, Giorgio Grillo, Alessandro Barge and Giancarlo Cravotto
Dipartimento di Scienza e Tecnologia del Farmaco, University of Turin, Via P. Giuria 9, 10125 Turin, Italy

Pietro Allegrini and Daniele Ciceri
INDENA S.p.A., Viale Ortles, 12, 20139 Milan, Italy

Thanh-Thien Tran-Lam, Yen Hai Dao, Duong Thanh Nguyen, Trung Quoc Pham and Giang Truong Le
Institute of Chemistry, Vietnam Academy of Science and Technology (VAST), 18 Hoang Quoc Viet, Cau Giay, Hanoi 100000, Vietnam

Hoi Kim Ma
University of Science, Vietnam National University HCMC, Linh Trung Ward, Thu Duc District, Ho Chi Minh City 720400, Vietnam

Emmanouil Tsochatzis
Department of Chemical Engineering, Aristotle University of Thessaloniki, GR-54124 Thessaloniki, Greece

Maria Papageorgiou
Department of Food Science and Technology, International Hellenic University, Sindos Campus, GR-57400 Thessaloniki, Greece

Stavros Kalogiannis
Department of Nutritional Sciences and Dietetics, International Hellenic University, Sindos Campus, GR-57400 Thessaloniki, Greece

Index

pliance